Graduate Texts in Mathematics   231

Editorial Board
S. Axler   F.W. Gehring   K.A. Ribet

T0236686

Anders Bjorner
Francesco Brenti

Combinatorics of Coxeter Groups

With 81 Illustrations

 Springer

Anders Bjorner
Department of Mathematics
Royal Institute of Technology
Stockholm 100 44
Sweden
bjorner@math.kth.se

Francesco Brenti
Dipartimento di Matematica
Universita di Roma
Via della Ricerca Scientifica, 1
Roma 00133
Italy
brenti@mat.uniroma2.it

Editorial Board:

S. Axler
Mathematics Department
San Francisco State
  University
San Francisco, CA 94132
USA
axler@sfsu.edu

F.W. Gehring
Mathematics Department
East Hall
University of Michigan
Ann Arbor, MI 48109
USA
fgehring@math.lsa.umich.edu

K.A. Ribet
Mathematics Department
University of California,
  Berkeley
Berkeley, CA 94720-3840
USA
ribet@math.berkeley.edu

Mathematics Subject Classification (2000): 20F55, 05C25

Printed on acid-free paper.

ISBN-13: 978-3-642-07922-1   e-ISBN-13: 978-3-540-27596-1

© 2010 Springer Science+Business Media, Inc.
All rights reserved. This work may not be translated or copied in whole or in part without the
written permission of the publisher (Springer Science+Business Media, Inc., 233 Spring Street, New
York, NY 10013, USA), except for brief excerpts in connection with reviews or scholarly analysis.
Use in connection with any form of information storage and retrieval, electronic adaptation, computer software, or by similar or dissimilar methodology now known or hereafter developed is for-
bidden.
The use in this publication of trade names, trademarks, service marks, and similar terms, even if
they are not identified as such, is not to be taken as an expression of opinion as to whether or not
they are subject to proprietary rights.

Printed in the United States of America.   (EB)

9 8 7 6 5 4 3 2 1

springeronline.com

To Annamaria and Christine

Contents

Foreword

Coxeter groups arise in a multitude of ways in several areas of mathematics. They are studied in algebra, geometry, and combinatorics, and certain aspects are of importance also in other fields of mathematics. The theory of Coxeter groups has been exposited from algebraic and geometric points of view in several places, also in book form. The purpose of this work is to present its core combinatorial aspects.

By "combinatorics of Coxeter groups" we have in mind the mathematics that has to do with reduced expressions, partial order of group elements, enumeration, associated graphs and combinatorial cell complexes, and connections with combinatorial representation theory. There are some other topics that could also be included under this general heading (e.g., combinatorial properties of reflection hyperplane arrangements on the geometric side and deeper connections with root systems and representation theory on the algebraic side). However, with the stated aim, there is already more than plenty of material to fill one volume, so with this "disclaimer" we limit ourselves to the chosen core topics.

It is often the case that phenomena of Coxeter groups can be understood in several ways, using either an algebraic, a geometric, or a combinatorial approach. The interplay between these aspects provides the theory with much of its richness and depth. When alternate approaches are possible, we usually choose a combinatorial one, since it is our task to tell this side of the story. For a more complete understanding of the subject, the reader is urged to study also its algebraic and geometric aspects. The notes at the end of each chapter provide references and hints for further study.

The book is divided into two parts. The first part, comprising Chapters 1 – 4, gives a self-contained introduction to combinatorial Coxeter group theory. We treat the combinatorics of reduced decompositions, Bruhat order, weak order, and some aspects of root systems. The second part consists of four independent chapters dealing with certain more advanced topics. In Chapters 5 – 7, some external references are necessary, but we have tried to minimize reliance on other sources. Chapter 8, which is elementary, discusses permutation representations of the most important finite and affine Coxeter groups.

Exercises are provided to all chapters — both easier exercises, meant to test understanding of the material, and more difficult ones representing results from the research literature. Open problems are marked with an asterisk. Thus, the book is meant to have a dual character as both graduate textbook (particularly Part I) and as research monograph (particularly Part II).

Acknowledgments: Work on this book has taken place at highly irregular intervals during the years 1993–2004. An essentially complete and final version was ready in 1999, but publication was delayed due to unfortunate circumstances. During the time of writing we have enjoyed the support of the Volkswagen-Stiftung (RiP-program at Oberwolfach), of the Fondazione San Michele, and of EC grants Nos. CHRX-CT93-0400 and HPRN-CT-2001-00272 (Algebraic Combinatorics in Europe).

Several people have offered helpful comments and suggestions. We particularly thank Sergey Fomin and Victor Reiner, who used preliminary versions of the book as course material at MIT and University of Minnesota and provided invaluable feedback. Useful suggestions have been given also by Christos Athanasiadis, Henrik Eriksson, Axel Hultman, and Federico Incitti. Günter Ziegler provided much needed help with the mysteries of LATEX. Special thanks go to Annamaria Brenti and Siv Sandvik, who did much of the original typing of text, and to Federico Incitti, who helped us create many of the figures and improve some of the ones created by us. Figure 1.1 was provided by Frank Lutz and Figure 1.3 by Jürgen Richter-Gebert.

Stockholm and Rome, September 2004

Anders Björner and Francesco Brenti

Notation

We collect here some notation that is adhered to throughout the book.

$\mathbb{Z}$ the integers

$\mathbb{N}$ the non-negative integers

$\mathbb{P}$ the positive integers

$\mathbb{Q}, \mathbb{R}, \mathbb{C}$ the rational, real, and complex numbers

$[n]$ the set $\{1, 2, \ldots, n\}$ $(n \in \mathbb{N})$, in particular $[0] = \emptyset$

$[a, b]$ the set $\{n \in \mathbb{Z} : a \leq n \leq b\}$ $(a, b \in \mathbb{Z})$

$[\pm n]$ the set $[-n, n] \setminus \{0\}$

$\{a_1, \ldots, a_n\}_<$ the set $\{a_1, \ldots, a_n\}$ with total order $a_1 < \cdots < a_n$

$\lfloor a \rfloor$ the largest integer $\leq a$ $(a \in \mathbb{R})$

$\lceil a \rceil$ the smallest integer $\geq a$ $(a \in \mathbb{R})$

$\text{sgn}(a)$ the sign of a real number: $\text{sgn}(a) \overset{\text{def}}{=} \begin{cases} 1, & \text{if } a > 0, \\ 0, & \text{if } a = 0, \\ -1, & \text{if } a < 0. \end{cases}$

δ_{ij} or $\delta(i, j)$ the Kronecker delta: $\delta_{ij} \overset{\text{def}}{=} \begin{cases} 1, & \text{if } i = j, \\ 0, & \text{if } i \neq j. \end{cases}$

$|A|, \#A,$
$\quad$ or $\text{card}(A)$ the cardinality of a set A

$A \uplus B$ the union of two disjoint sets

$A \Delta B$ the symmetric difference $A \cup B \setminus (A \cap B)$

2^A the family of all subsets of a finite set A

$\binom{A}{k}$ the family of all k-element subsets of a finite set A

A^* the set of all words with letters from an alphabet A

Each result (theorem, corollary, proposition, or lemma) is numbered consecutively within sections. So, for example, Theorem 2.3.3 is the third result in the third section of Chapter 2 (i.e., in Section 2.3). The symbol □ denotes the end of a proof or an example. A □ appearing at the end of the statement of a result signifies that the result should be obvious at that stage of reading, or else that a reference to a proof is given.

1

The basics

Coxeter groups are defined in a simple way by generators and relations. A key example is the symmetric group S_n, which can be realized as permutations (combinatorics), as symmetries of a regular $(n-1)$-dimensional simplex (geometry), or as the Weyl group of the type A_{n-1} root system or of the general linear group (algebra). The general theory of Coxeter groups expands and interweaves the many mathematical themes and aspects suggested by this example.

In this chapter, we give the basic definitions, present some examples, and derive the most elementary combinatorial facts underlying the rest of the book. Readers who already know the fundamentals of the theory can skim or skip this chapter.

1.1 Coxeter systems

Let S be a set. A matrix $m : S \times S \rightarrow \{1, 2, \ldots, \infty\}$ is called a *Coxeter matrix* if it satisfies

$$m(s, s') = m(s', s); \tag{1.1}$$
$$m(s, s') = 1 \Leftrightarrow s = s'. \tag{1.2}$$

Equivalently, m can be represented by a *Coxeter graph* (or *Coxeter diagram*) whose node set is S and whose edges are the unordered pairs $\{s, s'\}$ such that $m(s, s') \geq 3$. The edges with $m(s, s') \geq 4$ are labeled by that

number. For instance,

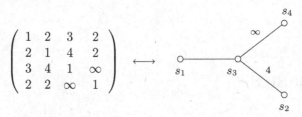

$$\begin{pmatrix} 1 & 2 & 3 & 2 \\ 2 & 1 & 4 & 2 \\ 3 & 4 & 1 & \infty \\ 2 & 2 & \infty & 1 \end{pmatrix}$$

Let $S^2_{\text{fin}} = \{(s, s') \in S^2 : m(s, s') \neq \infty\}$. A Coxeter matrix m determines a group W with the presentation

$$\begin{cases} \text{Generators: } S \\ \text{Relations: } (ss')^{m(s,s')} = e, \quad \text{for all } (s, s') \in S^2_{\text{fin}}. \end{cases} \tag{1.3}$$

Here, and in the sequel, "e" denotes the identity element of any group under consideration. Since $m(s, s) = 1$, we have that

$$s^2 = e, \quad \text{for all } s \in S, \tag{1.4}$$

which, in turn, shows that the relation $(ss')^{m(s,s')} = e$ is equivalent to

$$\underbrace{s\, s'\, s\, s'\, s \ldots}_{m(s,s')} = \underbrace{s'\, s\, s'\, s\, s' \ldots}_{m(s,s')}. \tag{1.5}$$

In particular, $m(s, s') = 2$ (i.e., two distinct nodes s and s' are not neighbors in the Coxeter graph) if and only if s and s' commute.

For instance, the group determined by the above Coxeter diagram is generated by s_1, s_2, s_3, and s_4 subject to the relations

$$\begin{cases} s_1^2 = s_2^2 = s_3^2 = s_4^2 = e \\ s_1 s_2 = s_2 s_1 \\ s_1 s_3 s_1 = s_3 s_1 s_3 \\ s_1 s_4 = s_4 s_1 \\ s_2 s_3 s_2 s_3 = s_3 s_2 s_3 s_2 \\ s_2 s_4 = s_4 s_2. \end{cases}$$

If a group W has a presentation such as (1.3), then the pair (W, S) is called a *Coxeter system*. The group W is the *Coxeter group* and S is the set of *Coxeter generators*. The cardinality of S is called the *rank* of (W, S). Most groups of interest will be of finite rank. The system is *irreducible* if its Coxeter graph is connected.

When referring to an abstract group as a Coxeter group, one usually has in mind not only W but the pair (W, S), with a specific generating set S tacitly understood. Some caution is necessary in such cases, since the isomorphism type of (W, S) is not determined by the group W alone; see Exercise 2.

The following three statements are equivalent and make explicit what it means for W to be determined by m via the presentation (1.3):

1. (Universality Property) If G is a group and $f : S \to G$ is a mapping such that

$$(f(s)f(s'))^{m(s,s')} = e$$

for all $(s, s') \in S_{\mathrm{fin}}^2$, then there is a unique extension of f to a group homomorphism $f : W \to G$.

2. $W \cong F/N$, where F is the free group generated by S and N is the normal subgroup generated by $\{(ss')^{m(s,s')} : (s, s') \in S_{\mathrm{fin}}^2\}$.

3. Let S^* be the free monoid generated by S (i.e., the set of words in the alphabet S with concatenation as product). Let $\equiv$ be the equivalence relation generated by allowing insertion or deletion of any word of the form

$$(ss')^{m(s,s')} = \underbrace{s\,s'\,s\,s'\,s\ldots s'\,s\,s'}_{2m(s,s')}$$

for $(s, s') \in S_{\mathrm{fin}}^2$. Then, $S^*/\equiv$ forms a group isomorphic to W.

It might seem that to be precise we should use different symbols for the elements of S and for their images in $W \cong S^*/\equiv$ under the surjection

$$\varphi : S^* \to W. \tag{1.6}$$

However, this is needlessly pedantic since, in practice, the possibility of confusion is negligible. It will be shown (Proposition 1.1.1) that $s \neq s'$ in S implies $\varphi(s) \neq \varphi(s')$ in W and (Corollary 1.4.8) that S is a minimal generating system for W.

Let (W, S) be a Coxeter system. Definition (1.3) leaves some uncertainty about the orders of pairwise products ss' as elements of W $(s, s' \in S)$. All that immediately follows is that the order of ss' divides $m(s, s')$ if $m(s, s')$ is finite. This leaves open the possibility that distinct Coxeter graphs might determine isomorphic Coxeter systems. However, this is not the case.

Proposition 1.1.1 *Let (W, S) be the Coxeter system determined by a Coxeter matrix m. Let s and s' be distinct elements of S. Then, the following hold:*

(i) (The classes of) s and s' are distinct in W.

(ii) The order of ss' in W is $m(s, s')$.

The proof is postponed to Section 4.1, where it is obtained for free as a by-product of some other material. Section 4.1 makes no use of (or even mention of) any material in the intermediate sections, so it is possible for a systematic reader, who wants to see a proof for Proposition 1.1.1 at this stage of reading, to go directly from here to Section 4.1.

It is a consequence of Proposition 1.1.1 that the Coxeter matrix $(m(s, s'))_{s,s' \in S}$ can be fully reconstructed from the group W and the generating set S. This leads to an important conclusion.

Theorem 1.1.2 *Up to isomorphism there is a one-to-one correspondence between Coxeter matrices and Coxeter systems.* □

The finite irreducible Coxeter systems, as well as certain classes of infinite ones, have been classified. See Appendix A1 for the classification of the finite and so-called affine groups and [306] for additional information. From now on, we will every now and then refer to these Coxeter groups by their conventional names mentioned in Appendix A1, but the classification as such will not play any significant role in the book. There is no essential restriction in confining attention to the irreducible case, since reducible Coxeter groups decompose uniquely as a product of irreducible ones (see Exercise 2.3).

The finite Coxeter groups for which $m(s, s') \in \{2, 3, 4, 6\}$ for all $(s, s') \in S^2, s \neq s'$ are called *Weyl groups*, a name motivated by Lie theory (see Example 1.2.10). The Coxeter groups for which $m(s, s') \in \{2, 3\}$ for all $(s, s') \in S^2, s \neq s'$ are called *simply-laced*.

1.2 Examples

Let us now look at a few examples. The following list is not intended to be systematic — the aim is merely to acquaint the reader with some of the groups that play an important role in the combinatorial theory of Coxeter groups and to exemplify some of the diverse ways in which Coxeter groups arise. More examples can be found in Chapter 8.

Example 1.2.1 The graph with n isolated vertices (no edges) is the Coxeter graph of the group $\mathbb{Z}_2 \times \mathbb{Z}_2 \times \cdots \times \mathbb{Z}_2$ of order 2^n. □

Example 1.2.2 The *universal Coxeter group* U_n of rank n is defined by the complete graph with all $\binom{n}{2}$ edges marked by "∞." Equivalently, it is the group having n generators of order 2 and no other relations. Each group element can be uniquely expressed as a word in the alphabet of generators, and these words are precisely the ones where no adjacent letters are equal. □

Example 1.2.3 The path

$$s_1 \qquad s_2 \qquad s_3 \qquad\qquad s_{n-2} \qquad s_{n-1}$$

is the Coxeter graph of the symmetric group S_n with respect to the generating system of adjacent transpositions $s_i = (i, i+1)$, $1 \leq i \leq n-1$. This is proved in Proposition 1.5.4. An understanding of this particular example is very valuable, both because of the importance of the symmetric group as such and its role as the most accessible nontrivial example of a Coxeter group. We will frequently return to S_n in order to concretely illustrate various general concepts and constructions. □

Example 1.2.4 The graph

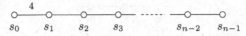

is the Coxeter graph of the group S_n^B of all signed permutations of the set $[n] = \{1, 2, \ldots, n\}$. See Section 8.1 for a detailed description of this group. It can be thought of in terms of the following combinatorial model. Suppose that we have a deck of n cards, such that the j-th card has "$+j$" written on one side and "$-j$" on the other. The elements of S_n^B can then be identified with the possible rearrangements of stacks of cards; that is, a group element is a permutation of $[n] = \{1, 2, \ldots, n\}$ (the order of the cards in the stack) together with the sign information $[n] \to \{+, -\}$ (telling which side of each card is up). The Coxeter generators s_i, $1 \le i \le n - 1$, interchange the card in position i with that in position $i + 1$ in the stack (preserving orientation), and s_0 flips card 1 (the top card).

The group S_n^B has a subgroup, denoted S_n^D, with Coxeter graph

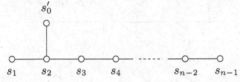

Here, $s_0' = s_0 s_1 s_0$. In terms of the card model this group consists of the stacks with an even number of turned-over cards (i.e., with minus side up). The generators s_i, $1 \le i \le n - 1$, are adjacent interchanges as before, and s_0 flips cards 1 and 2 together (as a package). See Section 8.2 for more about this group. □

Example 1.2.5 The circuit

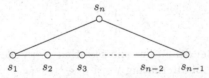

is the Coxeter graph of the group $\widetilde{S}_n$ of *affine permutations* of the integers. This is the group of all permutations x of the set $\mathbb{Z}$ such that

$$x(j + n) = x(j) + n, \quad \text{for all } j \in \mathbb{Z},$$

and

$$\sum_{i=1}^{n} x(i) = \binom{n+1}{2},$$

with composition as group operation. The Coxeter generators are the periodic adjacent transpositions $\widetilde{s}_i = \prod_{j \in \mathbb{Z}} (i + jn, i + 1 + jn)$ for $i = 1, \ldots, n$. See Section 8.3 for more about these infinite permutation groups. □

Example 1.2.6 The one-way infinite path

$$\underset{s_1}{\circ}\!\!\!-\!\!\!-\!\!\!-\!\!\!\underset{s_2}{\circ}\!\!\!-\!\!\!-\!\!\!-\!\!\!\underset{s_3}{\circ}\!\!\!-\!\!\!-\!\!\!-\!\!\!\underset{s_4}{\circ}\!\!\!-\!\!\!-\!\!\!-\cdots$$

is the Coxeter graph of the group of permutations with finite support of the positive integers (i.e., permutations that leave all but a finite subset fixed). The generators are the adjacent transpositions $s_i = (i, i+1)$, $1 \leq i$. $\square$

Example 1.2.7 *Dihedral groups.* Let L_1 and L_2 be straight lines through the origin of the Euclidean plane $\mathbb{E}^2$. Assume that the angle between them is $\frac{\pi}{m}$, for some integer $m \geq 2$. Let r_1 be the orthogonal reflection through L_1, and similarly for r_2. Then, $r_1 r_2$ is a rotation of the plane through the angle $\frac{2\pi}{m}$ and, hence, $(r_1 r_2)^m = e$.

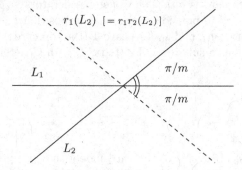

Let G_m be the group generated by r_1 and r_2. Simple geometric considerations show that G_m consists of the m rotations of the plane through angles $\frac{2\pi k}{m}$, $0 \leq k < m$, and these m rotations followed by the reflection r_1. Hence, $|G_m| = 2m$.

Now, define $I_2(m)$ to be the Coxeter group given by the Coxeter graph

$$\underset{s_1}{\circ}\!\!\!-\!\!\!-\!\!\!\overset{m}{-}\!\!\!-\!\!\!-\!\!\!\underset{s_2}{\circ}$$

Directly from the definition, one sees that every element of $I_2(m)$ can be represented as an alternating word $s_1 s_2 s_1 s_2 s_1 \ldots$ or $s_2 s_1 s_2 s_1 s_2 \ldots$ of length $\leq m$. (This includes the identity element represented by the empty word.) Since there are two such words of each positive length and the two words of length m represent the same group element, it follows that $|I_2(m)| \leq 2m$.

Since $r_1^2 = r_2^2 = (r_1 r_2)^m = e$, there is a surjective homomorphism $f : I_2(m) \rightarrow G_m$ extending $s_i \mapsto r_i$, $i = 1, 2$. We have seen that $|I_2(m)| \leq |G_m|$. Consequently, f must be an isomorphism.

The group $I_2(m)$ is called the *dihedral group* of order $2m$. Similarly, the group $I_2(\infty)$ (which is easily seen to be of infinite order) is called the *infinite dihedral group*. It arises as the group generated by orthogonal reflections r_1 and r_2 in lines whose angle is a nonrational multiple of π. $\square$

Example 1.2.8 *Symmetry groups of regular polytopes.* The symmetries of a regular m-gon in the plane are the m rotations and the m orthogonal reflections through a line of symmetry. Thus, the symmetry group is the dihedral group $I_2(m)$ discussed in the previous example.

The 2-dimensional regular polygons have their counterparts in higher dimensions among the regular polytopes. A d-dimensional convex polytope is *regular* if given two nested sequences of faces $F_0 \subseteq F_1 \subseteq \cdots \subseteq F_{d-1}$ $(\dim F_i = i)$, there is some isometry of d-space that maps the polytope onto itself and maps the first sequence of faces to the other one. It turns out that the symmetry groups of regular polytopes are always Coxeter groups.

The 3-dimensional regular polytopes are known since antiquity. They are the five Platonic solids. The higher-dimensional regular poytopes were classified by Schläfli in the mid-1800s. The full classification, with corresponding Coxeter groups as symmetry groups, is as follows:

Dimension	Regular Polytope	Coxeter group
d	simplex	A_d
d	cube	B_d
d	hyperoctahedron	B_d
2	m-gon	$I_2(m)$
3	dodecahedron	H_3
3	icosahedron	H_3
4	24-cell	F_4
4	120-cell	H_4
4	600-cell	H_4

Certain of these polytopes appear in pairs of dual polytopes that share the same symmetry group. The rest are self-dual.

Let us illustrate the link between polytope and group by having a look at the geometry of the dodecahedron. As illustrated in Figure 1.1, the dodecahedron has 15 planes of symmetry, and these planes subdivide its boundary into 120 congruent triangles. The orthogonal reflections through the planes generate the full symmetry group W, and this group acts simply transitively on the triangles.

Seen from this geometric perspective, what are the Coxeter generators? Fix any one of the 120 triangles and call this the "fundamental region." Take as S the three reflections in the "walls" of this triangle. Then, (W, S) is a Coxeter system.

The Coxeter matrix can be read from the geometry in the following way. Notice in Figure 1.1 that the dihedral angles in the corners of any triangle (in particular, of the fundamental region) are $\pi/2, \pi/3$, and $\pi/5$. The denominators are the defining numbers $m(s, s')$ of the Coxeter system of type H_3. $\square$

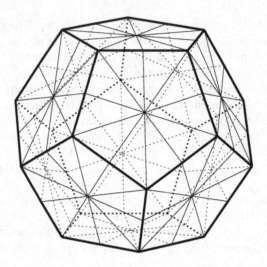

Figure 1.1. Symmetries of the dodecahedron.

Example 1.2.9 *Reflection groups.* The example of the dodecahedron shows how a certain finite Coxeter group can be realized as a group of geometric transformations generated by reflections. This is, in fact, true of *all* finite Coxeter groups, not only the ones related to regular polytopes. It is also true for the infinite Coxeter groups, although here one may need to relax the concept of reflection.

The two most important classes of infinite Coxeter groups are defined in terms of their realizations as reflection groups. These are the *affine* and *hyperbolic* Coxeter groups. We will not discuss the precise definitions here; suffice it to say that they arise from suitably defined reflections in affine (resp. hyperbolic) space. The irreducible groups of both types have been classified.

Here are a few low-dimensional examples that should convey the general idea. There are three affine irreducible Coxeter systems of rank 3: $\widetilde{A}_2$, $\widetilde{C}_2$, and $\widetilde{G}_2$ (cf. Appendix A1). The corresponding arrangements of reflecting lines are shown in Figure 1.2. There are infinitely many hyperbolic irreducible Coxeter systems of rank 3 (but only finitely many in higher ranks); the system of reflecting lines for one of them is shown in Figure 1.3.

Just as for the dodecahedron, the Coxeter generators for these affine and hyperbolic groups can be taken to be the reflections in the three lines that border a fundamental region. Furthermore, the Coxeter matrix of the group can be read off from the angles at which these lines pairwise meet. For instance, these angles are, in the case of Figure 1.3, respectively $\pi/2$, $\pi/3$, and $0 = \pi/\infty$. Again, the denominators are the edge labels of the Coxeter diagram ∘—∘$\overset{\infty}{—}$∘. □

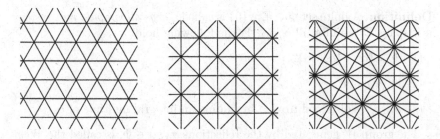

Figure 1.2. The $\widetilde{A}_2$, $\widetilde{C}_2$, and $\widetilde{G}_2$ tesselations of the affine plane.

Figure 1.3. The ○——○—$\overset{\infty}{}$—○ tesselation of the hyperbolic plane.

Example 1.2.10 *Weyl groups of root systems.* This example concerns a special class of groups generated by reflections, which is of great importance in the theory of semisimple Lie algebras. In that context, the following finite vector systems in Euclidean space $\mathbb{E}^d$ naturally arise. (Recall that $\mathbb{E}^d$ is the same as $\mathbb{R}^d$ endowed with a positive definite symmetric bilinear form.)

For $\alpha \in \mathbb{E}^d \backslash \{0\}$, let σ_α denote the orthogonal reflection in the hyperplane orthogonal to α. In particular, $\sigma_\alpha(\alpha) = -\alpha$.

Definition. A finite set $\Phi \subset \mathbb{E}^d \setminus \{0\}$ is called a *crystallographic root system* if it spans $\mathbb{E}^d$ and for all $\alpha, \beta \in \Phi$, the following hold:

(1) $\Phi \cap \mathbb{R}\alpha = \{\alpha, -\alpha\}$.

(2) $\sigma_\alpha(\Phi) = \Phi$.

(3) $\sigma_\alpha(\beta)$ is obtained from β by adding an integral multiple of α.

The group W generated by the reflections σ_α, $\alpha \in \Phi$, is called the *Weyl group* of Φ. It is (with a natural choice of generators) a Coxeter group. It is known that every finite irreducible Coxeter group, with the exception of H_3, H_4, and $I_2(m)$ for $m \neq 2, 3, 4, 6$, can appear as the Weyl group of a crystallographic root system. The classification of semisimple Lie algebras proceeds via the classification of their root systems and is thus closely linked to the classification of finite Coxeter groups.

In Chapter 4, we consider a more general concept of root system, available for *every* Coxeter group. The more restrictive crystallographic root systems will not reappear in this book. □

Example 1.2.11 *Matrix groups and BN-pairs.* Coxeter groups can arise as groups of matrices. For instance, $\circ\!\!-\!\!\circ\overset{\infty}{\text{---}}\!\circ \cong \mathrm{PGL}_2(\mathbb{Z})$, the projective general linear group (consisting of invertible 2×2 matrices with integer entries where a matrix and its negative are identified).

However, the classical matrix groups over fields are not themselves Coxeter groups. Nevertheless, Coxeter groups play an important role in their theory. In a precise sense, there sits "inside" such a matrix group G a certain Coxeter group, called its "Weyl group," which controls important features of the structure of G.

We now sketch the connection with matrix groups, via the axiomatization as "groups with BN-pair," due to Tits.

Definition. A pair B, N of subgroups of a group G is called a *BN-pair* (or *Tits system*) if the following hold:

(1) $B \cup N$ generates G, and $B \cap N$ is normal in N.

(2) $W \overset{\text{def}}{=} N/(B \cap N)$ is generated by some set S of involutions.

(3) $s \in S, w \in W \Rightarrow BsB \cdot BwB \subseteq BswB \cup BwB$.

(4) $s \in S \Rightarrow BsB \cdot BsB \neq B$.

It can be shown to follow from these axioms that the set S is uniquely determined and that the pair (W, S) is a Coxeter system. The group W is called the *Weyl group* and the number $|S|$ is the *rank* of the BN-pair $(G; B, N)$.

Notice that the double coset BwB is well-defined by the coset $w \in N/(B \cap N)$. Axioms (3) and (4) suggest the possibility of an induced algebraic structure on the set $\{BwB\}_{w \in W}$. This leads to the so-called "Hecke algebra" of (W, S), underlying Chapters 5 and 6.

The simplest example of a BN-pair is that of a group G acting doubly transitively on a set E of size ≥ 3. Let $x \neq y$ in E, and let $B \overset{\text{def}}{=} \text{Stab}(\{x\})$ and $N \overset{\text{def}}{=} \text{Stab}(\{x, y\})$. This is a BN-pair of rank 1. Conversely, given a BN-pair of rank 1, one can show that G acts doubly transitively on G/B.

A more instructive example is that of the general linear group $GL_n(\mathbb{F})$, consisting of invertible $n \times n$ matrices over a field $\mathbb{F}$. Here, there is a "canonical" BN-pair consisting of the group B of upper-triangular matrices and the group N of monomial matrices (having exactly one nonzero element in each row and each column). In this case, as is easy to see, the Weyl group is the group of permutation matrices. Hence, $GL_n(\mathbb{F})$ has a BN-pair of type A and of rank $n - 1$. The other classical matrix groups (orthogonal, symplectic, etc.) have BN-pairs, as do all groups "of Lie type." Hence, they can be classified according to the type of their Weyl group.

It is known that every finite irreducible Coxeter group, with the exception of H_3, H_4, and $I_2(m)$ for $m \neq 2, 3, 4, 6, 8$, can appear as the Weyl group of a finite group with a BN-pair. The finite groups with BN-pairs of rank ≥ 3 have been classified by Tits. They include all of the finite simple groups except the cyclic groups of prime order, the alternating groups A_n ($n \geq 5$), and the 26 sporadic groups.

One of the most important features of a group G with a BN-pair is the following:

Bruhat decomposition: $G = \uplus_{w \in W} BwB$.

Thus, the Weyl group W acts as indexing set for a partition of the group G into pairwise disjoint subsets (double cosets w.r.t. the subgroup B). In the cases of classical matrix groups, this partition induces a partial order on the set W. This partial order is the topic of Chapter 2. $\square$

1.3 A permutation representation

We now return to the program of deriving the basics of combinatorial Coxeter group theory "from scratch," and we continue the discussion where we left off in Section 1.1. Our immediate goal is to get as quickly as possible to the core combinatorial properties of a Coxeter group, such as the exchange property that is discussed in the next section. It turns out that the description of the group via its defining presentation (1.3) is ill suited for this purpose — one needs the added structure coming from some suitable concrete realization of the group.

This section describes a realization as a permutation group that leads quite quickly to the goal. This permutation representation is introduced here for the sole purpose of proving Theorem 1.4.3 of the following section;

it will not reappear in this form after that. (However, it will reappear in the guise of permutations of the root system of (W, S) in Section 4.4, for the connection see Exercise 4.7.)

Throughout this section, (W, S) denotes a Coxeter system. Let $T \overset{\text{def}}{=} \{wsw^{-1} : s \in S, w \in W\}$. The elements of T (i.e., the elements conjugate to some Coxeter generator) are called *reflections*. The definition shows that $S \subseteq T$ and that

$$t^2 = e, \quad \text{for all } t \in T. \tag{1.7}$$

The elements of S are sometimes called *simple reflections*.

Given a word $s_1 s_2 \ldots s_k \in S^*$, define

$$t_i \overset{\text{def}}{=} s_1 s_2 \ldots s_{i-1} s_i s_{i-1} \ldots s_2 s_1, \quad \text{for } 1 \le i \le k, \tag{1.8}$$

and the ordered k-tuple

$$\widehat{T}(s_1 s_2 \ldots s_k) \overset{\text{def}}{=} (t_1, t_2, \ldots, t_k). \tag{1.9}$$

For instance,

$$\widehat{T}(abca) = (a, aba, abcba, abcacba).$$

As stated earlier, we consider words in S^* also as elements in W (reading them as a product) without change of notation. Note that

$$t_i = (s_1 \ldots s_{i-1}) s_i (s_1 \ldots s_{i-1})^{-1} \in T, \tag{1.10}$$

$$t_i s_1 s_2 \ldots s_k = s_1 \ldots \widehat{s}_i \ldots s_k \quad (s_i \text{ omitted}), \tag{1.11}$$

and

$$s_1 s_2 \ldots s_i = t_i t_{i-1} \ldots t_1. \tag{1.12}$$

Lemma 1.3.1 *If* $w = s_1 s_2 \ldots s_k$, *with* k *minimal, then* $t_i \ne t_j$ *for all* $1 \le i < j \le k$.

Proof. If $t_i = t_j$ for some $i < j$ then $w = t_i t_j s_1 s_2 \ldots s_k = s_1 \ldots \widehat{s}_i \ldots \widehat{s}_j \ldots s_k$ (i.e., s_i and s_j deleted), which contradicts the minimality of k. $\square$

For $s_1 s_2 \ldots s_k \in S^*$ and $t \in T$, let

$$n(s_1 s_2 \ldots s_k; t) \overset{\text{def}}{=} \text{the number of times } t \text{ appears in } \widehat{T}(s_1 s_2 \ldots s_k). \tag{1.13}$$

Furthermore, for $s \in S$ and $t \in T$, let

$$\eta(s\,;t) \overset{\text{def}}{=} \begin{cases} -1, & \text{if } s = t, \\ +1, & \text{if } s \ne t. \end{cases} \tag{1.14}$$

Note that

$$(-1)^{n(s_1 s_2 \ldots s_k; t)} = \prod_{i=1}^{k} \eta(s_i\,; s_{i-1} \ldots s_1 t s_1 \ldots s_{i-1}). \tag{1.15}$$

We will consider the group $S(R)$ of all permutations of the set

$$R = T \times \{+1, -1\}.$$

For $s \in S$, define a mapping π_s of R to itself by

$$\pi_s(t, \varepsilon) \overset{\text{def}}{=} (sts, \varepsilon \, \eta(s\,;t)).$$

The computation

$$\pi_s^2(t, \varepsilon) = \pi_s(sts, \varepsilon \, \eta(s\,;t)) = (sstss, \varepsilon \, \eta(s\,;t) \, \eta(s\,;sts)) = (t, \varepsilon)$$

shows that $\pi_s \in S(R)$.

Theorem 1.3.2   *(i) The mapping $s \mapsto \pi_s$ extends uniquely to an injective homomorphism $w \mapsto \pi_w$ from W to $S(R)$.*

(ii) $\pi_t(t, \varepsilon) = (t, -\varepsilon)$, for all $t \in T$.

Proof. We verify the assertions in several steps.

(1) It was already shown that $\pi_s^2 = \text{id}_R$.

(2) Let $s, s' \in S$ and $m(s, s') = p \neq \infty$. We claim that

$$(\pi_s \, \pi_{s'})^p = \text{id}_R.$$

To prove this, let

$$s_i = \begin{cases} s', & \text{if } i \text{ is odd,} \\ s, & \text{if } i \text{ is even} \end{cases}$$

and let $\mathbf{s}$ denote the word $s_1 s_2 \ldots s_{2p} = s'ss'ss' \ldots s' s$. Let $\widehat{T}(\mathbf{s}) = (t_1, t_2, \ldots, t_{2p})$; that is,

$$t_i = s_1 s_2 \ldots s_i \ldots s_2 s_1 = (s's)^{i-1}s', \quad 1 \le i \le 2p.$$

Since $(s's)^p = e$, we have that

$$t_{p+i} = t_i, \quad 1 \le i \le p,$$

which implies that $n(\mathbf{s}\,;t)$ is even for all $t \in T$. Let

$$(t', \varepsilon') = (\pi_s \, \pi_{s'})^p(t, \varepsilon) = \pi_{s_{2p}} \, \pi_{s_{2p-1}} \ldots \pi_{s_1}(t, \varepsilon).$$

Then, $t' = s_{2p} \ldots s_1 \, t \, s_1 \ldots s_{2p} = t$, since $s_1 s_2 \ldots s_{2p} = (s's)^p = e$. Furthermore, using (1.15), we get

$$\varepsilon' = \varepsilon \prod_{i=1}^{2p} \eta(s_i\,;s_{i-1} \ldots s_1 \, t \, s_1 \ldots s_{i-1}) = \varepsilon \, (-1)^{n(\mathbf{s};t)} = \varepsilon.$$

So, the claim is proved.

(3) By the universality property and what has just been shown, the mapping $s \mapsto \pi_s$ extends to a homomorphism $w \mapsto \pi_w$ of W. If $w =$

$s_k\, s_{k-1} \ldots s_1$, we compute

$$\pi_w(t,\varepsilon) = \pi_{s_k}\pi_{s_{k-1}}\ldots\pi_{s_1}(t,\varepsilon)$$

$$= \left(s_k \ldots s_1\, t s_1 \ldots s_k, \varepsilon \prod_{i=1}^{k} \eta(s_i\,;\, s_{i-1}\ldots s_1\, t\, s_1 \ldots s_{i-1}) \right)$$

$$= (w\, t\, w^{-1}, \varepsilon\,(-1)^{n(s_1 s_2 \ldots s_k\,;\, t)}). \tag{1.16}$$

In particular, the parity of $n(s_1 s_2 \ldots s_k; t)$ only depends on w and t.

(4) Suppose that $w \neq e$. Choose an expression $w = s_k\, s_{k-1} \ldots s_1$ with k minimal, and let $\widehat{T}(s_1 s_2 \ldots s_k) = (t_1, t_2, \ldots, t_k)$. By Lemma 1.3.1, all t_i's are distinct, so $n(s_1 s_2 \ldots s_k; t_i) = 1$. Therefore, $\pi_w(t_i, \varepsilon) = (w t_i w^{-1}, -\varepsilon)$ for $1 \leq i \leq k$ by equation (1.16), which shows that $\pi_w \neq \mathrm{id}_R$. Hence, the homomorphism is injective.

(5) We show part (ii) of the theorem by induction on the size of a symmetric expression for t. Let

$$t = s_1 s_2 \ldots s_p \ldots s_2 s_1, \qquad s_i \in S.$$

The case $p = 1$ is clear by definition. Then, by induction,

$$\pi_{s_1 \ldots s_p \ldots s_1}(s_1 \ldots s_p \ldots s_1, \varepsilon)$$

$$= \pi_{s_1}\pi_{s_2 \ldots s_p \ldots s_2}(s_2 \ldots s_p \ldots s_2, \varepsilon\,\eta(s_1\,;\, s_1 \ldots s_p \ldots s_1))$$

$$= \pi_{s_1}(s_2 \ldots s_p \ldots s_2, -\varepsilon\,\eta(s_1\,;\, s_2 \ldots s_p \ldots s_2))$$

$$= (s_1 \ldots s_p \ldots s_1, -\varepsilon\,\eta^2(s_1\,;\, s_2 \ldots s_p \ldots s_2))$$

$$= (t, -\varepsilon).$$

$\square$

For $w \in W$ and $t \in T$, let

$$\eta(w\,;t) \overset{\text{def}}{=} (-1)^{n(s_1 s_2 \ldots s_k; t)}, \tag{1.17}$$

where $w = s_1 s_2 \ldots s_k$ is an arbitrary expression, $s_i \in S$. Step (3) of the proof shows that this is well defined. This definition extends that of $\eta(s\,;t)$ for $s \in S$ given in equation (1.14) and makes it possible to rewrite equation (1.16) as follows:

$$\pi_w(t,\varepsilon) = (w\, t\, w^{-1}, \varepsilon\,\eta(w^{-1}\,;t)). \tag{1.18}$$

1.4   Reduced words and the exchange property

In this section, we prove some fundamental combinatorial properties of the system of words representing any given element of a Coxeter group.

Let (W, S) be a Coxeter system. Each element $w \in W$ can be written as a product of generators:

$$w = s_1 s_2 \ldots s_k, \quad s_i \in S.$$

If k is minimal among all such expressions for w, then k is called the *length* of w (written $\ell(w) = k$) and the word $s_1 s_2 \ldots s_k$ is called a *reduced word* (or *reduced decomposition* or *reduced expression*) for w. As discussed in Section 1.1, we let "$s_1 s_2 \ldots s_k$" denote both the product of these generators (an element of W) and the word formed by listing them in this order (an element of the free monoid S^*).

The following is an immediate consequence of the Universality Property.

Lemma 1.4.1 *The map $\varepsilon : s \mapsto -1$, for all $s \in S$, extends to a group homomorphism $\varepsilon : W \to \{+1, -1\}$.* $\square$

Here are some basic properties of the length function.

Proposition 1.4.2 *For all $u, w \in W$:*

(i) $\varepsilon(w) = (-1)^{\ell(w)}$,

(ii) $\ell(uw) \equiv \ell(u) + \ell(w) \pmod{2}$,

(iii) $\ell(sw) = \ell(w) \pm 1$, for all $s \in S$,

(iv) $\ell(w^{-1}) = \ell(w)$,

(v) $|\ell(u) - \ell(w)| \leq \ell(uw) \leq \ell(u) + \ell(w)$,

(vi) $\ell(uw^{-1})$ is a metric on W.

Proof. Parts (i) – (iii) follow from Lemma 1.4.1. We leave the rest as exercises. $\square$

It is a consequence of Lemma 1.4.1 that the elements of even length form a subgroup of W of index 2. This is called the *alternating subgroup* (following the terminology of the symmetric group) or the *rotation subgroup* (following the terminology of finite reflection groups) of W.

We now come to the so-called "exchange property," which is a fundamental combinatorial property of a Coxeter group. In its basic version, appearing in the following section, the condition $t \in T$ in the statement below is weakened to $t \in S$, hence the adjective "strong" for the version given here.

Theorem 1.4.3 (Strong Exchange Property) *Suppose $w = s_1 s_2 \ldots s_k$ ($s_i \in S$) and $t \in T$. If $\ell(tw) < \ell(w)$, then $tw = s_1 \ldots \widehat{s_i} \ldots s_k$ for some $i \in [k]$.*

Proof. Recall the number $\eta(w; t) \in \{+1, -1\}$ defined in definition (1.17). We prove the equivalence of these two conditions:

(a) $\ell(tw) < \ell(w)$,

(b) $\eta(w;t) = -1$.

First, assume that $\eta(w;t) = -1$, and choose a reduced expression $w = s'_1 s'_2 \ldots s'_d$. Since $n(s'_1 s'_2 \ldots s'_d; t)$ is odd, we deduce that $t = s'_1 s'_2 \ldots s'_i \ldots s'_2 s'_1$ for some $1 \leq i \leq d$. Hence,

$$\ell(tw) = \ell(s'_1 \ldots \widehat{s'_i} \ldots s'_d) < d = \ell(w).$$

Second, assume that $\eta(w;t) = 1$. Using equation (1.18), we get

$$\begin{aligned} \pi_{(tw)^{-1}}(t, \varepsilon) = \pi_{w^{-1}} \pi_t(t, \varepsilon) &= \pi_{w^{-1}}(t, -\varepsilon) \\ &= (w^{-1}tw, -\varepsilon\, \eta(w;t)) = (w^{-1}tw, -\varepsilon). \end{aligned}$$

This means that $\eta(tw\,;t) = -1$, which by the implication (b) $\Rightarrow$ (a) already proved shows that $\ell(t\,tw) < \ell(tw)$, that is, $\ell(tw) > \ell(w)$.

The implication (a) $\Rightarrow$ (b) now concludes the proof. Suppose that $\ell(tw) < \ell(w)$. Then, since $\eta(w\,;t) = (-1)^{n(s_1 s_2 \ldots s_k; t)}$, we deduce that $n(s_1 s_2 \ldots s_k\,;t)$ is odd, and hence that $t = s_1 s_2 \ldots s_i \ldots s_2 s_1$ for some i. Therefore, $tw = s_1 \ldots \widehat{s_i} \ldots s_k$. $\square$

Corollary 1.4.4 *If $w = s_1 s_2 \ldots s_k$ is reduced and $t \in T$, then the follwing are equivalent:*

(a) $\ell(tw) < \ell(w)$,

(b) $tw = s_1 \ldots \widehat{s_i} \ldots s_k$, *for some $i \in [k]$,*

(c) $t = s_1 s_2 \ldots s_i \ldots s_2 s_1$, *for some $i \in [k]$.*

Furthermore, the index "i" appearing in (b) and (c) is uniquely determined.

Proof. The equivalence (b) $\Leftrightarrow$ (c) is easy to see (and does not require the hypothesis that $s_1 s_2 \ldots s_k$ is reduced). Uniqueness is provided by Lemma 1.3.1.

Theorem 1.4.3 shows that (a) $\Rightarrow$ (b), and the converse is obvious. $\square$

We now have the following definitions:

$$\begin{aligned} T_L(w) &\stackrel{\text{def}}{=} \{t \in T : \ell(tw) < \ell(w)\}, \\ T_R(w) &\stackrel{\text{def}}{=} \{t \in T : \ell(wt) < \ell(w)\}. \end{aligned} \tag{1.19}$$

In this notation "L" and "R" are mnemonic for "left" and "right." $T_L(w)$ is called the set of *left associated reflections* to w, and similarly for $T_R(w)$. Corollary 1.4.4 gives some useful characterizations of the set $T_L(w)$. Applying these to w^{-1}, we get the corresponding "mirrored" statements for $T_R(w)$, since clearly

$$T_R(w) = T_L(w^{-1}).$$

Corollary 1.4.5 $|T_L(w)| = \ell(w)$.

Proof. Let $w = s_1 s_2 \ldots s_k$, $k = \ell(w)$. Then, $T_L(w) = \{s_1 s_2 \ldots s_i \ldots s_2 s_1 : 1 \le i \le k\}$ by Corollary 1.4.4, and these elements are all distinct by Lemma 1.3.1. $\square$

We will quite often need to refer to the associated *simple* reflections, for which we introduce the following special notation and terminology:

$$D_L(w) \stackrel{\text{def}}{=} T_L(w) \cap S,$$
$$D_R(w) \stackrel{\text{def}}{=} T_R(w) \cap S. \tag{1.20}$$

$D_R(w)$ is called the *right descent set*, and similarly for $D_L(w)$. Their elements are called right (resp. left) *descents*. The reason for this terminology will become clear in Proposition 1.5.3, where we specialize to the symmetric groups, and even more so in Chapter 8. Note that, by symmetry, $D_R(w) = D_L(w^{-1})$.

Corollary 1.4.6 *For all $s \in S$ and $w \in W$, the following hold:*

(i) $s \in D_L(w)$ *if and only if some reduced expression for w begins with the letter s.*

(ii) $s \in D_R(w)$ *if and only if some reduced expression for w ends with the letter s.*

Proof. The "if" direction is clear. The opposite direction follows easily from Corollary 1.4.4 (or directly from Proposition 1.4.2(iii)). $\square$

We now come to another important consequence of the exchange property.

Proposition 1.4.7 (Deletion Property) *If $w = s_1 s_2 \ldots s_k$ and $\ell(w) < k$, then $w = s_1 \ldots \widehat{s_i} \ldots \widehat{s_j} \ldots s_k$ for some $1 \le i < j \le k$.*

Proof. Choose i maximal so that $s_i s_{i+1} \ldots s_k$ is not reduced. Then, $\ell(s_i s_{i+1} \ldots s_k) < \ell(s_{i+1} \ldots s_k)$ and hence, by Theorem 1.4.3,

$$s_i s_{i+1} \ldots s_k = s_{i+1} \ldots \widehat{s_j} \ldots s_k,$$

for some $i < j \le k$. Now multiply on the left by $s_1 s_2 \ldots s_{i-1}$. $\square$

Corollary 1.4.8 (i) *Any expression $w = s_1 s_2 \ldots s_k$ contains a reduced expression for w as a subword, obtainable by deleting an even number of letters.*

(ii) *Suppose $w = s_1 s_2 \ldots s_k = s_1' s_2' \ldots s_k'$ are two reduced expressions. Then, the set of letters appearing in the word $s_1 s_2 \ldots s_k$ equals the set of letters appearing in $s_1' s_2' \ldots s_k'$.*

(iii) *S is a minimal generating set for W; that is, no Coxeter generator can be expressed in terms of the others.*

Proof. Part (i) is a direct consequence of the deletion property.

To prove part (ii), suppose that $s_j \notin I \overset{\text{def}}{=} \{s'_1, \ldots, s'_k\}$ and that j is chosen to be minimal with this property. Then by Corollary 1.4.4,

$$s_1 s_2 \ldots s_j \ldots s_2 s_1 = s'_1 s'_2 \ldots s'_i \ldots s'_2 s'_1$$

for some i and, hence,

$$s_j = s_{j-1} \ldots s_1 s'_1 \ldots s'_i \ldots s'_1 s_1 \ldots s_{j-1} .$$

All letters in the word on the right-hand side belong to I, so taking a reduced subword, we get that $s_j \in I$, contradicting the hypothesis.

Part (iii) follows from (ii). $\square$

1.5 A characterization

The statement that the Exchange Property is fundamental in the combinatorial theory of Coxeter groups can be made very precise: It characterizes such groups! This is often a convenient way to prove that a given group is a Coxeter group, as will be exemplified in the case of the symmetric groups at the end of this section and for some other cases in Chapters 2 and 8.

Throughout this section we assume that W is an arbitrary group and that $S \subseteq W$ is a generating subset such that $s^2 = e$ for all $s \in S$. The concepts of *length* $\ell(w)$, $w \in W$, and *reduced expression* $s_1 s_2 \ldots s_k$, $s_i \in S$, can be defined just as earlier. Note however that properties (i) and (v) of Proposition 1.4.2 are no longer necessarily true; all we can say is that $|\ell(sw) - \ell(w)| \leq 1$, since $\ell(sw) = \ell(w)$ is now also a possibility. To say that such a pair (W, S) has the "Exchange Property" means the following:

Exchange Property. Let $w = s_1 s_2 \ldots s_k$ be a reduced expression and $s \in S$. If $\ell(sw) \leq \ell(w)$, then $sw = s_1 \ldots \widehat{s_i} \ldots s_k$ for some $i \in [k]$.

Similarly, we say that (W, S) has the "Deletion Property" if the statement of Proposition 1.4.7 is valid.

Theorem 1.5.1 *Let W be a group and S a set of generators of order 2. Then the following are equivalent:*

(i) (W, S) is a Coxeter system.

(ii) (W, S) has the Exchange Property.

(iii) (W, S) has the Deletion Property.

Proof. (i) $\Rightarrow$ (ii) This is a special case of Theorem 1.4.3.

(ii) $\Rightarrow$ (iii) The proof of Proposition 1.4.7 goes through to prove this implication, even if (W, S) is not (a priori) a Coxeter system.

(iii) $\Rightarrow$ (ii) Suppose $\ell(s s_1 \ldots s_k) \leq \ell(s_1 \ldots s_k) = k$. Then, by the Deletion Property, two letters can be deleted from $s s_1 \ldots s_k$, giving a new expression for sw. If s is not one of these letters, then $ss_1 \ldots s_k = ss_1 \ldots \widehat{s}_i \ldots \widehat{s}_j \ldots s_k$, would give $\ell(w) = \ell(s_1 \ldots \widehat{s}_i \ldots \widehat{s}_j \ldots s_k) < k$, a contradiction. Hence, s must be one of the deleted letters and we obtain $sw = ss_1 \ldots s_k = s_1 \ldots \widehat{s}_j \ldots s_k$.

(ii) $\Rightarrow$ (i) Let $s_1 s_2 \ldots s_r = e$ be a relation in a group with the Exchange Property. Then, r must be even, say $r = 2k$; this follows from the already established Deletion Property. So we can write our relation on the form

$$s_1 s_2 \ldots s_k = s_1' s_2' \ldots s_k'. \tag{1.21}$$

We must now prove that (1.21) is a consequence of the pairwise relations $(ss')^{m(s,s')} = e$, where $m(s, s')$ is defined as the order of the product ss' whenever this is finite. We frequently omit mention of the trivial relations $s^2 = e$, as we did when restating our relation on the form (1.21).

The proof is by induction on k, the case $k = 1$ being trivially correct. For simplicity, we will say that a given relation is "fine" if it can be derived from the relations $(ss')^{m(s,s')} = e$. Thus, we now assume that all relations of length less that $2k$ are fine.

Case 1: $s_1 s_2 \ldots s_k$ is not reduced. Then, there exists a position $1 \leq i < k$ such that $s_{i+1} s_{i+2} \ldots s_k$ is reduced but $s_i s_{i+1} s_{i+2} \ldots s_k$ is not. By the Exchange Property, we then have that

$$s_{i+1} s_{i+2} \ldots s_k = s_i s_{i+1} \ldots \widehat{s}_j \ldots s_k$$

for some $i < j \leq k$. This relation is of length $< 2k$ and hence fine. Plugging it into equation (1.21) yields

$$s_1 \ldots s_i s_i s_{i+1} \ldots \widehat{s}_j \ldots s_k = s_1' s_2' \ldots s_k'.$$

Hence, the factor $s_i s_i$ can be deleted, leaving a relation of length $< 2k$. Hence, equation (1.21) is fine.

Case 2: $s_1 s_2 \ldots s_k$ is reduced. We may assume that $s_1 \neq s_1'$, since otherwise equation (1.21) is trivially equivalent to a shorter relation. The Exchange Property shows that

$$s_1 s_2 \ldots s_i = s_1' s_1 \ldots s_{i-1} \tag{1.22}$$

for some $1 \leq i \leq k$. From equations (1.21) and (1.22), we conclude that

$$s_1 \ldots \widehat{s}_i \ldots s_k = s_2' s_3' \ldots s_k'.$$

Being of length $< 2k$, this relation is fine, and hence so is

$$s_1' s_1 \ldots \widehat{s}_i \ldots s_k = s_1' s_2' \ldots s_k'. \tag{1.23}$$

If $i < k$, then also equation (1.22) is fine, and we are done since equation (1.21) is obtained by substituting equation (1.22) in equation (1.23).

If $i = k$, we have to work a little harder. The relation (1.23), which we know is fine, now reads

$$s_1' s_1 \ldots s_{k-1} = s_1' s_2' \ldots s_k'.$$

Thus, it will suffice to show that

$$s_1's_1\ldots s_{k-1} = s_1s_2\ldots s_k \tag{1.24}$$

is fine. Let equation (1.24) take the role of equation (1.21) and repeat the whole argument of Case 2. If not settled along the way, the question will now (when we again reach "stage (1.24)") be reduced to whether

$$s_1's_1\ldots s_{k-1} = s_1s_1's_1\ldots s_{k-2}$$

is fine. Another iteration will then reduce to the relation

$$s_1s_1's_1\ldots s_{k-2} = s_1's_1s_1's_1\ldots s_{k-3},$$

and so on. Thus, in the end, the question will be reduced to the relation

$$s_1s_1's_1s_1'\ldots = s_1's_1s_1's_1\ldots,$$

which is, of course, implied by $(s_1s_1')^{m(s_1,s_1')} = e$. $\square$

The Exchange Property is stated above in its "left" version, since we are acting with s on the left of w. There is, of course, a "right" version (replace sw by ws), which is equivalent as a consequence of Theorem 1.5.1.

The rest of this section is devoted to a brief discussion of the symmetric groups from a Coxeter group point of view. The elements of S_n are permutations of the set $[n]$. See Appendix A3 for definitions and notational conventions pertaining to permutations.

As a set of generators for S_n we take $S = \{s_1,\ldots,s_{n-1}\}$, where $s_i \overset{\text{def}}{=} (i, i+1)$ for $i = 1,\ldots,n-1$. The effect of multiplying an element $x \in S_n$ on the right by the transposition s_i is that of interchanging the places of $x(i)$ and $x(i+1)$ in the complete notation of x. For instance, $31524 \cdot s_3 = 31254$. This makes it clear that $s_1,\ldots,s_{n-1}$ generate S_n.

Define the *inversion number* of $x \in S_n$ as the number of its inversions; that is,

$$\text{inv}(x) \overset{\text{def}}{=} \text{card}\{(i,j) : i < j,\ x(i) > x(j)\}. \tag{1.25}$$

Note that

$$\text{inv}(xs_i) = \begin{cases} \text{inv}(x) + 1, & \text{if } x(i) < x(i+1), \\ \text{inv}(x) - 1, & \text{if } x(i) > x(i+1). \end{cases} \tag{1.26}$$

Let $\ell_A(\cdot)$ denote the length function of S_n with respect to S.

Proposition 1.5.2 *Let $x \in S_n$. Then,*

$$\ell_A(x) = \text{inv}(x). \tag{1.27}$$

Proof. Since $\text{inv}(e) = \ell_A(e) = 0$, relation (1.26) implies that $\text{inv}(x) \leq \ell_A(x)$. The opposite inequality will be proved by induction on $\text{inv}(x)$.

If $\text{inv}(x) = 0$, then $x = 12\ldots n = e$ and equation (1.27) clearly holds. So let $x \in S_n$ and $k \in \mathbb{N}$ be such that $\text{inv}(x) = k + 1$. Then, $x \neq e$

and hence there exists $s \in S$ such that $\mathrm{inv}(x\,s) = k$ (otherwise relation (1.26) would imply that $x(1) < x(2) < \ldots < x(n)$ and hence that $x = e$). This, by the induction hypothesis, implies that $\ell_A(x\,s) \leq k$ and hence that $\ell_A(x) \leq k + 1$. Therefore, $\ell_A(x) \leq \mathrm{inv}(x)$. $\square$

As a consequence we obtain the following combinatorial description of the right descent set of an element of S_n. Note that this agrees with the definition of right descent set of a permutation given in Appendix A3.4.

Proposition 1.5.3 *Let* $x \in S_n$. *Then,*

$$D_R(x) = \{s_i \in S : x(i) > x(i+1)\}. \tag{1.28}$$

Proof. By the definitions and Proposition 1.5.2, we have that

$$D_R(x) = \{s \in S : \mathrm{inv}(x\,s) < \mathrm{inv}(x)\},$$

so (1.28) follows from (1.26). $\square$

In the classification of finite irreducible Coxeter groups, the Coxeter system determined by the graph

is denoted by A_{n-1} (cf. Appendix A1).

Proposition 1.5.4 (S_n, S) *is a Coxeter system of type* A_{n-1}.

Proof. We show that the pair (S_n, S) has the Exchange Property (in its "right" version), and this, by Theorem 1.5.1, implies that (S_n, S) is a Coxeter system. Since

$$\begin{cases} s_i s_j = s_j s_i, & \text{if } |i - j| \geq 2 \\ s_i s_j s_i = s_j s_i s_j, & \text{if } |i - j| = 1, \end{cases} \tag{1.29}$$

one sees that the type of (S_n, S) is A_{n-1}.

Let $i, i_1, \ldots, i_p \in [n-1]$ and suppose that

$$\ell_A(s_{i_1} \ldots s_{i_p} s_i) < \ell_A(s_{i_1} \ldots s_{i_p}). \tag{1.30}$$

We want to show that there exists a $j \in [p]$ such that

$$s_{i_1} \ldots s_{i_p} s_i = s_{i_1} \ldots \widehat{s}_{i_j} \ldots s_{i_p}. \tag{1.31}$$

Let $x = s_{i_1} \ldots s_{i_p}$, $b = x(i)$, and $a = x(i+1)$. By Proposition 1.5.2, we know that relation (1.30) means that $b > a$. Therefore, a is to the left of b in the complete notation of the identity, but is to the right of b in that of x. Hence, there exists $j \in [p]$ such that a is to the left of b in $s_{i_1} \ldots s_{i_{j-1}}$ but a is to the right of b in $s_{i_1} \ldots s_{i_j}$. Hence, the complete notation of $s_{i_1} \ldots \widehat{s}_{i_j} \ldots s_{i_p}$ is the same as that of $s_{i_1} \ldots s_{i_p}$, except that a and b are interchanged. This, by the definitions of x, a, and b, implies equation (1.31). $\square$

There are, of course, other ways of proving that (S_n, S) is a Coxeter system of type A_{n-1}; see, for example, Exercises 1.5 and 4.2.

Exercises

1. The relation $bcbcacababcacbcabacbabacbc = e$ holds in the Coxeter group

Determine x.

2. Show that there exist Coxeter systems (W, S) and (W', S') with $|S| \neq |S'|$ such that W is isomorphic to W' as abstract groups.
[*Hint:* Consider the dihedral group of order 12.]

3. Let $a_1 = (12)(34)$, $a_2 = (12)(45)$, and $a_3 = (14)(23)$ be elements of S_5. Compute the orders of all products $a_i a_j$. Conclude the existence of a surjective homomorphism of H_3 onto the alternating subgroup of S_5. Then, prove that the alternating subgroups of H_3 and S_5 are isomorphic as abstract groups.

4. Show with a direct and elementary argument that the Coxeter diagram D of a finite irreducible Coxeter group must satisfy the following graph-theoretic requirements:

   (a) D is a tree.
   (b) D has at most one vertex of degree 3 and none of higher degree.
   (c) D has at most one marked (i.e., label ≥ 4) edge.
   (d) If D has a degree 3 vertex, then all edges are unmarked.

   [*Hint:* In the presence of any violation, exhibit an element of infinite order.]

5. For $n \geq 2$, let $(A_{n-1}, \{\bar{s}_1, \ldots, \bar{s}_{n-1}\})$ be a Coxeter system of type A_{n-1} (so $A_1 \subseteq A_2 \subseteq \cdots$).

   (a) Show that there is a unique group homomorphism $f : A_{n-1} \to S_n$ such that $f(\bar{s}_i) = s_i$, for $i \in [n-1]$, and that f is surjective.
   (b) For $x \in A_{n-1} \setminus A_{n-2}$, let $p \overset{\text{def}}{=} \min\{\ell(y) : y \in A_{n-2}x\}$, and $\bar{s}_{j_1} \ldots \bar{s}_{j_p} \in A_{n-2}x$. Show that $(j_1, \ldots, j_p) = (n-1, n-2, \ldots, n-p)$.
   (c) Deduce from (b) that there are at most n right cosets of A_{n-2} in A_{n-1} and hence, by induction on n, that f is injective.

6. Identify the set of transpositions $T = \{(i, j) : 1 \leq i < j \leq n\}$ in S_n with the edges of the complete graph K_n.

   (a) Show that $A \subseteq T$ is a minimal generating set for S_n if and only if A is a spanning tree of K_n.
   (b) show that $A \subseteq T$ is a system of Coxeter generators for S_n if and only if the corresponding tree is linear.

(c) Is it generally true for Coxeter groups (W, S) that every minimal generating set $A \subseteq T$ is of the same cardinality as S? [*Hint*: Consider the dihedral groups.]

7. Prove the statement made in Example 1.2.6.

8. Prove the converse of Lemma 1.3.1: If $t_i \neq t_j$ for all $i \neq j$, then $s_1 s_2 \ldots s_k$ is reduced.

9. Complete the proof of Proposition 1.4.2.

10. Show that every $t \in T$ has a palindromic reduced expression; that is, one can write $t = s_1 s_2 \ldots s_k$ with $k = \ell(t)$ and $s_i = s_{k-i}$ for $1 \leq i \leq k$, $s_i \in S$.

11. Show that $u \neq w \Rightarrow T_L(u) \neq T_L(w)$, for all $u, w \in W$.

12. Prove that $T_R(uw) = T_R(w) \,\triangle\, w^{-1} T_R(u) w$.

13. Show that the following conditions on $u, w \in W$ are equivalent:

    (a) $\ell(uw) = \ell(u) + \ell(w)$,
    (b) $T_R(u) \cap T_L(w) = \varnothing$,
    (c) $T_R(uw) = T_R(w) \,\cup\, w^{-1} T_R(u) w$,
    (d) $T_R(uw) = T_R(w) \,\uplus\, w^{-1} T_R(u) w$.

14. Let W be a group and S a set of generators of W of order 2. Show that (W, S) is a Coxeter system if and only if $\ell(sw) = \ell(ws') = \ell(w) + 1 \geq \ell(sws')$ implies $sw = ws'$, for all $s, s' \in S$ and $w \in W$.

15. Let (W, S) be a Coxeter group with $S = \{s_1, \ldots, s_n\}$. Show that the alternating subgroup of W is generated by the elements $\{s_i s_n\}_{i=1}^{n-1}$.

16. Let (W, S) be a Coxeter group, and let $K_1, \ldots, K_k$ be the connected components of the subgraph of W's Coxeter diagram obtained if all evenly labeled edges (i.e., with $m(s, s')$ even) are removed. Furthermore, let A and C be the alternating subgroup and the commutator subgroup of W, respectively, and let W_{K_i} be the parabolic subgroup generated by K_i (defined in Section 2.4). Show the following:

    (a) Two generators $s, s' \in S$ are conjugate (as elements of W) if and only if they belong to the same component K_i.
    (b) $C \subseteq A$.
    (c) $A_i \subseteq C$, where A_i is the alternating subgroup of W_{K_i}.
    (d) $C = A$ if and only if $k = 1$.

17. Preserve the notation from Exercise 16, and let $\varphi : W \to \mathbb{Z}_2^k$ be the mapping $\varphi(w) = (\varepsilon_1, \ldots, \varepsilon_k)$ defined by letting ε_i be the (mod 2)-number of occurences of elements from K_i in some expression $w = s_1 s_2 \ldots s_q$, $s_j \in S$. Show the following:

    (a) φ is well defined (not dependent on choice of expression $s_1 s_2 \ldots s_q$).

(b) φ is a group homomorphism.

(c) $C = \operatorname{Ker}\varphi$.

18. Let (W, S) be a Coxeter system, and let W' be a subgroup of W generated by a some subset of T (such subgroups are called *reflection subgroups*). Let

$$\Sigma(W') \overset{\text{def}}{=} \{t \in T \cap W' \,:\, \ell(t't) > \ell(t), \text{ for all } t' \in T \cap W' \setminus \{t\}\}.$$

Show that $(W', \Sigma(W'))$ is a Coxeter system.

19. Let (W, S) be a Weyl group. Define a directed graph Γ with vertex set W and edges $w \to sws$ for all $w \in W$ and $s \in S$ such that $\ell(w) \geq \ell(sws)$. (In case of equal length, the edge will be directed both ways.) Call $w \in W$ *conjugacy-reduced* if $\ell(w') = \ell(w)$ for all w' that can be reached on a directed path from w.

Let C be a conjugacy class of W, and put $\ell_C \overset{\text{def}}{=} \min\{\ell(w) : w \in C\}$.

(a) Show that if $w \in C$ and w is conjugacy-reduced, then $\ell(w) = \ell_C$.

(b)* Is the same true for general Coxeter groups?

Notes

For the basics of combinatorial group theory, see, e.g., the books by Coxeter and Moser [166] and Stillwell [518]. The equivalence of properties 2 and 3 in Section 1.1 is a special case of Dyck's theorem [518, p. 45].

The study of Coxeter groups can be said to go back to Greek antiquity, the symmetries and classification of regular polytopes being part of the theory. Finite Coxeter groups in the sense of definition (1.3) were first studied and classified by Coxeter [162, 163] and Witt [555]. The main references for the core algebraic and geometric aspects of Coxeter groups are the books by Bourbaki [79] and Humphreys [306]. The newcomer to this area is especially recommended to study [306], where a good and accessible account of the classification and the various geometric realizations as reflection groups is given.

The examples sketchily given in Section 1.2 unfortunately cannot convey a fair idea of the immense mathematical territory where Coxeter groups arise. More information can be found in the books by Billey and Lakshmibai [47], Brown [106], Carter [110, 111], Fulton [248], Hiller [295], Humphreys [304, 305], Kac [317], Kumar [334], McMullen and Schulte [396], and Tits [538] and the papers by Hazewinkel et al. [289] and Vinberg [546, 547].

The permutation representation as a tool for quick access to the exchange property appears in Bourbaki [79]. The strong exchange property was introduced by Verma [545]; it can also be found somewhat earlier in geometric form (for Coxeter complexes) in the work of Tits [534]. The fact

that Coxeter groups are characterized by the exchange property appears
to be due to Matsumoto [392].

Exercise 14 is from Deodhar [182], where one can also find several
    characterizations of Coxeter systems.
Exercise 18. See Deodhar [185] and Dyer [202].
Exercise 19(a). See Geck and Pfeiffer [261]. A positive answer to part (b)
    has been conjectured by A. Cohen. See also Gill [266].

2
Bruhat order

One of the most remarkable aspects of Coxeter groups, from a combinatorial point of view, is the crucial role that is played in their theory by a certain partial order structure. This partial order arises in a multitude of ways in algebra and geometry — for instance, from cell decompositions of certain varieties.

Although order structure is used in some other parts of algebra, the role of Bruhat order for the study of Coxeter groups, and the deep combinatorial and geometric properties of this order relation, are unique. In this chapter we introduce Bruhat order and derive its basic combinatorial properties.

2.1 Definition and first examples

Let (W, S) be a Coxeter system and $T = \{wsw^{-1} : w \in W, s \in S\}$ its set of reflections. The poset terminology that we use is reviewed in Appendix A2.2.

Definition 2.1.1 *Let $u, w \in W$. Then*

(i) $u \xrightarrow{t} w$ means that $u^{-1}w = t \in T$ and $\ell(u) < \ell(w)$.

(ii) $u \rightarrow w$ means that $u \xrightarrow{t} w$ for some $t \in T$.

(iii) $u \leq w$ means that there exist $w_i \in W$ such that

$$u = u_0 \rightarrow u_1 \rightarrow \cdots \rightarrow u_{k-1} \rightarrow u_k = w.$$

The Bruhat graph *is the directed graph whose nodes are the elements of* W *and whose edges are given by (ii).* Bruhat order *is the partial order relation on the set* W *defined by (iii).*

The following observations are immediate:

(i) $u < w$ implies $\ell(u) < \ell(w)$.

(ii) $u < ut \Leftrightarrow \ell(u) < \ell(ut)$, for all $u \in W$ and $t \in T$.

(iii) The identity element e satisfies $e \leq w$ for all $w \in W$ (any reduced decomposition $w = s_1 \ldots s_q$ induces $e \to s_1 \to s_1 s_2 \to \cdots \to s_1 \ldots s_q = w$).

Since Bruhat order is the transitive closure of the primitive relations $u \xrightarrow{t} ut$, it might seem at this stage that the concept favors multiplication on the right-hand side. However, this impression is false; see Exercise 1.

Example 2.1.2 Consider the dihedral group $I_2(4) \cong B_2$ with Coxeter graph

$$\overset{4}{\underset{a \qquad b}{\circ\!\!-\!\!-\!\!-\!\!-\!\!\circ}}$$

Then, $T = \{a, b, aba, bab\}$ and the group has the following diagram under Bruhat order:

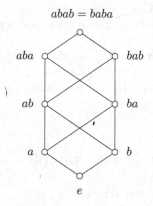

Figure 2.1. Bruhat order of B_2.

To obtain the Bruhat graph of B_2, direct all edges of Figure 2.1 upward and add the edges $e \to aba$, $e \to bab$, $a \to abab$, and $b \to baba$ (cf. Figure 8.3). □

Bruhat order of a general dihedral group $I_2(m)$ has the same structure: a graded poset of length m with two elements on each rank level except the top and bottom, and with all order relations between successive rank levels. Bruhat order of the symmetric groups is discussed later in this section and that of B_n and several other Coxeter groups in Chapter 8. Figure 2.2 is the diagram of B_3. (The pattern of solid and dashed/dotted edges will be explained following Corollary 2.4.5.)

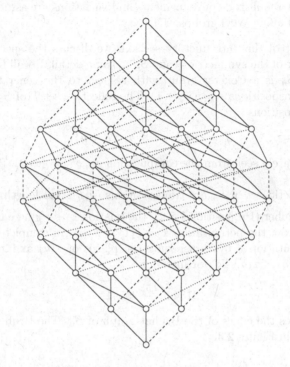

Figure 2.2. Bruhat order of B_3.

Example 2.1.3 *Inclusion order of Bruhat cells.* Here we sketch the algebraic–geometric origin of Bruhat order in a central case — that of cell decompositions of flag manifolds.

Let $G = GL_n(\mathbb{C})$, and let B be the subgroup of upper-triangular matrices. It turns out that the quotient G/B has the structure of a smooth projective algebraic variety, called the *flag variety*. Because of the Bruhat decomposition (see Example 1.2.11), we have an induced decomposition

$$G/B = \uplus_{w \in W} BwB/B, \qquad (2.1)$$

where $W = S_n$. The pieces $C_w \overset{\text{def}}{=} BwB/B$ of this decomposition are called *Bruhat cells* (or *Schubert cells*). The decomposition (2.1) is actually a cell decomposition in the sense of topology (a CW complex) and we may speak of the topological closure $\overline{C_w}$ of a Bruhat cell. How are these closed cells arranged? The elegant answer is

$$\overline{C_u} \subseteq \overline{C_w} \quad \Leftrightarrow \quad u \leq w;$$

that is, the combinatorial pattern of inclusion of Bruhat cells determines Bruhat order on S_n.

This example is only the tip of an iceberg. Similar decompositions of varieties and a multitude of refinements and variations are associated with all finite and affine Weyl groups. □

In the rest of this introductory section, we discuss the special case of Bruhat order of the symmetric group S_n in some detail. Recall from Chapter 1 that S_n is a Coxeter group with respect to the generating set of adjacent transpositions $s_i = (i, i+1)$. The reflection set T of S_n is the set of all transpositions

$$T = \{(a, b) : 1 \le a < b \le n\},$$

as is immediately seen from the computation $x s_i x^{-1} = (x(i), x(i+1))$, for $x \in S_n$.

Since reflections t in S_n are transpositions (a, b), and length equals the inversion number (Proposition 1.5.2), the relation $x \xrightarrow{(a,b)} y$ here means that one moves from the permutation $x = x(1)\ldots x(n)$ (in complete notation) to the permutation y by transposing the places of $x(a)$ and $x(b)$, where $a < b$ and $x(a) < x(b)$. So, for instance,

$$21543 \xrightarrow{(2,5)} 23541.$$

This describes the edges of the Bruhat graph of S_n. The Bruhat graph of S_3 is shown in Figure 2.3.

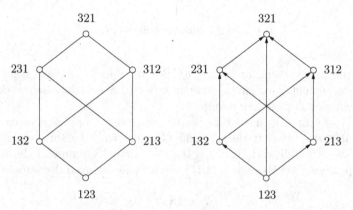

Figure 2.3. Bruhat order and Bruhat graph of S_3.

Lemma 2.1.4 *Let $x, y \in S_n$. Then, x is covered by y in Bruhat order if and only if $y = x \cdot (a, b)$ for some $a < b$ such that $x(a) < x(b)$ and there does not exist any c such that $a < c < b$, $x(a) < x(c) < x(b)$.*

Proof. If $y = x \cdot (a, b)$ with the stated properties, then $\mathrm{inv}(y) = \mathrm{inv}(x) + 1$; hence, we have a Bruhat covering. Suppose conversely that $y = x \cdot (a, b)$, $a < b$, and $\mathrm{inv}(y) > \mathrm{inv}(x)$. Then, $x(a) < x(b)$. If $x(a) < x(c) < x(b)$ for some $a < c < b$, then $x < x \cdot (a, c) < y$, so $x < y$ is not a covering. $\square$

Using Lemma 2.1.4, one can easily work out small cases, such as that of $n = 4$ shown in Figure 2.4. However, for larger values of n, it is compu-

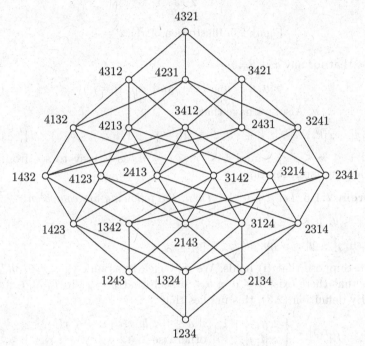

Figure 2.4. Bruhat order of S_4.

tationally hard to see from the definition (or from Lemma 2.1.4) whether two permutations are comparable in Bruhat order. For example,

$$\textit{are } x = 368475912 \textit{ and } y = 694287531 \textit{ comparable?} \qquad (2.2)$$

Fortunately, there exist effective criteria. We now present one such rule; for another one see Theorem 2.6.3.

For $x \in S_n$, let

$$x[i, j] \overset{\mathrm{def}}{=} |\{a \in [i] : x(a) \geq j\}| \qquad (2.3)$$

for $i, j = 1, \ldots, n$. One may interpret this function as follows. Represent the permutation x by placing dots at the points with coordinates $(a, x(a))$ in the plane, for $1 \leq a \leq n$. Then, $x[i, j]$ counts the number of dots in the northwest corner above the point with coordinates (i, j). So, for example, if $x = 31524$, then $x[1, 3] = 1$, $x[3, 3] = 2$, and $x[4, 2] = 3$. See Figure 2.5.

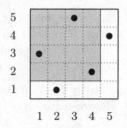

Figure 2.5. Illustration of $x[4, 2] = 3$.

Note that for any $x \in S_n$,

$$x[n, i] = n + 1 - i \quad \text{and} \quad x[i, 1] = i \tag{2.4}$$

for $i = 1, \ldots, n$. Also, we have that

$$x[i, j] - x[k, j] - x[i, l] + x[k, l] = |\{a \in [k + 1, i] : j \le x(a) < l\}| \tag{2.5}$$

for all $1 \le k \le i \le n$ and $1 \le j \le l \le n$, as is easy to see from the dot-counting interpretation.

Theorem 2.1.5 *Let $x, y \in S_n$. Then, the following are equivalent:*

(i) $x \le y$.

(ii) $x[i, j] \le y[i, j]$, *for all* $i, j \in [n]$.

Proof. Suppose that (i) holds. We may clearly assume that $x \to y$. This means that there exist $1 \le a < b \le n$ such that $y = x \cdot (a, b)$ and $x(a) < x(b)$. By definition (2.3), this implies that

$$y[i, j] = \begin{cases} x[i, j] + 1, & \text{if } a \le i < b, \, x(a) < j \le x(b), \\ x[i, j], & \text{otherwise} \end{cases} \tag{2.6}$$

and (ii) follows.

Assume now that (ii) holds. For brevity, let $M(i, j) \stackrel{\text{def}}{=} y[i, j] - x[i, j]$ for all $i, j \in [n]$. If $M(i, j) = 0$ for all $i, j \in [n]$, then $x = y$. Let $(a_1, b_1) \in [n]^2$ be such that $M(a_1, b_1) > 0$ and $M(i, j) = 0$ for all $(i, j) \in [1, a_1] \times [b_1, n] \setminus \{(a_1, b_1)\}$. Then, $y(a_1) = b_1$ and $x(a_1) < b_1$. Now, let $(a_2, b_2) \in [n]^2$ be the bottom right corner of a maximal positive connected submatrix of M having (a_1, b_1) as the upper left corner. It follows from equation (2.4) that $a_2 < n$ and $b_2 > 1$. Because of maximality, there exist $c \in [a_1, a_2]$ and $d \in [b_2, b_1]$ such that $M(c, b_2 - 1) = 0$ and $M(a_2 + 1, d) = 0$. Hence,

$$M(a_2 + 1, b_2 - 1) - M(c, b_2 - 1) - M(a_2 + 1, d) + M(c, d) > 0. \tag{2.7}$$

This, by (2.5), implies that

$$|\{e \in [c + 1, a_2 + 1] : y(e) \in [b_2 - 1, d - 1]\}| > 0.$$

So let $(a_0, b_0) \in [c + 1, a_2 + 1] \times [b_2 - 1, d - 1]$ be such that $y(a_0) = b_0$. Then, $a_1 < a_0$ and $y(a_1) = b_1 > b_0 = y(a_0)$. It follows that $z \to y$, where

$z \stackrel{\text{def}}{=} y \cdot (a_1, a_0)$. However, $x[i, j] \leq z[i, j]$ for all $i, j \in [n]$ by equation (2.6) and our choice of (a_2, b_2). Hence, by induction, we get that $x \leq z$ and, therefore, $x \leq y$. □

Let us illustrate the preceding theorem by answering question (2.2). Since in that example $x[1, 6] < y[1, 6]$ and $x[4, 3] > y[4, 3]$, we find that x and y are incomparable in Bruhat order.

2.2 Basic properties

Throughout this section, let (W, S) be a Coxeter system. Here, we will establish two fundamental properties of Bruhat order: the "subword property" and the "chain property." They are consequences of the following lemma. By a *subword* of a word $s_1 s_2 \ldots s_q$ we mean a word of the form $s_{i_1} s_{i_2} \ldots s_{i_k}$, where $1 \leq i_1 < \cdots < i_k \leq q$.

Lemma 2.2.1 *For $u, w \in W$, $u \neq w$, let $w = s_1 s_2 \ldots s_q$ be reduced, and suppose that some reduced expression for u is a subword of $s_1 s_2 \ldots s_q$. Then, there exists $v \in W$ such that the following hold:*

(i) $v > u$.

(ii) $\ell(v) = \ell(u) + 1$.

(iii) *Some reduced expression for v is a subword of $s_1 s_2 \ldots s_q$.*

Proof. Of all reduced subword expressions

$$u = s_1 \ldots \widehat{s_{i_1}} \ldots \widehat{s_{i_k}} \ldots s_q, \qquad 1 \leq i_1 < \cdots < i_k \leq q,$$

choose one such that i_k is minimal. Let

$$t = s_q s_{q-1} \ldots s_{i_k} \ldots s_{q-1} s_q.$$

Then, $ut = s_1 \ldots \widehat{s_{i_1}} \ldots \widehat{s_{i_{k-1}}} \ldots s_{i_k} \ldots s_q$, so $\ell(ut) \leq \ell(u) + 1$. We claim that, in fact, $ut > u$. If so, $v = ut$ satisfies (i) – (iii), and we are done.

Suppose on the contrary that $ut < u$. Then, by the Strong Exchange Property, either

$$t = s_q s_{q-1} \ldots s_p \ldots s_{q-1} s_q, \qquad \text{for some } p > i_k \tag{2.8}$$

or

$$t = s_q \ldots \widehat{s_{i_k}} \ldots \widehat{s_{i_d}} \ldots s_r \ldots \widehat{s_{i_d}} \ldots \widehat{s_{i_k}} \ldots s_q, \qquad \text{for some } r < i_k, r \neq i_j. \tag{2.9}$$

In the first case,

$$\begin{aligned} w &= w t^2 \\ &= (s_1 s_2 \ldots s_q)(s_q \ldots s_{i_k} \ldots s_q)(s_q \ldots s_p \ldots s_q) \\ &= s_1 \ldots \widehat{s_{i_k}} \ldots \widehat{s_p} \ldots s_q, \end{aligned}$$

which contradicts $\ell(w) = q$. Similarly, in the second case,

$$
\begin{aligned}
u &= ut^2 \\
&= (s_1 \ldots \widehat{s}_{i_1} \ldots \widehat{s}_{i_k} \ldots s_q)(s_q \ldots \widehat{s}_{i_k} \ldots s_r \ldots \widehat{s}_{i_k} \ldots s_q)(s_q \ldots s_{i_k} \ldots s_q) \\
&= s_1 \ldots \widehat{s}_{i_1} \ldots \widehat{s}_r \ldots s_{i_k} \ldots s_q,
\end{aligned}
$$

which contradicts the minimality of i_k.

[*Remark*: The notation in equation (2.9) and on the preceding line may seem to indicate that $r > i_1$; however, $r < i_1$ is also possible as should be clear from the context.] $\square$

Theorem 2.2.2 (Subword Property) *Let* $w = s_1 s_2 \ldots s_q$ *be a reduced expression. Then,*

$$u \leq w \quad \Leftrightarrow \quad \text{there exists a reduced expression}$$
$$u = s_{i_1} s_{i_2} \ldots s_{i_k}, \ \ 1 \leq i_1 < \ldots < i_k \leq q.$$

Proof. ($\Rightarrow$) Suppose that $u = x_0 \overset{t_1}{\to} x_1 \overset{t_2}{\to} \ldots \overset{t_m}{\to} x_m = w$. Then, $x_{m-1} = wt_m = s_1 \ldots \widehat{s}_i \ldots s_q$ for some i by the Strong Exchange Property and, similarly, $x_{m-2} = x_{m-1}t_{m-1} = s_1 \ldots \widehat{s}_i \ldots \widehat{s}_j \ldots s_q$, and so on for $x_{m-3}, x_{m-4}, \ldots$. Finally, we obtain an expression for u that is a subword of $s_1 s_2 \ldots s_q$ (with m deleted letters). By the Deletion Property (Proposition 1.4.7), this contains a reduced subword, which is the sought-after expression for u.

($\Leftarrow$) This direction follows from Lemma 2.2.1 via induction on $\ell(w) - \ell(u)$. $\square$

Corollary 2.2.3 *For* $u, w \in W$, *the following are equivalent:*

(i) $u \leq w$.

(ii) *Every reduced expression for* w *has a subword that is a reduced expression for* u.

(iii) *Some reduced expression for* w *has a subword that is a reduced expression for* u. $\square$

Corollary 2.2.4 *Bruhat intervals* $[u, w]$ *are finite (even if S is infinite). In fact,* $\text{card}[u, w] \leq 2^{\ell(w)}$.

Proof. There are $2^{\ell(w)}$ subwords of any reduced expression for w, and there is an injective map from $[u, w]$ into the set of those subwords. $\square$

Corollary 2.2.5 *The mapping* $w \mapsto w^{-1}$ *is an automorphism of Bruhat order (i.e.,* $u \leq w \Leftrightarrow u^{-1} \leq w^{-1}$).

Proof. The subword relation is unaffected by reversing all expressions. (*Remark*: The result is also easy to derive directly from Definition 2.1.1, see Exercise 1). $\square$

Theorem 2.2.6 (Chain Property) *If $u < w$, there exists a chain $u = x_0 < x_1 < \cdots < x_k = w$ such that $\ell(x_i) = \ell(u) + i$, for $1 \leq i \leq k$.*

Proof. This follows directly from Lemma 2.2.1 and the Subword Property. $\square$

We will use the symbol "$u \lhd w$" or "$w \rhd u$" to denote a covering in Bruhat order. Thus, by the Chain Property, $u \lhd w$ means that $u < w$ and $\ell(u) + 1 = \ell(w)$. The Chain Property shows that Bruhat order is a graded poset whose rank function is the length function. The same is true of every Bruhat interval $[u, w]$.

The following simple technical tool, allowing Bruhat relations to be "lifted" (see Figure 2.6), is very useful. In fact, it can be shown to characterize Bruhat order (see Exercise 14).

Proposition 2.2.7 (Lifting Property) *Suppose $u < w$ and $s \in D_L(w) \setminus D_L(u)$. Then, $u \leq sw$ and $su \leq w$.*

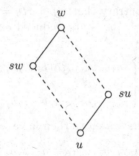

Figure 2.6. The lifting property.

Proof. Let $\alpha \prec \beta$ here denote the subword relation between a word β and a subword α. Choose a reduced decomposition $sw = s_1 s_2 \ldots s_q$. Then, $w = s\, s_1 s_2 \ldots s_q$ is also reduced, and there exists a reduced subword

$$u = s_{i_1} s_{i_2} \ldots s_{i_k} \prec s\, s_1 s_2 \ldots s_q.$$

Now, $s_{i_1} \neq s$ since $su > u$; hence,

$$s_{i_1} s_{i_2} \ldots s_{i_k} \prec s_1 s_2 \ldots s_q \Rightarrow u \leq sw$$

and

$$s\, s_{i_1} s_{i_2} \ldots s_{i_k} \prec s s_1 s_2 \ldots s_q \Rightarrow su \leq w.$$

$\square$

The Lifting Property has the following corollaries about local configurations in Bruhat order:

Corollary 2.2.8   (i) *For $s \in S$, $t \in T$, $s \neq t$: $w \lhd sw, tw \Rightarrow sw, tw \lhd stw$.*

(ii) For $s, s' \in S$: $w \lhd sw, ws' \Rightarrow$ either $sw, ws' \lhd sws'$ or $w = sws'$. □

Recall that a poset P is said to be *directed* if for any $u, w \in P$, there exists $z \in P$ such that $u, w \leq z$.

Proposition 2.2.9 *Bruhat order is a directed poset.*

Proof. We will use induction on $\ell(u) + \ell(w)$, the $\ell(u) + \ell(w) = 0$ case being trivially correct. Choose $s \in S$ so that $su < u$ (we may assume that $\ell(u) > 0$). By induction, there exists $x \in W$ such that $su, w \leq x$. By the Lifting Property, $sx < x \Rightarrow u \leq x$ and $sx > x \Rightarrow u \leq sx$. Hence, in the first case, $\{u, w\}$ has the upper bound x and, in the second case, it has the upper bound sx. □

We end this section with a technical lemma that is needed later.

Lemma 2.2.10 *Suppose that $x < xt$ and $y < ty$, for $x, y \in W$, $t \in T$. Then, $xy < xty$.*

Proof. Suppose to the contrary that $xy > xty = t'xy$, where $t' = xtx^{-1}$. Let $x = s_1 \ldots s_k$ and $y = s'_1 \ldots s'_q$ be reduced expressions. Then, by the Strong Exchange Property,

$$t'xy = \begin{cases} a_1 \ldots \widehat{a_i} \ldots a_k b_1 \ldots b_j \ldots b_q \\ \text{or} \\ a_1 \ldots a_i \ldots a_k b_1 \ldots \widehat{b_j} \ldots b_q \end{cases}$$

for some i and j. In the first case, we then have

$$xt = t'x = a_1 \ldots \widehat{a_i} \ldots a_k < a_1 \ldots a_i \ldots a_k = x,$$

and in the second, $xty = t'xy = xb_1 \ldots \widehat{b_j} \ldots b_q$, and hence

$$ty = b_1 \ldots \widehat{b_j} \ldots b_q < b_1 \ldots b_j \ldots b_q = y.$$

Thus, in both cases, we reach a contradiction. □

2.3   The finite case

If W is finite, directedness (Proposition 2.2.9) just says that W has a greatest element. This unique element of maximal length is customarily denoted "w_0." In this section, we derive some of its basic properties. We also discuss automorphisms of Bruhat order.

Proposition 2.3.1 *(i) If W is finite, there exists an element $w_0 \in W$ such that $w \leq w_0$ for all $w \in W$.*

(ii) Conversely, suppose that (W, S) has an element x such that $D_L(x) = S$. Then, W is finite and $x = w_0$.

Proof. Existence and uniqueness were already motivated.

For part (ii), we prove that $u \leq x$ for all $u \in W$ by induction on length. If $u \neq e$, we can find $s \in S$ such that $su < u$. By induction, $su \leq x$ and this can be lifted (Proposition 2.2.7) to $u \leq x$. Thus, $W = [e, x]$, which is finite. $\square$

Proposition 2.3.2 *The top element w_0 of a finite group has the following properties:*

(i) $w_0^2 = e$.

(ii) $\ell(ww_0) = \ell(w_0) - \ell(w)$, *for all $w \in W$.*

(iii) $T_L(ww_0) = T \setminus T_L(w)$, *for all $w \in W$.*

(iv) $\ell(w_0) = |T|$.

Proof. (i) Since $\ell(w_0^{-1}) = \ell(w_0)$, uniqueness of w_0 implies that $w_0^{-1} = w_0$.

(ii) The inequality $\geq$ follows from $\ell(w^{-1}) + \ell(ww_0) \geq \ell(w_0)$. For the opposite inequality, we will use induction on $\ell(w_0) - \ell(w)$, starting with $w = w_0$. For $w < w_0$, choose $s \in S$ such that $w < sw$. This is possible according to Proposition 2.3.1(ii). Then,

$$\ell(ww_0) \leq \ell(sww_0) + 1 \leq \ell(w_0) - \ell(sw) + 1$$
$$= \ell(w_0) - (\ell(w) + 1) + 1 = \ell(w_0) - \ell(w).$$

(iii) A consequence of (ii) is that for every $t \in T$ and $w \in W$: $tw < w \Leftrightarrow tww_0 > ww_0$.

(iv) Putting $w = e$ in equation (iii) and using Corollary 1.4.5, we get $\ell(w_0) = |T_L(w_0)| = |T|$. $\square$

Corollary 2.3.3   (i) $\ell(w_0w) = \ell(w_0) - \ell(w)$, *for all $w \in W$.*

(ii) $\ell(w_0ww_0) = \ell(w)$, *for all $w \in W$.*

Proof. $\ell(w_0w) = \ell(w^{-1}w_0) = \ell(w_0) - \ell(w^{-1}) = \ell(w_0) - \ell(w)$. $\square$

Translation and conjugation by the top element w_0 induce (anti)automorphisms of Bruhat order, as can be seen from Proposition 2.3.2(ii) and Corollary 2.3.3.

Proposition 2.3.4 *For Bruhat order on a finite Coxeter group, the following hold:*

(i) $w \mapsto ww_0$ *and* $w \mapsto w_0w$ *are antiautomorphisms.*

(ii) $w \mapsto w_0ww_0$ *is an automorphism.* $\square$

The top element w_0 in the symmetric group S_n is the "reversal permutation" $i \mapsto n + 1 - i$. Hence, the effects of the mappings of Proposition

2.3.4 in S_5 are exemplified by

$$41523 \longrightarrow \begin{cases} (ww_0) & 32514, & \text{reverse the places,} \\ (w_0w) & 25143, & \text{reverse the values,} \\ (w_0ww_0) & 34152, & \text{reverse places and values.} \end{cases}$$

The mapping $x \mapsto w_0xw_0$ is an inner group automorphism of W, and $w_0Sw_0 = S$ by Corollary 2.3.3(ii). Hence, $x \mapsto w_0xw_0$ preserves *all* Coxeter group structure, in particular its action on S induces an automorphism of the Coxeter diagram. Conversely, every such *diagram automorphism* (there may be others) amounts to a renaming of the Coxeter generators preserving all relations and therefore induces an automorphism of Bruhat order, also in the infinite case.

Temporarily dropping the assumption of this section that W be finite, let us ask for a description of *all* automorphisms of Bruhat order. For rank 2, the answer is nontypical and quite special (see Exercise 2). For rank > 2 however, the following result of van den Hombergh [298] and Waterhouse [551] provides the answer.

Theorem 2.3.5 *Suppose that (W, S) is irreducible and $|S| \geq 3$. If $\varphi : W \to W$ is an automorphism of Bruhat order and $\varphi(s) = s$ for all $s \in S$, then either $\varphi(x) = x$ for all $x \in W$ or $\varphi(x) = x^{-1}$ for all $x \in W$.* □

Corollary 2.3.6 *If (W, S) is irreducible and $|S| \geq 3$, then the automorphism group of Bruhat order is generated by the diagram automorphisms and the mapping $x \mapsto x^{-1}$.* □

For instance, the diagram of type A_n, $n \geq 2$, has a unique nontrivial automorphism, and this, in fact, induces the mapping $x \mapsto w_0xw_0$. Hence, the automorphism group of Bruhat order of the symmetric group S_n, $n \geq 4$, is the dihedral group of order 4 generated by $x \mapsto w_0xw_0$ and $x \mapsto x^{-1}$. In fact, it follows from Corollary 2.3.6 that the automorphism group of Bruhat order of a finite irreducible Coxeter group of rank ≥ 3 (other than D_4) is always either $\mathbb{Z}_2$ (no nontrivial diagram automorphism) or $\mathbb{Z}_2 \times \mathbb{Z}_2$.

The mapping $x \mapsto w_0xw_0$ is an inner group automorphism of order ≤ 2 and also a Bruhat order automorphism. When is it the identity? Equivalently, when does w_0 belong to the center of W? For the answer, see Exercise 4.10.

2.4 Parabolic subgroups and quotients

For $J \subseteq S$, let W_J be the subgroup of W generated by the set J. Subgroups of Coxeter groups (W, S) of this form are called *parabolic*. In this section, we will describe their basic combinatorial properties. Subscripts "J" appended to familiar symbols will always refer to such a subgroup; for example, "$\ell_J(\cdot)$" refers to the length function of W_J with respect to the system J of involutory generators.

Proposition 2.4.1   (i) (W_J, J) is a Coxeter group.

(ii) $\ell_J(w) = \ell(w)$, for all $w \in W_J$.

(iii) $W_I \cap W_J = W_{I \cap J}$.

(iv) $\langle W_I \cup W_J \rangle = W_{I \cup J}$.

(v) $W_I = W_J \Rightarrow I = J$.

Proof. Let $w \in W_J$. By definition, $w = s_1 s_2 \ldots s_q$, for some $s_i \in J$, and by the Deletion Condition, we may assume that this is reduced in W, and hence in W_J. This proves (ii). Since $\ell_J(w) = \ell(w)$, the Exchange Property holds in (W_J, J) as a special case of the Exchange Property in (W, S). Hence, (i) follows from Theorem 1.5.1.

Statements (iii) and (v) are implied by Corollary 1.4.8, and (iv) is elementary. $\square$

It is a consequence that the parabolic subgroups form a sublattice of W's subgroup lattice that is isomorphic to the Boolean lattice 2^S. The Coxeter diagram for (W_J, J) is obtained by removing all nodes in $S \setminus J$ and their incident edges from the diagram for (W, S).

If W_J is finite it has a top element (Proposition 2.3.2), which will be denoted as follows:

$$w_0(J) \overset{\text{def}}{=} \text{ top element of } W_J. \tag{2.10}$$

Thus for instance, $w_0(\emptyset) = e$ and $w_0(S) = w_0$ (if W is finite). One sees from Corollary 1.4.8(ii) that $w_0(I) \neq w_0(J)$ if $I \neq J$.

Parabolic subgroups have complete systems of combinatorially distinguished coset representatives; namely, each coset has a unique member of shortest length. To discuss these and other systems of coset representatives, we need the following concepts.

Definition 2.4.2 For $I \subseteq J \subseteq S$, let

$$\mathcal{D}_I^J \overset{\text{def}}{=} \{w \in W : I \subseteq D_R(w) \subseteq J\},$$
$$W^J \overset{\text{def}}{=} \mathcal{D}_{\emptyset}^{S \setminus J},$$
$$\mathcal{D}_I \overset{\text{def}}{=} \mathcal{D}_I^I.$$

Sets of the form $\mathcal{D}_I^J$ are called (right) *descent classes*. The special descent classes $W^J = \{w \in W : ws > w \text{ for all } s \in J\}$ are called *quotients* for a reason that will soon become clear.

Lemma 2.4.3 *An element w belongs to W^J if and only if no reduced expression for w ends with a letter from J.*

Proof. This follows from Corollary 1.4.6. $\square$

Proposition 2.4.4 *Let $J \subseteq S$. Then, the following hold:*

(i) *Every $w \in W$ has a unique factorization $w = w^J \cdot w_J$ such that $w^J \in W^J$ and $w_J \in W_J$.*

(ii) *For this factorization, $\ell(w) = \ell(w^J) + \ell(w_J)$.*

Proof. (Existence) Choose $s_1 \in J$ so that $ws_1 < w$, if such s_1 exists. Continue choosing $s_i \in J$ so that $ws_1 \ldots s_i < ws_1 \ldots s_{i-1}$ as long as such s_i can be found. The process must end after at most $\ell(w)$ steps. If it ends with $w_k = ws_1 \ldots s_k$, then $w_k s > w_k$ for all $s \in J$; that is, $w_k \in W^J$. Now, let $v = s_k s_{k-1} \ldots s_1 \in W_J$. We have that $w = w_k v$, and, by construction, $\ell(w) = \ell(w_k) + k$.

(Uniqueness) Suppose that $w = uv = xy$, with $u, x \in W^J$ and $v, y \in W_J$. Let $u = s_1 s_2 \ldots s_k$ and $vy^{-1} = s'_1 s'_2 \ldots s'_q$ with the first expression reduced, $s_i \in S$, $s'_j \in J$. Then,

$$x = uvy^{-1} = s_1 s_2 \ldots s_k s'_1 s'_2 \ldots s'_q.$$

From this, we can extract a reduced subword for x. It cannot end in some letter s'_j, since $x \in W^J$. Hence, it is a subword of $s_1 s_2 \ldots s_k$, and $x \leq u$ follows. By symmetry, $u \leq x$. So, $u = x$ and $v = y$ follow. $\square$

The following statements are immediate consequences of Proposition 2.4.4:

Corollary 2.4.5   (i) *Each left coset wW_J has a unique representative of minimal length. The system of such minimal coset representatives is $W^J = \mathcal{D}_\emptyset^{S \backslash J}$.*

(ii) *If W_J is finite, then each left coset wW_J has a unique representative of maximal length. The system of such maximal coset representatives is $\mathcal{D}_J^S$.* $\square$

Proposition 2.4.4 and its corollary are illustrated in Figure 2.2, where $W = B_3$ and $W_J = B_2$. The six cosets of B_2 are drawn with solid lines, the eight translates of the six-element chain W^J (including $\mathcal{D}_J^S$) are drawn with dashed lines, and the additional Bruhat edges are drawn with dotted lines. Notice how the direct product poset $W^J \times W_J$ (i.e., the poset induced by the solid and dashed lines only) sits as a scaffolding inside the poset W.

Suppose that $S = \{s_1, s_2, \ldots, s_n\}$, and for $i = 1, \ldots, n$, let

$$Q_i = (W_{\{s_1, \ldots, s_i\}})^{\{s_1, \ldots, s_{i-1}\}}.$$

Thus, $Q_1 = W_{\{s_1\}} = \{e, s_1\}$, and Q_i for $i \geq 2$ is the system of minimal left coset representatives of $W_{\{s_1,\ldots,s_{i-1}\}}$ in $W_{\{s_1,\ldots,s_i\}}$. Repeated application of Proposition 2.4.4 then gives the following:

Corollary 2.4.6 *The product map $Q_1 \times \cdots \times Q_n \to W$, defined by*

$$(q_1, q_2, \ldots, q_n) \mapsto q_n q_{n-1} \cdots q_1,$$

is a bijection satisfying $\ell(q_n q_{n-1} \cdots q_1) = \ell(q_1) + \ell(q_2) + \cdots + \ell(q_n)$. $\square$

The preceding constructions can of course be mirrored. There is a complete system

$$^J W \overset{\text{def}}{=} \{w \in W : D_L(w) \subseteq S \setminus J\} = (W^J)^{-1} \tag{2.11}$$

of minimal length representatives of right cosets $W_J w$. Every $w \in W$ can be uniquely factorized,

$$w = w_J \cdot {}^J w, \quad \text{where } w_J \in W_J \text{ and } {}^J w \in {}^J W, \tag{2.12}$$

and then

$$\ell(w) = \ell(w_J) + \ell({}^J w).$$

Furthermore, an element w belongs to $^J W$ if and only if no reduced expression for w begins with a letter from J.

Let us now exemplify the preceding in the case of the symmetric groups. The parabolic subgroups of S_n are often called *Young subgroups* in the literature. Since there is no essential loss of generality, we describe for notational simplicity only the *maximal* parabolic subgroups and their quotients. All permutations $x \in S_n$ will be denoted here in complete notation as $x = x_1 x_2 \ldots x_n$ where $x_i = x(i)$. For $k \in [n-1]$, let

$$S_n^{(k)} \overset{\text{def}}{=} \{x \in S_n : x_1 < \cdots < x_k \text{ and } x_{k+1} < \cdots < x_n\}. \tag{2.13}$$

Recall our convention in the case of S_n to let s_i denote the adjacent transposition $(i, i+1)$. The following is then clear from the definitions.

Lemma 2.4.7 *Let $J \overset{\text{def}}{=} S \setminus \{s_k\}$. Then,*

$$(S_n)_J = \mathrm{Stab}([k]) \cong S_k \times S_{n-k} \quad \text{and} \quad (S_n)^J = S_n^{(k)}. \quad \square$$

The reader will have no trouble figuring out what the corresponding statements are for a general subset $J \subseteq S$. The following is an example that should make the general situation clear. To simplify notation, we identify s_i with i. Say $n = 6$ and $J = \{2, 3, 5\}$. Then, $(S_n)^J = \{x \in S_6 : x_2 < x_3 < x_4,\ x_5 < x_6\}$.

Again, let $J = S \setminus \{s_k\}$. The map $x \mapsto x^J$ from S_n to $S_n^{(k)}$ is easy to describe; namely, x^J is obtained from x by first rearranging the values $x_1, \ldots, x_k$ so that they appear in increasing order in the places $1, \ldots, k$, and then similarly for $x_{k+1}, \ldots, x_n$.

Again, the general case should be quite clear. We illustrate with an example: say $n = 7$, $x = 7125346$, and $J = \{1, 2, 4, 6\}$. Then, x^J is obtained from x by rearranging in increasing order the numbers $\{7, 1, 2\}$, $\{5, 3\}$, and $\{4, 6\}$; hence, $x^J = 1273546$. Similarly (see Exercise 4), $^J x$ is obtained by permuting the elements $\{1, 2, 3\}$, $\{4, 5\}$, and $\{6, 7\}$ so that they form increasing subsequences; hence, $^J x = 6124357$.

Bruhat order restricted to the quotient $S_n^{(k)}$ has a simple description.

Proposition 2.4.8 *For $x, y \in S_n^{(k)}$, the following are equivalent:*

(i) $x \leq y$.

(ii) $x_i \leq y_i$, *for $1 \leq i \leq k$.*

(iii) $x_i \geq y_i$, *for $k + 1 \leq i \leq n$.*

Proof. (i) $\Rightarrow$ (ii). This is an immediate consequence of Theorem 2.1.5.

(ii) $\Rightarrow$ (i). Suppose that $x_j < y_j$ for some $1 \leq j \leq k$ and $x_i = y_i$ for all $j + 1 \leq i \leq k$. Then, $x_j + 1 = x_m$ for some $m > k$ (since $x_j + 1 \leq y_j < y_{j+1} = x_{j+1}$ if $j < k$). Let $x' = (x_j, x_j + 1) \cdot x = x \cdot (j, m)$. Then, $x_i' \leq y_i$ for $1 \leq i \leq k$ and $x \lhd x'$, so we are done by induction on $\sum_{i=1}^{k} (y_i - x_i)$.

The equivalence of (ii) and (iii) is left to the reader. $\square$

It is clear from definition (2.13) that an element $x \in S_n^{(k)}$ is determined by the *set* $\{x_1, x_2, \ldots, x_k\}$, so we can make the identification

$$S_n^{(k)} \leftrightarrow \binom{[n]}{k}.$$

Thus, Proposition 2.4.8 shows that the maximal parabolic quotient $S_n^{(k)}$ under Bruhat order can be identified with the family of k-subsets of $[n]$ under product order of k-tuples. This latter poset is sometimes denoted $L(k, n - k)$ in the literature. It is isomorphic to the lattice of Ferrers diagrams that fit into a $k \times (n - k)$ box, ordered by inclusion. The Bruhat poset $S_6^{(3)} \cong L(3, 3)$ is depicted in Figure 2.7.

2.5   Bruhat order on quotients

Quotients W^J (and, more generally, descent classes) have interesting poset structure under Bruhat order. See Figures 2.7 and 2.8 for examples. The latter depicts W^J for $(W, S) = E_6$ and $J \subset S$ such that $(W_J, J) = D_5$.

Much of the structure found in Bruhat order on all of W is inherited when restricting to the subposet W^J. This can to some extent be understood as transfer of structure via the projection maps defined as follows.

Let $J \subseteq S$. Adhering to the notation used in Proposition 2.4.4, define a mapping

$$P^J : W \to W^J \tag{2.14}$$

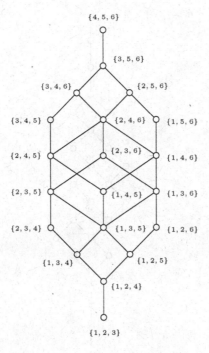

Figure 2.7. The Bruhat poset $S_6^{(3)}$.

by $P^J(w) \overset{\text{def}}{=} w^J$. In other words, the *projection map* P^J sends w to its minimal coset representative modulo W_J.

Proposition 2.5.1 *The map P^J is order-preserving.*

Proof. Suppose that $w_1 \leq w_2$ in W. We will show that $w_1^J \leq w_2^J$ by induction on $\ell(w_2)$.

To begin with, note that $w_1^J \leq w_1 \leq w_2$. Hence, if $w_2^J = w_2$, we are done. If not, then there exists some $s \in J$ such that $w_2 s < w_2$. The relation $w_1^J \leq w_2$ can then be lifted (Proposition 2.2.7) to $w_1^J \leq w_2 s$. By induction, $w_1^J \leq (w_2 s)^J = w_2^J$. $\square$

Corollary 2.5.2 *Suppose $u \in W^J$, $w \in W$ and $u \lhd w$. Then, either $w = us$, for some $s \in J$, or $w \in W^J$.*

Proof. If $w \notin W^J$, then $u \leq P^J(w) < w$. $\square$

Corollary 2.5.3 *W^J is a directed poset.*

Proof. This follows from Propositions 2.2.9 and 2.5.1. $\square$

In particular, if W^J is finite, then it has a unique maximal element, which will be denoted

$$w_0^J \overset{\text{def}}{=} \text{top element of } W^J. \tag{2.15}$$

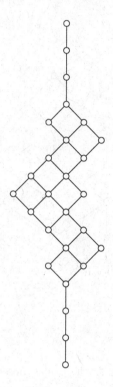

Figure 2.8. The Bruhat poset E_6 modulo D_5.

Thus, in that case, $w \leq w_0^J$ for all $w \in W^J$.

If W is finite, we have the following relation between the various top elements:

$$w_0 = w_0^J \, w_0(J), \quad \ell(w_0) = \ell(w_0^J) + \ell(w_0(J)), \tag{2.16}$$

or, equivalently, $w_0^J = (w_0)^J$ and $w_0(J) = (w_0)_J$. Since w_0 and $w_0(J)$ are involutions, it follows that w_0^J is an involution if and only if w_0 and $w_0(J)$ commute. In particular,

$$w_0^J \text{ is an involution if } w_0 \in \text{center } (W). \tag{2.17}$$

This sufficient (but not necessary) condition is often fulfilled; see Exercise 4.10.

Quotients in finite groups have a remarkable combinatorial symmetry.

Proposition 2.5.4 *Let (W, S) be finite, $J \subseteq S$. Then,*

$$\alpha : x \mapsto w_0 x w_0(J)$$

defines an antiautomorphism $\alpha : W^J \to W^J$ of Bruhat order (that is, $x \leq y \Leftrightarrow \alpha(x) \geq \alpha(y)$).

Proof. We have that $x \in W^J \Rightarrow xw_0(J) \in \mathcal{D}_J^S \Rightarrow w_0xw_0(J) \in W^J$, by Proposition 2.4.4 and Corollary 2.3.3. Furthermore, if $x, y \in W^J$, then $x \leq y \Rightarrow xw_0(J) \leq yw_0(J) \Rightarrow w_0xw_0(J) \geq w_0yw_0(J)$. $\square$

The following is a stronger version of Theorem 2.2.6 (the $J = \emptyset$ case). It can be further generalized to descent classes $\mathcal{D}_I^J$ (see Exercise 23).

Theorem 2.5.5 (Chain Property) *If $u < w$ in W^J, then there exist elements $w_i \in W^J$, $\ell(w_i) = \ell(u) + i$, for $0 \leq i \leq k$, such that $u = w_0 < w_1 < \cdots < w_k = w$.*

Proof. It suffices to construct w_1; the rest follows via induction. Let $w = s_1 s_2 \ldots s_q$, and take a reduced subword expression

$$u = s_1 \ldots \widehat{s}_{i_1} \ldots \widehat{s}_{i_k} \ldots s_q, \qquad 1 \leq i_1 < \cdots < i_k \leq q,$$

such that i_k is minimal. Let $w_1 = s_1 \ldots \widehat{s}_{i_1} \ldots \widehat{s}_{i_{k-1}} \ldots s_{i_k} \ldots s_q$. The proof of Lemma 2.2.1 shows that $u \lhd w_1 \leq w$. It remains only to check that $w_1 \in W^J$. If not, then, by Corollary 2.5.2, $w_1 = us$, where $s = s_q s_{q-1} \ldots s_{i_k} \ldots s_{q-1} s_q \in J$. However, then $ws = s_1 \ldots \widehat{s}_{i_k} \ldots s_q < w$, contradicting that $w \in W^J$. $\square$

Corollary 2.5.6 *All maximal chains from u to w in W^J have the same length.* $\square$

We will later make use of the notation

$$[u, w]^J \overset{\text{def}}{=} [u, w] \cap W^J \tag{2.18}$$

for intervals in W^J. The corollary shows that such intervals are graded posets. In particular, if W^J is finite, then $W^J = [e, w_0^J]^J$ is a graded poset.

2.6   A criterion

Bruhat order on proper quotients W^J is induced by the order on W. It is an interesting and useful fact that the converse is also true: If one knows the order relation of sufficiently many projections of two group elements onto quotients, then one can deduce the original order relation of the elements.

Theorem 2.6.1 *Let $J_i \subseteq S$, $i \in E$, be a family of subsets and $I \overset{\text{def}}{=} \bigcap_{i \in E} J_i$. Let $u \in W^I$ and $w \in W$. Then,*

$$u \leq w \quad \Longleftrightarrow \quad P^{J_i}(u) \leq P^{J_i}(w), \text{ for all } i \in E.$$

Proof. The forward direction is known from Proposition 2.5.1. The backward direction will be proved by induction on $\ell(w)$. If $\ell(w) = 0$, then $P^{J_i}(u) = e$ for all $i \in E$, which implies that $u \in \bigcap_{i \in E} W_{J_i} = W_I$. Since $W_I \cap W^I = \{e\}$, we deduce that $u = e$, so $u = w$ in this case.

Assume now that $\ell(w) > 0$ and choose $s \in S$ such that $sw < w$. We make the following claim:

Claim: $P^J(u) \leq P^J(w) \Rightarrow \begin{cases} P^J(su) \leq P^J(sw), & \text{if } su < u, \\ P^J(u) \leq P^J(sw), & \text{if } su > u. \end{cases}$

Here, $J \subseteq S$ is arbitrary. Based on this claim, the proof can be concluded as follows.

If $su < u$, then $P^{J_i}(su) \leq P^{J_i}(sw)$ for all i, which by induction (since $\ell(sw) < \ell(w)$ and $su \in W^I$) implies that $su < sw$. However, this can be lifted (Proposition 2.2.7) to $u \leq w$. If on the other hand $su > u$, then $P^{J_i}(u) \leq P^{J_i}(sw)$ for all i, which in the same way via induction gives $u \leq sw < w$.

So, it remains to prove the claim. For this, we need to have a handle on what the relevant projections are in various cases. We begin by assembling this information in statements (2.19) and (2.20). From now on, we write $x^J = P^J(x)$, etc., to simplify notation.

$$sx < x \text{ and } sx^J < x^J \quad \Rightarrow \quad (sx)^J = sx^J. \qquad (2.19)$$

If sx^J had a reduced expression ending in a letter from J, then so would x^J; hence, $sx^J \in W^J$. Furthermore, $sx = sx^J x_J$ with $x_J \in W_J$, and statement (2.19) follows.

$$sx < x \text{ and } sx^J > x^J \quad \Rightarrow \quad (sx)^J = x^J. \qquad (2.20)$$

We have that $sx = sx^J x_J$, but $\ell(sx) < \ell(x) = \ell(x^J) + \ell(x_J) < \ell(sx^J) + \ell(x_J)$. Hence, $sx^J \notin W^J$, so (by Corollary 2.5.2) $sx^J = x^J s'$ for some $s' \in J$. We conclude that $sx = sx^J x_J = x^J s' x_J$ with $s' x_J \in W_J$; hence, statement (2.20) holds.

We are now ready to prove the claim. There will be four cases, all tacitly referring to statements (2.19) and (2.20).

Case 1: $su < u$, $sw^J < w^J$. We have to prove that $su^J \leq sw^J$ (if $su^J < u^J$) or that $u^J \leq sw^J$ (if $su^J > u^J$). This follows from $u^J \leq w^J$ by lifting.

Case 2: $su < u$, $sw^J > w^J$. This time we want $su^J \leq w^J$ (if $su^J < u^J$) or $u^J \leq w^J$ (if $su^J > u^J$), which follows from $u^J \leq w^J$ by transitivity.

Case 3: $su > u$, $sw^J < w^J$. We want $u^J \leq sw^J$. Since $\ell(su^J) + \ell(u_J) \geq \ell(su^J u_J) = \ell(su) > \ell(u) = \ell(u^J) + \ell(u_J)$, it follows that $su^J > u^J$. Hence, we get $u^J \leq sw^J$ from $u^J \leq w^J$ by lifting.

Case 4: $su > u$, $sw^J > w^J$. We want $u^J \leq w^J$, as assumed. $\square$

The quotients W^J that are of greatest interest in connection with Theorem 2.6.1 are those of maximal parabolic subgroups (i.e., for $|J| = |S| - 1$). These quotients are the smallest in size, and Bruhat order sometimes admits an easy explicit characterization when restricted to them (see, e.g., Proposition 2.4.8). Theorem 2.6.1 shows that the size of the right descent set $D_R(u)$ determines how many projections of this kind are needed to test for a Bruhat relation $u \leq w$. Let us formally state this important specialization.

Corollary 2.6.2 *Let* $u, w \in W$. *Then,*

$$u \leq w \iff P^{S \setminus \{s\}}(u) \leq P^{S \setminus \{s\}}(w), \quad \text{for all } s \in D_R(u). \quad \square$$

If (W, S) is finite, one gets via antiautomorphism the following alternative criterion:

$$u \leq w \iff P^{S \setminus \{s\}}(w_0 w) \leq P^{S \setminus \{s\}}(w_0 u), \quad \text{for all } s \in S \setminus D_R(w). \quad (2.21)$$

We now return to the topic of describing Bruhat order for the symmetric groups S_n, continuing the discussion from Section 2.1. Recall that a combinatorial procedure for deciding Bruhat relations in S_n (the "dot criterion") was given in Theorem 2.1.5.

Theorem 2.6.1 implies, because of the simplicity of Bruhat order on quotients $S_n^{(k)}$, a very efficient procedure for deciding when two permutations are comparable. This is the so-called "tableau criterion," in practice often the most convenient algorithm.

Theorem 2.6.3 (Tableau Criterion) *For* $x, y \in S_n$, *let* $x_{i,k}$ *be the i-th element in the increasing rearrangement of* $x_1, x_2, \ldots, x_k$, *and similarly define* $y_{i,k}$. *Then, the following are equivalent:*

(i) $x \leq y$.

(ii) $x_{i,k} \leq y_{i,k}$, *for all* $k \in D_R(x)$ *and* $1 \leq i \leq k$.

(iii) $x_{i,k} \leq y_{i,k}$, *for all* $k \in [n-1] \setminus D_R(y)$ *and* $1 \leq i \leq k$.

Proof. Condition (ii) can, as shown by Proposition 2.4.8, be restated as saying that $P^{S \setminus \{k\}}(x) \leq P^{S \setminus \{k\}}(y)$ for all $k \in D_R(x)$. Similarly, condition (iii) says that $P^{S \setminus \{k\}}(w_0 y) \leq P^{S \setminus \{k\}}(w_0 x)$ for all $k \in D_R(w_0 y)$. The result therefore follows from Corollary 2.6.2 and its alternative version (2.21). $\square$

For example, let us check whether $x = 368475912 \overset{?}{<} y = 694287531$. Since $D_R(x) = \{3, 5, 7\}$, we generate the three-line arrays of increasing rearrangements of initial segments of lengths 3, 5, and 7:

$$x \qquad\qquad\qquad\qquad y$$

3	4	5	6	7	8	9
3	4	6	7	8		
3	6	8				

2	4	5	6	7	8	9
2	4	6	8	9		
4	6	9				

Comparing entry by entry, we find two violations $(3 > 2)$ in the upper left corner, so we conclude that $x \not\leq y$. Since $[8] \setminus D_R(y) = \{1, 4\}$, it is quicker to use the alternative version (iii) of the criterion, which requires comparing the smaller arrays

$$x \qquad\qquad\qquad y$$

3	4	6	8
3			

2	4	6	9
6			

The fact that the arrays used are tableaux (increasing along rows and columns) explains the name of the criterion. To reduce the size of a calculation based on this criterion (the size of the tableaux), it may be worth having a preprocessing step to determine which of the sets $D_R(x)$, $D_L(x)$, $[n-1] \setminus D_R(y)$, and $[n-1] \setminus D_L(y)$ has the smallest size. If it is $D_L(x) = D_R(x^{-1})$, one uses that $x \leq y \Leftrightarrow x^{-1} \leq y^{-1}$ and applies the criterion to x^{-1} and y^{-1}, and similarly for $D_L(y) = D_R(y^{-1})$.

2.7   Interval structure

What can be said about the combinatorial structure of intervals $[u, w]$ in Bruhat order, and more generally of such intervals $[u, w]^J$ in quotients W^J? One result in this direction is Corollary 2.5.6, which implies that $[u, w]^J$ is a graded poset with rank function $\rho(x) = \ell(x) - \ell(u)$. In this section, we will probe deeper into this question. It turns out that some of the results are best expressed using concepts from topology.

Throughout the section, (W, S) denotes a general Coxeter system, $J \subseteq S$, $u, w \in W^J$, and $u < w$. We already introduced the notation $[u, w]^J = [u, w] \cap W^J$ for the closed interval spanned by u and w in W^J. We use the corresponding notation for open intervals: $(u, w)^J = (u, w) \cap W^J$. Also, we write

$$\ell(u, w) \stackrel{\text{def}}{=} \ell(w) - \ell(u).$$

Let $\mathcal{M}(u, w)$ denote the set of maximal chains in the Bruhat interval $[u, w]$, and let $w = s_1 s_2 \ldots s_q$ be a reduced expression. We are going to associate with each $\mathbf{m} \in \mathcal{M}(u, w)$ a string of integers $\lambda(\mathbf{m}) = (\lambda_1(\mathbf{m}), \lambda_2(\mathbf{m}), \ldots, \lambda_k(\mathbf{m}))$, where $k = \ell(u, w)$. The k-tuple $\lambda(\mathbf{m})$ is induced by the given reduced word as follows.

Suppose that $\mathbf{m}$ is the chain $w = x_0 \rhd x_1 \rhd \cdots \rhd x_k = u$. By the Strong Exchange Property (see Corollary 1.4.4), $x_1 = x_0 t_1 = s_1 s_2 \ldots \widehat{s_i} \ldots s_q$, where the deleted generator s_i is uniquely determined. Let $\lambda_1(\mathbf{m}) = i$. Now repeat the process. After f steps, we have reached x_f, and after f deletions, we have obtained a uniquely determined reduced subword expression $x_f = s_{j_1} s_{j_2} \ldots s_{j_{q-f}}$, $1 \leq j_1 < j_2 < \cdots < j_{q-f} \leq q$. Again, $x_{f+1} = x_f t_{f+1} = s_{j_1} s_{j_2} \ldots \widehat{s_{j_i}} \ldots s_{j_{q-f}}$ where the deleted generator s_{j_i} is uniquely determined. Let $\lambda_{f+1}(\mathbf{m}) = j_i$. Hence, the idea is to label by the positions of the generators which are successively deleted from the chosen reduced expression for w as we go down the maximal chain $\mathbf{m}$ from w to u.

Example 2.7.1 Let W be the dihedral group of order 6 on two generators $S = \{a, b\}$ (or equivalently, the symmetric group S_3). Its Bruhat order is depicted in Figure 2.9.

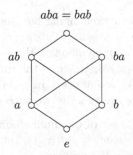

$$aba = bab$$

Figure 2.9. Bruhat order of S_3.

Choosing "aba" as reduced expression for the top element, the induced labels of the four maximal chains are

$$\lambda(aba \vartriangleright ba \vartriangleright a \vartriangleright \emptyset) = (1, 2, 3),$$
$$\lambda(aba \vartriangleright ba \vartriangleright b \vartriangleright \emptyset) = (1, 3, 2),$$
$$\lambda(aba \vartriangleright ab \vartriangleright b \vartriangleright \emptyset) = (3, 1, 2),$$
$$\lambda(aba \vartriangleright ab \vartriangleright a \vartriangleright \emptyset) = (3, 2, 1). \quad \square$$

Lemma 2.7.2 *There is at most one chain* $\mathbf{m} \in \mathcal{M}(u, v)$ *for which* $\lambda(\mathbf{m})$ *is increasing (meaning that* $\lambda_1(\mathbf{m}) < \lambda_2(\mathbf{m}) < \cdots < \lambda_k(\mathbf{m})$*).*

Proof. The statement is clear for intervals of length 1, so we may inductively suppose that it has been shown for length $k - 1$. Suppose that there are two maximal chains $\mathbf{m} : w = x_0 \vartriangleright x_1 \vartriangleright \cdots \vartriangleright x_k = u$ and $\mathbf{m}' : w = x_0' \vartriangleright x_1' \vartriangleright \cdots \vartriangleright x_k' = u$ with increasing labels $\lambda(\mathbf{m}) = (i_1, i_2, \ldots, i_k)$ and $\lambda(\mathbf{m}') = (j_1, j_2, \ldots, j_k)$. Then, $u = s_1 \ldots \widehat{s}_{i_1} \ldots \widehat{s}_{i_2} \ldots \widehat{s}_{i_k} \ldots s_q = s_1 \ldots \widehat{s}_{j_1} \ldots \widehat{s}_{j_2} \ldots \widehat{s}_{j_k} \ldots s_q$.

Assume that $i_k < j_k$, and let $t_{j_k} = s_q s_{q-1} \ldots s_{j_k} \ldots s_{q-1} s_q$. Then, $x_{k-1}' = u t_{j_k} = s_1 \ldots \widehat{s}_{i_1} \ldots \widehat{s}_{i_2} \ldots \widehat{s}_{i_k} \ldots \widehat{s}_{j_k} \ldots s_q$, so $\ell(x_{k-1}') \leq \ell(u) - 1$, contradicting that $x_{k-1}' \vartriangleright u$. Hence, $j_k \leq i_k$, and by symmetry, $i_k \leq j_k$. The equality $i_k = j_k$ implies that $x_{k-1} = x_{k-1}'$. Since the interval $[x_{k-1}, w]$ by the induction assumption does not admit two distinct maximal chains with increasing labels, we conclude that $\mathbf{m} = \mathbf{m}'$. $\square$

From this, we can deduce the structure of intervals of length 2.

Lemma 2.7.3 *If* $\ell(u, w) = 2$*, then* $[u, w] \cong$ ⬦ *.*

Proof. Among all reduced expressions for u which are subwords of $s_1 s_2 \ldots s_q$, choose $u = s_1 \ldots \widehat{s}_i \ldots \widehat{s}_j \ldots s_q$ so that $i < j$ and j is *minimal*. Let $y = s_1 \ldots \widehat{s}_i \ldots s_j \ldots s_q$. The proof of Theorem 2.5.5 shows that $w \vartriangleright y \vartriangleright u$, and the label (i, j) of this chain is increasing. It is thus, by Lemma 2.7.2, the unique chain in $[u, w]$ with increasing label.

Now, mirror the situation, exchanging left and right (i.e., choose a reduced subword expression $u = s_1 \ldots \widehat{s}_m \ldots \widehat{s}_p \ldots s_q$ so that $m < p$ and m is *maximal*, and so on $\ldots$). By symmetry, the same argument produces a chain in $[u, w]$ with label (p, m), which is the unique one with decreasing label. The label $\lambda(\mathbf{m}) = (\lambda_1, \lambda_2)$ of every $\mathbf{m} \in \mathcal{M}(u, w)$ is either increasing or decreasing. Hence, there are exactly two such chains. $\square$

Let $\mathcal{M}^J(u, w)$ denote the set of maximal chains in the quotient Bruhat interval $[u, w]^J$. Since $\mathcal{M}^J(u, w) \subseteq \mathcal{M}(u, w)$, the injective mapping $\mathbf{m} \mapsto \lambda(\mathbf{m})$ restricts to $\mathcal{M}^J(u, w)$. We write $(a_1, \ldots, a_k) \prec (b_1, \ldots, b_k)$ for the lexicographic order relation of distinct integer strings. This means that $a_i < b_i$ for the minimal index i such that $a_i \neq b_i$.

Lemma 2.7.4   *(i) There is a unique chain $\mathbf{m}_0 \in \mathcal{M}^J(u, w)$ such that $\lambda(\mathbf{m}_0)$ is increasing.*

(ii) $\lambda(\mathbf{m}_0) \prec \lambda(\mathbf{m})$ for all $\mathbf{m} \neq \mathbf{m}_0$ in $\mathcal{M}^J(u, w)$.

Proof. A chain $\mathbf{m}$ with increasing $\lambda(\mathbf{m})$ is constructed in Theorem 2.5.5, and it is unique by Lemma 2.7.2. This proves part (i).

Part (ii) will be proved by induction on length. For the $\ell(u, w) = 2$ case, we refer to the proof of Lemma 2.7.3. It shows that there are at most two maximal chains in $[u, w]^J$, namely the increasing one with label (i, j) and (perhaps) a decreasing one with label (p, m), where, by construction, $i < j \leq p$. Since $(i, j) \prec (p, m)$, we are done with this case.

Suppose that $\ell(u, w) > 2$, and let $\mathbf{m} : w = x_0 \rhd x_1 \rhd \cdots \rhd x_k = u$ be the maximal chain in $[u, w]^J$ whose label $\lambda(\mathbf{m}) = (\lambda_1(\mathbf{m}), \ldots, \lambda_k(\mathbf{m}))$ is lexicographically minimal. We want to show that $\lambda(\mathbf{m})$ is increasing. The simple key observation is as follows (with the labelings induced by the reduced expressions $w = s_1 s_2 \ldots s_q$ and $x_1 = s_1 s_2 \ldots \widehat{s}_{\lambda_1(\mathbf{m})} \ldots s_q$, respectively):

- The chain $w = x_0 \rhd x_1 \rhd \cdots \rhd x_{k-1}$ has lexicographically minimal label in the interval $[x_{k-1}, w]^J$.

- The chain $x_1 \rhd x_2 \rhd \cdots \rhd x_k = u$ has lexicographically minimal label in the interval $[u, x_1]^J$.

Hence, by induction, the two strings

$$(\lambda_1(\mathbf{m}), \ldots, \lambda_{k-1}(\mathbf{m})) \quad \text{and} \quad (\lambda_2(\mathbf{m}), \ldots, \lambda_k(\mathbf{m}))$$

are increasing. Since they overlap, we obtain that $\lambda(\mathbf{m}) = (\lambda_1(\mathbf{m}), \ldots, \lambda_k(\mathbf{m}))$ is increasing. $\square$

We now discuss the order complex $\Delta((u, w)^J)$, whose faces are the chains of the open interval $(u, w)^J$. By Corollary 2.5.6, this complex is pure $(\ell(u, w) - 2)$-dimensional. See Appendices A2.3 and A2.4 for all relevant definitions, results, and references concerning these notions and the concept of shellability. The following result provides the main technical tool of this section.

Theorem 2.7.5 *The order complex of $(u, w)^J$ is shellable.*

Proof. It is notationally more convenient to prove the statement for the order complex of the closed interval $[u, w]^J$, whose facets are the maximal chains in $\mathcal{M}^J(u, w)$. Since $\Delta([u, w]^J)$ is the double cone over $\Delta((u, w)^J)$ (with cone points u and w), shellability for one is equivalent to shellability of the other; see statement (A2.4).

For $\mathbf{m}', \mathbf{m} \in \mathcal{M}^J(u, w)$, define $\mathbf{m}' \prec \mathbf{m}$ to mean that $\lambda(\mathbf{m}') \prec \lambda(\mathbf{m})$. Thus, we are letting the lexicographic order of labels $\lambda(\mathbf{m})$ induce a linear order on the set $\mathcal{M}^J(u, w)$ (i.e., on the set of facets of $\Delta([u, w]^J)$). We will show that this is a shelling order. We have to prove (cf. Definition A2.4.1) that if $\mathbf{m}' \prec \mathbf{m}$, then there exists $\mathbf{k} \in \mathcal{M}^J(u, w)$ such that $\mathbf{k} \prec \mathbf{m}$, $\mathbf{m}' \cap \mathbf{m} \subseteq \mathbf{k} \cap \mathbf{m}$, and $|\mathbf{k} \cap \mathbf{m}| = |\mathbf{m}| - 1$.

Consider two maximal chains $\mathbf{m} : w = x_0 \rhd x_1 \rhd \cdots \rhd x_k = u$ and $\mathbf{m}' : w = x_0' \rhd x_1' \rhd \cdots \rhd x_k' = u$, and suppose that $\mathbf{m}' \prec \mathbf{m}$. Let d be the greatest integer such that $x_i = x_i'$ for $i = 0, 1, \ldots, d$, and let g be the least integer such that $g > d$ and $x_g = x_g'$. Then, $g - d \geq 2$, and $d < i < g$ implies that $x_i \neq x_i'$.

Now, focus attention on the subinterval $[x_g, x_d]^J$ and the labeling of its maximal chains that is induced by the reduced word for x_d that one gets by following the chain $w = x_0 \rhd x_1 \rhd \cdots \rhd x_d$ (or, equivalently, the chain $w = x_0' \rhd x_1' \rhd \cdots \rhd x_d'$) and making the corresponding deletions in the given reduced expression $w = s_1 s_2 \ldots s_q$. (To be precise, the letters $s_{\lambda_i(\mathbf{m})}$ for $i = 1, \ldots, d$ are the deleted ones in the reduced subword for x_d.)

The chain $x_d \rhd x_{d+1} \rhd \cdots \rhd x_g$ cannot be the unique maximal chain of $[x_g, x_d]^J$ with increasing label, because then the property stated in Lemma 2.7.4 would force $\lambda(\mathbf{m}) \prec \lambda(\mathbf{m}')$, contrary to the assumption that $\mathbf{m}' \prec \mathbf{m}$. Consequently, the label $\lambda(\mathbf{m})$ must have a descent $\lambda_e(\mathbf{m}) > \lambda_{e+1}(\mathbf{m})$ for some e with $d < e < g$. Then, in the sub-subinterval $[x_{e+1}, x_{e-1}]^J$, with its induced labeling of maximal chains, the chain $x_{e-1} \rhd x_e \rhd x_{e+1}$ has a decreasing label. So, again by Lemma 2.7.4, there is a chain $x_{e-1} \rhd y \rhd x_{e+1}$ whose label is increasing and comes earlier in lexicographic order. If we define $\mathbf{k}$ to be the chain $w \rhd x_1 \rhd \cdots \rhd x_{e-1} \rhd y \rhd x_{e+1} \rhd x_{e+2} \rhd \cdots \rhd u$, it follows that $\lambda(\mathbf{k}) \prec \lambda(\mathbf{m})$. Furthermore, the construction shows that $\mathbf{k} \cap \mathbf{m} = \mathbf{m} - \{w_e\} \supseteq \mathbf{m}' \cap \mathbf{m}$. $\square$

Corollary 2.7.6 *The order complex of $(u, w)^J$ is Cohen-Macaulay.*

A closed interval $[u, w]^J$ in the quotient poset W^J is said to be *full* if $[u, w]^J = [u, w]$, and similarly for open intervals.

Theorem 2.7.7 *The order complex of $(u, w)^J$ is PL homeomorphic to*

(i) *the sphere $\mathbb{S}^{\ell(u,w)-2}$, if $(u, w)^J$ is full;*

(ii) *the ball $\mathbb{B}^{\ell(u,w)-2}$, otherwise.*

Proof. The complex $\Delta\left((u,w)^J\right)$ is pure $(\ell(u,w)-2)$-dimensional, and Lemma 2.7.3 implies that it is thin if $(u,w)^J$ is full and subthin otherwise. Hence, Theorem 2.7.5 and Fact A2.4.3 force the conclusion. $\square$

The theorem implies the following characterization of full intervals of length 3. Call a poset a *k-crown* if it is isomorphic to the poset depicted in Figure 2.10. For instance, Bruhat order of S_3 is a 2-crown; see Figure 2.9.

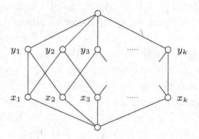

Figure 2.10. A *k*-crown.

Corollary 2.7.8 *Suppose that $\ell(u,w)=3$. Then, the closed interval $[u,w]$ is a k-crown, for some $k\geq 2$.*

Proof. The order complex of (u,w) triangulates the circle $\mathbb{S}^1$. The k-crown clearly has the only possible isomorphism type. $\square$

The length 3 intervals in Bruhat order of S_4 provide examples of 2-, 3-, and 4-crowns; see Figure 2.4. It is a nontrivial fact that these are the *only* types that can occur in a finite Weyl group (H_3 and H_4 also contain 5-crowns); see Section 2.8. In contrast, arbitrarily large k-crowns can exist in infinite Coxeter groups, as we now show.

Example 2.7.9 Let (W,S) be given by the Coxeter diagram

Let $u=(abc)^{n-1}$ and $w=(abc)^n$. Then, $u<w$, $\ell(u,w)=3$ and the interval $[u,w]$ is a $3n$-crown. To see this, observe that each of the $3n$ letters in the reduced word $w=abc\,abc\ldots abc$ can be deleted, creating a reduced subword of length $3n-1$, which uniquely represents an element of $[u,w]$. $\square$

Let $\mu^J(u,w)$ denote the Möbius function computed on the quotient Bruhat poset W^J; see Appendix A2.2 for the definition. Since, by Fact A2.3.1, $\mu^J(u,w)$ is the reduced Euler characteristic of $\Delta((u,w)^J)$, the following is a direct consequence of Theorem 2.7.7.

Corollary 2.7.10 $\mu^J(u,w) = \begin{cases} (-1)^{\ell(u,w)}, & \text{if } [u,w]^J \text{ is full,} \\ 0, & \text{otherwise.} \end{cases}$ □

Computing on the full Bruhat order of W, the corollary specializes to say that $\mu(u,w) = (-1)^{\ell(u,w)}$ for all $u \leq w$. From definition (A2.1), one sees that this is equivalent to the following:

Corollary 2.7.11 *In a closed interval* $[u,w]$, $u < w$, *the number of elements of odd length equals the number of elements of even length.* □

The last two corollaries can, of course, be given direct combinatorial proofs. If there exists some $s \in D_L(w) \setminus D_L(u)$, then the mapping $x \mapsto sx$ matches the odd-length and the even-length elements of $[u,w]$, as is easily seen from Proposition 2.2.7. In general, a more detailed combinatorial argument is needed; see Exercise 13.

Theorem 2.7.7 shows that the order complex of an open interval (u,w) triangulates a sphere. This triangulation is actually the barycentric subdivision of a more intrinsic cell decomposition of the sphere. We refer to Appendix A2.5 for explanations of the concepts used.

Theorem 2.7.12 *Suppose that* $\ell(u,w) \geq 2$. *Then, there exists a regular CW complex* $\Gamma_{u,w}$, *uniquely determined up to cellular homeomorphism, whose cell poset is isomorphic to* (u,w) *and such that* $\|\Gamma_{u,w}\| \cong \mathbb{S}^{\ell(u,w)-2}$.

Proof. Let $X = \|\Delta((u,w))\|$, the geometric realization of the order complex of (u,w). Then, by Theorem 2.7.7, $X \cong \mathbb{S}^{\ell(u,w)-2}$. For each $z \in (u,w)$, let $\sigma_z = \|\Delta((u,z))\|$. This subspace of X is a cone (with cone point z) over the sphere $\|\Delta((u,z))\|$; hence, it is a ball. Our claim is that $\Gamma_{u,w} = \{\sigma_z : z \in (u,w)\}$ is a regular CW decomposition of X.

The two conditions of Definition A2.5.1 are easily verified. A point $p \in X$ lies in the interior of σ_z if and only if the support of its barycentric coordinates is a chain in which z is the greatest element. Hence, the interiors $\overset{\circ}{\sigma}_z$ partition X. The boundary of σ_z is $\|\Delta((u,z))\|$, which is the union of σ_y for all $u < y < z$. □

Example 2.7.13 Consider an interval of length 3 (i.e., a k-crown as in Figure 2.10). The CW complex $\Gamma_{u,w}$ has vertices $x_1, \ldots, x_k$ and edges $y_1, \ldots, y_k$, forming the boundary of a k-gon. Its barycentric subdivision, the order complex of (u,w), has vertices $x_1, \ldots, x_k$ and $y_1, \ldots, y_k$ and $2k$ edges.

For another example, take the open interval $(1234, 3241)$ in S_4, shown to the left in Figure 2.11. The corresponding regular CW decomposition of $\mathbb{S}^2$ appears to the right. It has three vertices, four edges, and three 2-cells (one of the triangular 2-cells fills the outer region of the compactified plane). □

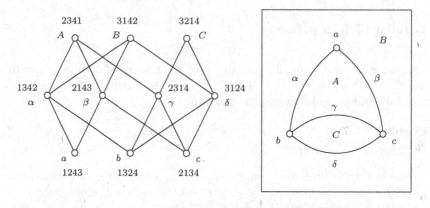

Figure 2.11. Regular CW interpretation of a Bruhat interval.

Theorem 2.7.12 has a combinatorial corollary concerning labelings of the edges of the poset diagram of $[u, w]$ by $+1$ and -1. Let

$$\mathrm{Cov}[u, w] \overset{\text{def}}{=} \{(v, z) \in W^2 : u \leq v \lhd z \leq w\},$$

and say that a mapping $\mathrm{sg} : \mathrm{Cov}[u, w] \to \{+1, -1\}$ is *balanced* if for all $x < y$ in $[u, w]$ such that $\ell(x, y) = 2$:

$$\mathrm{sg}(x \lhd a)\mathrm{sg}(a \lhd y) + \mathrm{sg}(x \lhd b)\mathrm{sg}(b \lhd y) = 0, \qquad (2.22)$$

where a and b are the two "middle" elements of $[x, y]$ (cf. Lemma 2.7.3). Equivalently, the number of "(-1)"s assigned to the four edges of $[x, y]$ is odd.

For each $i \in [\ell(u), \ell(w)]$, let $C_i[u, w]$ be the free Abelian group generated by the set of elements $z \in [u, w]$ such that $\ell(z) = i$. Assume that we have a signature $\mathrm{sg} : \mathrm{Cov}[u, w] \to \{+1, -1\}$, and for $i \in [\ell(u) + 1, \ell(w)]$, define a homomorphism $d_i : C_i[u, w] \to C_{i-1}[u, w]$ by linear extension of

$$d_i(z) = \sum_{u \leq x \lhd z} \mathrm{sg}(x, z)\, x.$$

Thus, we have a sequence of successive maps d_i:

$$0 \to C_{\ell(w)}[u, w] \to C_{\ell(w)-1}[u, w] \to \cdots \to C_{\ell(u)}[u, w] \to 0. \qquad (2.23)$$

One easily sees that sg is balanced if and only if $\mathrm{Im}\, d_i \subseteq \mathrm{Ker}\, d_{i-1}$, for all i.

Corollary 2.7.14 *Suppose that $\ell(u, w) \geq 2$. Then, there exists a signature* $\mathrm{sg} : \mathrm{Cov}[u, w] \to \{+1, -1\}$ *such that the following hold:*

(i) sg *is balanced.*

(ii) $\mathrm{Im}\, d_i = \mathrm{Ker}\, d_{i-1}$, *for all $i \in [\ell(u), \ell(w)]$.*

Proof. Extend the cell complex $\Gamma_{u,w}$ by attaching a $(\ell(u, w) - 1)$-cell via some homeomorphism of its boundary onto $\|\Gamma_{u,w}\|$. This gives a regular

CW complex $\widehat{\Gamma}_{u,w}$ which decomposes the ball $\mathbb{B}^{\ell(u,w)-1}$ and whose cell poset is isomorphic to the half-open interval $(u, w]$.

Now, consider the cellular chain complex of $\widehat{\Gamma}_{u,w}$ (see Appendix A2.5). The incidence numbers of cells $[\sigma : \tau] \in \{+1, -1\}$ can, via the poset isomorphism, be transferred to the coverings $v \lhd z$ such that $u < v \lhd z \leq w$. This induces a partial labeling, which extends to a complete labeling $\mathrm{sg} : \mathrm{Cov}[u, w] \to \{+1, -1\}$ by putting $\mathrm{sg}(u \lhd z) = +1$ for the "bottom edges" (i.e., for all elements z that cover u in $[u, w]$). This signature sg is balanced, since $d_{i-1} \circ d_i = 0$ in the cellular chain complex. Furthermore, via transfer of structure, the chain complex (2.23) can be identified with the cellular chain complex of $\widehat{\Gamma}_{u,w}$. Therefore, part (ii) is a consequence of the fact that reduced cellular homology of the ball $\mathbb{B}^{\ell(u,w)-1}$ vanishes in all dimensions. $\square$

2.8   Complement: Short intervals

When studying Bruhat intervals of length m in finite Coxeter groups it is sufficient to consider groups of rank m. The reason for this surprising fact is made precise by the following theorem. Recall that a *reflection subgroup* is a subgroup generated by some subset of the set T of reflections. It is known (see Exercise 1.18) that such subgroups are themselves Coxeter groups.

Theorem 2.8.1 *Suppose that (W, S) is finite, and let $[u, w]$ be a Bruhat interval in W with $\ell(u, w) = m$. Then, there exists a reflection subgroup (W', S') of rank $|S'| \leq m$ and a Bruhat interval $[u', w']$ in W' such that $[u, w] \cong [u', w']$.* $\square$

This theorem, due to Dyer [203], has the following consequence, since by the classification there are only finitely many finite Coxeter groups of each rank.

Corollary 2.8.2 *Up to isomorphism, only finitely many posets of each length $m \geq 0$ occur as intervals in Bruhat order of finite Coxeter groups.* $\square$

What are the possible types? Or, at least, how many are there? Here is what is known about this question for three important classes of finite groups.

	$m = 2$	$m = 3$	$m = 4$	$m = 5$
Symmetric groups	1	3	7	25
Simply-laced groups	1	3	10	
Weyl groups	1	3	24	

Number of length m intervals

That all length 3 intervals in a finite Weyl group are k-crowns with $2 \leq k \leq 4$ was shown by Janzen [316]. Since there are only two irreducible Weyl groups of rank 3 (A_3 and B_3), and the rank 2 and reducible cases are easy, this follows from Theorem 2.8.1 by checking A_3 and B_3. The group H_3 also contains some 5-crowns.

The classification of length 4 intervals, and for symmetric groups also length 5, was done by Hultman [301, 302]. The 24 types of intervals that occur for $m = 4$ are shown in Figure 2.12. The pictures show the regular CW decompositions of $\mathbb{S}^2$ that the respective intervals determine. For instance, the poset diagram of an interval of type 1 is shown in Figure 2.11.

Referring to Figure 2.12, the following is also shown in [302]:

> *Only intervals of type 1–7 appear in the symmetric groups.*
> *Only intervals of type 1–10 appear in the simply-laced groups.*

It is an interesting fact that *all* 24 types of length four intervals are represented within the group F_4.

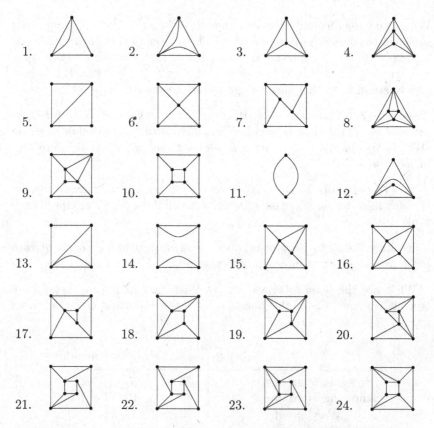

Figure 2.12. All length 4 intervals that appear in Weyl groups.

Exercises

1. Define "left Bruhat order" by exchanging "$w = ut$" for "$w = tu$" in Definition 2.1.1. Show (without using the Subword Property) that it coincides with the "right Bruhat order" of Definition 2.1.1.

2. Show that the automorphism group of Bruhat order of the dihedral group $I_2(p)$ is isomorphic to $\mathbb{Z}_2^{p-1}$.

3. Let $J_1, \ldots, J_k$ be the node sets of the connected components of the Coxeter graph for (W, S). Show the following:

   (a) W is (group-theoretically) isomorphic to the direct product of the irreducible subgroups W_{J_i}.

   (b) Bruhat order on W is (order-theoretically) isomorphic to the direct product of Bruhat order on the respective subgroups W_{J_i}.

4. Let $k \in [n-1]$ and $J = S \setminus \{s_k\}$, and use complete notation for permutations.

   (a) Show that $^J(S_n)$ consists of the shuffles of the two sequences $1, \ldots, k$ and $k+1, \ldots, n$. [*Remark*: A *shuffle* of two sequences is any linear order of their set-theoretic union in which the elements of each sequence appear in their original order.]

   (b) Show that the projection map $x \mapsto {}^Jx \in {}^J(S_n)$ has the following description: ^Jx is obtained from x by permuting the values $1, \ldots, k$ so that they form an increasing subsequence on their original set of places, and then similarly for $k+1, \ldots, n$.

5. Let $x = 316725948$. Compute x^J and ^Jx in S_9, when $J = \{1, 7\}$, $\{1\}$, $\{7\}$, $\emptyset$, and $\{1, 2, 3, 4, 5, 6, 7, 8\}$, respectively.

6. Show that the set $\{x \in S_{2n} : |x(i) - i| \le n\}$ forms an interval in the Bruhat order of S_{2n}, and that this interval has $2n - 1$ atoms and n^2 coatoms.

7. For $W = S_n$, interpret concretely in terms of permutations the bijections of Corollary 2.4.6 and Proposition 2.5.4.

8. Prove the following version of the tableau criterion for S_n. Choose $k \in [n]$ and let $x, y \in S_n$. Then, $x \le y$ if and only if $\overline{x_1 \ldots x_i} \le \overline{y_1 \ldots y_i}$ for $1 \le i \le k-1$ and $\overline{x_j \ldots x_n} \ge \overline{y_j \ldots y_n}$ for $k+1 \le j \le n$. Here, the overline denotes increasing rearrangement, and comparison of strings is componentwise.

9. Deduce Theorem 2.6.3 from Theorem 2.1.5 by direct combinatorial reasoning.

10. Let W be finite and $w \in W$. Show the following:

    (a) $D_L(ww_0) = S \setminus D_L(w)$.

     (b) $D_L(w_0 w) = w_0(S \setminus D_L(w))w_0 = S \setminus w_0 D_L(w)w_0$.
     (c) $D_L(w_0 w w_0) = w_0 D_L(w)w_0$.
     (d) $T_L(w_0 w w_0) = w_0 T_L(w)w_0$.

11. Let $J \subseteq S$ and $u, w \in W_J$. Show that $u \leq w$ in W_J if and only if $u \leq w$ in W.

12. Suppose $u \leq w$, $J \subseteq S$, and that $u < us$ and $w > ws$ for all $s \in J$. Show that the Bruhat interval $[u, w]$ is a union of left cosets xW_J.

13. Prove Corollary 2.7.11 by a combinatorial matching argument.

14. For a Coxeter system (W, S) and a partial order $\leq$ on W, say that $\leq$ has the *lifting* property if for any $u, w \in W$ and $s \in S$ with $\ell(su) > \ell(u)$ and $\ell(sw) < \ell(w)$, one has that the following are equivalent:

     (i) $su \leq w$
     (ii) $u \leq w$
     (iii) $u \leq sw$.

Show that any ordering on W with the lifting property must coincide with the Bruhat ordering.

15. Let $I, J \subseteq S$.

     (a) Show that every double coset $W_I w W_J$, $w \in W$, has a unique element of minimal length.
     (b) Characterize the system $^I W^J$ of such minimal double coset representatives in terms of the quotients W^I and W^J.

16. Suppose that W_J is finite and define an *upper projection* operator $\overline{P}^J : W \to \mathcal{D}^S_J$ by $\overline{P}^J(w) = $ maximal length representative of the coset wW_J. Show that the following hold:

     (a) $\overline{P}^J$ is order preserving.
     (b) W^J and $\mathcal{D}^S_J$ are isomorphic as posets.
     (c) Let $u, w \in W$ and $J = D_R(w)$. Show that $u \leq w$ if and only if $\overline{P}^J(u) \leq w$.

17. Suppose that W is infinite and irreducible. Show that W^J is infinite for all proper subsets $J \subset S$.

18. Let $s_i = (i, i+1)$, $i = 1, \ldots, n-1$, be the standard Coxeter generators of the symmetric group S_n, and let $J \subset \{s_1, \ldots, s_{n-1}\}$. Show that the following conditions are equivalent:

     (a) $(S_n)^J$ is a lattice.
     (b) $(S_n)^J$ is a distributive lattice.
     (c) $\mathrm{card}(J) = n - 2$.

19. Let (W, S) be an irreducible finite Coxeter group. Show that if W^J is a lattice, then it is in fact a distributive lattice and $\mathrm{card}(J) = \mathrm{card}(S) - 1$.

20. Suppose $w \in W$ and $J \subseteq S$. Show that then $[e, w] \cap W_J = [e, u]$, for some $u \in W_J$.

21. Suppose that $\ell(xw) = \ell(x) + \ell(w)$ and $\ell(yw) = \ell(y) + \ell(w)$, for $x, y, w \in W$. Show that $xw < yw \Leftrightarrow x < y$.

22. Let (W, S) be a Coxeter group and $V \subseteq W$. Let

$$W/V \overset{\text{def}}{=} \{w \in W : \ell(wv) = \ell(w) + \ell(v) \text{ for all } v \in V\}.$$

Such subsets, called *generalized quotients*, are considered here as posets under the induced Bruhat order. Show the following:

(a) *Ordinary quotients*: If $J \subseteq S$, then $W/J = W^J$.
(b) W/V is a graded poset whose rank function is $r(w) = \ell(w)$.
(c) The order complex of any interval in W/V is shellable.
(d) If W is finite, then for every $V \subseteq W$, there exists an element $u \in W$ such that $W/V = W/\{u\}$.

23. Prove the following properties of descent classes $\mathcal{D}_I^J$ under Bruhat order, $I \subseteq J \subseteq S$:

(a) $\mathcal{D}_I^J \neq \emptyset$ if and only if W_I is finite. Suppose in the following parts that this is the case.
(b) $\mathcal{D}_I^J$ has a least element, namely $w_0(I)$.
(c) $\mathcal{D}_I^J$ is finite if and only if $W^{S \setminus J}$ is finite, and if so, $\mathcal{D}_I^J$ has a greatest element, namely $w_0^{S \setminus J}$.
(d) $\mathcal{D}_I^J$ is isomorphic (as a poset) to a generalized quotient, namely

$$\mathcal{D}_I^J \cong W/V, \text{ where } V \overset{\text{def}}{=} \{w_0(I) \cdot s : s \in S \setminus J\}.$$

(e) Conclude from part (d) and Exercise 22 that all maximal chains in an interval $[u, w]_I^J$ in $\mathcal{D}_I^J$ are of length equal to $\ell(w) - \ell(u)$, that the order complex of $[u, w]_I^J$ is shellable, and that the expression for the Möbius function in Corollary 2.7.10 generalizes.

24. The *order dimension* of a finite poset P, denoted $\text{odim}(P)$, is the least integer d such that P is isomorphic to an induced subposet of $\mathbb{N}^d$ (with product order). See [540] for a discussion of this concept. Consider now the finite irreducible Coxeter groups under Bruhat order. Show the following:

(a) $\text{odim}(I_2(m)) = 2$, for all $m \geq 2$,
(b) $\text{odim}(A_{n-1}) = \lfloor \frac{n^2}{4} \rfloor$, for all $n \geq 2$,
(c) $\text{odim}(B_n) = \binom{n}{2} + 1$, for all $n \geq 2$,
(d) $\text{odim}(H_3) = 6$,
(e) $\text{odim}(H_4) = 25$,
(f) $14 \leq \text{odim}(D_6) \leq 22$,
(g) $14 \leq \text{odim}(E_6) \leq 26$,
(h) $10 \leq \text{odim}(F_4) \leq 12$,

(i)* Determine the order dimension of Bruhat order in the remaining open cases.

25. Show that Bruhat order determines the Bruhat graph (i.e., knowledge of the edges that come from Bruhat order coverings determine knowledge of *all* edges of the Bruhat graph).

26. For any subset $A \subseteq W$, let $BG(A)$ denote the directed graph on the vertex set A induced by the Bruhat graph of (W, S) (i.e., $u \to w$ is an edge of $BG(A)$ if and only if $u \to w$ in W and $u, w \in A$). Let $J \subseteq S$. Show that the following hold:

   (a) The Bruhat graph of (W_J, J) coincides with $BG(W_J)$.
   (b) This graph is isomorphic to $BG(wW_J)$, for any $w \in W$.

27. Let (W, S) be a Coxeter group, and for any $i \geq 0$, let Γ_i be the (undirected) bipartite graph whose vertices are the group elements of lengths i and $i + 1$ and whose edges are pairs $x < y$. Show that Γ_i does not contain the complete bipartite graph $K_{2,3}$ as an induced subgraph.

28. Let (W, S) be a Coxeter group with $|S| \neq \infty$. Show that every antichain in Bruhat order of W is finite. (Equivalently, every linear extension of Bruhat order is well-ordered.)

29. Let $x, y \in S_n$, $n \geq 4$. Say that the ordered pair (x, y) satisfies "condition 4C" if the following hold:

   (i) x and y (written in complete notation) agree in all but four positions.
   (ii) In these positions, we have the patterns
   $$x = \ldots a \ldots c \ldots b \ldots d \ldots,$$
   $$y = \ldots c \ldots d \ldots a \ldots b \ldots$$
   for some numbers $1 \leq a < b < c < d \leq n$.
   (iii) No number e with $a < e < c$ appears in a position between a and c in x.
   (iv) No number f with $a < f < d$ appears in a position between c and b in x.
   (v) No number g with $b < g < d$ appears in a position between b and d in x.

   (a) Show that the interval $[x, y]$ is a 4-crown if and only if either (x, y) or (yw_0, xw_0) satisfies condition 4C.
   (b) Show by a direct argument that k-crowns for $k \geq 5$ cannot occur in the symmetric groups.

30. Suppose that $\ell(u, w) = 3$. Then, $[u, w]$ is a 2-crown if and only if $u \to w$ is an edge in the Bruhat graph.

31. Show that both the finite group H_3 and the affine group $\tilde{G}_2$ contain Bruhat intervals that are 5-crowns.

32. For any Coxeter group (W, S) and any $k \geq 0$, show that

$$\sum_{\ell(w) \leq k} (-1)^{k - \ell(w)} \geq 0,$$

with equality if and only if W is finite and $k \geq \ell(w_0)$.

33. Consider an open interval $(u, w)^J$, $\ell(u, w) = k \geq 2$. Suppose that $E \subseteq [k-1]$ and define the *length-selected subposet*

$$(u, w)_E^J \overset{\text{def}}{=} \left\{ x \in (u, w)^J : \ell(x) - \ell(u) \in E \right\}.$$

Show the following:

(a) The order complex of $(u, w)_E^J$ has the homotopy type of a wedge of $(|E| - 1)$-dimensional spheres.

(b) Suppose that the interval is full, and let h_E denote the number of spheres in the wedge. Then,

(i) $h_E \geq 1$, and

(ii) $h_E = h_{[k-1] \setminus E}$.

34. Let E be a finite set of positive integers and (W, S) a Coxeter group that is either infinite or finite with $\ell(w_0) > \max E$. Show that the order complex of the induced Bruhat poset on $\{w \in W : \ell(w) \in E\}$ has the homotopy type of a wedge of $(|E| - 1)$-dimensional spheres.

35. For a Coxeter group (W, S), let $\mathrm{Invol}(W)$ denote the subposet of involutions in W with induced Bruhat order. Figure 2.13 shows the set of involutions as a subset of Bruhat order of S_4 and Figure 2.14 shows the poset $\mathrm{Invol}(S_4)$ by itself.

The *absolute length* of $w \in W$ is defined by

$$a\ell(w) \overset{\text{def}}{=} \min\{k : w = t_1 t_2 \cdots t_k, \text{ for some } t_1, t_2, \ldots, t_k \in T\}.$$

Show the following:

(a) $\mathrm{Invol}(W)$ is a graded poset (i.e., every interval is pure).

(b) The rank function of $\mathrm{Invol}(W)$ is

$$r(w) = \frac{\ell(w) + a\ell(w)}{2}.$$

(c) The Möbius function of $\mathrm{Invol}(W)$ is

$$\mu(u, w) = (-1)^{r(w) - r(u)}.$$

(d) If W is finite of type A, B, or D, then every interval $[u, w]$ is shellable. Conclude that the order complex of (u, w) is homeomorphic to a sphere.

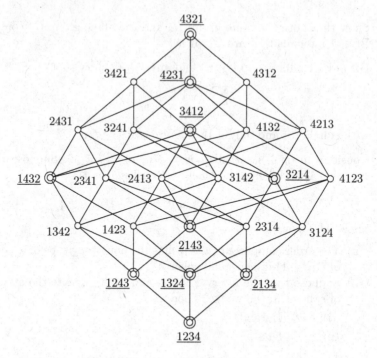

Figure 2.13. Bruhat order of S_4, with the involutions marked.

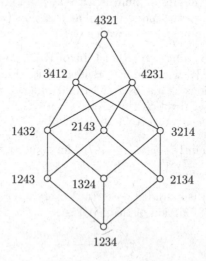

Figure 2.14. The induced subposet Invol(S_4).

(e)* Prove that every interval $[u, w]$ is shellable for general (W, S).

36.* The Robinson-Schensted correspondence specializes to a bijection $w \leftrightarrow (P, P)$ between involutions w and standard Young tableaux P.

Thus, Bruhat order on involutions induces a partial order on the set SYT_n. For $n = 4$, see Figure 2.14. Characterize directly in terms of the tableaux this partial order structure on SYT_n.

37.* Consider only Bruhat intervals occurring in finite Weyl groups. For $m = 1, 2, \ldots$, let $a(m)$ be the number of isomorphism types of length m intervals, $b(m)$ the maximum number of atoms of a length m interval, and $c(m)$ the maximum cardinality of a length m interval. Determine the order of growth (and, if possible, the generating functions) of the sequences

$$\{a(m)\}_{m=1}^{\infty} = \{1, 1, 3, 24, \ldots\},$$
$$\{b(m)\}_{m=1}^{\infty} = \{1, 2, 4, 8, \ldots\},$$
$$\{c(m)\}_{m=1}^{\infty} = \{2, 4, 10, 32, \ldots\}.$$

The same question can be asked about the corresponding numbers for intervals occurring in the symmetric groups.

Notes

Bruhat order was first considered in a geometric context, namely as describing the containment ordering of Schubert varieties in flag manifolds, Grassmannians, and other homogeneous spaces. In this form, Bruhat order was first considered probably by Ehresmann [220] and later in more general settings by Chevalley [134] and others. Since these beginnings, and because of its intimate relationship with naturally induced cell decompositions of certain varieties, Bruhat order has frequently figured in the vast literature on the geometry and representation theory of groups and algebras of Lie type.

The first step to a purely combinatorial study of Bruhat order seems to have been taken by Verma, who conjectured [543], and later proved [544], the formula for the Möbius function of a general Coxeter group (the $J = \emptyset$ case of Corollary 2.7.10). Further steps in this direction were later taken by Bernstein, Gelfand, and Gelfand [35]; and by Deodhar [176].

Incidentally, Verma [543] also seems to be responsible for coining the name "Bruhat order," presumably because of its connection with Bruhat decomposition of semisimple algebraic groups. The name has been questioned, and the historically more correct "Chevalley order" [72], or the more neutral "strong order" [54], has been proposed. However, the name "Bruhat order" seems by now to be firmly established and is undoubtedly here to stay.

The subword property (Theorem 2.2.2) is due to Chevalley [134]. The tableau criterion for the symmetric groups (Theorem 2.6.3) was known to Ehresmann [220], whereas the general version (Theorem 2.6.1) is due to

Deodhar [176]. Section 2.7 is based on the work of Björner and Wachs [65, 55].

Exercise 13. See Verma [544].

Exercise 14. See Deodhar [176].

Exercise 17. See Deodhar [179].

Exercise 18 and 19. See Proctor [424]. This also contains the classification of all cases in which W^J is a lattice.

Exercise 20. See van den Hombergh [298].

Exercises 21, 22, and 23. See Björner and Wachs [67].

Exercise 24. See Reading [430].

Exercises 25 and 26. See Dyer [203].

Exercise 27. See Brenti, Caselli, and Marietti [100].

Exercise 28. See Higman [294] and Björner [54].

Exercise 29(a). See Kerov [325].

Exercise 30 follows from a more general result of Dyer; see [206].

Exercise 33. See Björner and Wachs [65].

Exercise 35. See Richardson and Springer [445], Incitti [309, 310, 311, 312] and Hultman [303].

3

Weak order and reduced words

When working with a Coxeter group, one is sooner or later faced with problems concerning the combinatorics of reduced words. When do two such words represent the same group element? How should one best choose "distinguished" reduced words to represent the group elements? Such questions are discussed in this chapter and the following one.

We begin with a partial order structure on the group W that is intimately related to the language of reduced words. This partial order is, in fact, a semilattice (and in the finite case a lattice), which introduces additional algebraic structure.

3.1 Weak order

Let (W, S) be a Coxeter group, and let $u, w \in W$.

Definition 3.1.1 *(i)* $u \leq_R w$ *means that* $w = us_1s_2 \ldots s_k$, *for some* $s_i \in S$ *such that* $\ell(us_1s_2 \ldots s_i) = \ell(u) + i$, $0 \leq i \leq k$.

(ii) $u \leq_L w$ *means that* $w = s_ks_{k-1} \ldots s_1u$, *for some* $s_i \in S$ *such that* $\ell(s_is_{i-1} \ldots s_1u) = \ell(u) + i$, $0 \leq i \leq k$.

This defines the *right weak order* and the *left weak order*, respectively. Right and left weak order are distinct partial orderings of W; however, they are isomorphic via the map $w \mapsto w^{-1}$. We henceforth often drop the adjective "weak" and speak only of *right order* and *left order*. Also, results are usually stated for the right order only, although sometimes referred to for left order.

It is immediate from the definition that these orderings are strictly weaker than Bruhat order in the sense of having fewer relations:

$$u \leq_R w \text{ or } u \leq_L w \Rightarrow u \leq w. \tag{3.1}$$

We have the convention that all partial order notation refers to Bruhat order, unless given a subscript "R" or "L," in which case it refers to the respective weak orderings. For instance, $[u, w]_R = \{x \in W : u \leq_R x \leq_R w\}$ is a right order interval, and $u \lessdot_L w$ denotes a covering in left order.

The following are a few immediate observations.

Proposition 3.1.2   *(i) There is a one-to-one correspondence between reduced decompositions of w and maximal chains in the interval $[e, w]_R$.*

(ii) $u \leq_R w \iff \ell(u) + \ell(u^{-1}w) = \ell(w)$.

(iii) If W is finite, then $w \leq_R w_0$ for all $w \in W$.

(iv) Weak order satisfies the "prefix property":

$$u \leq_R w \quad \Leftrightarrow \quad \text{there exist reduced expressions}$$
$$u = s_1 s_2 \ldots s_k \text{ and } w = s_1 s_2 \ldots s_k s_1' s_2' \ldots s_q'.$$

(v) Weak order satisfies a "chain property" analogous to Theorem 2.2.6.

(vi) Suppose $s \in D_L(u) \cap D_L(w)$. Then, $u \leq_R w \iff su \leq_R sw$. $\square$

Property *(v)* shows that W under weak order is a graded poset ranked by the length function, and so is also every interval $[u, w]_R$.

The diagrams of the dihedral groups $I_2(4) \cong B_2$ and $I_2(\infty)$ under right order are shown in Figure 3.1.

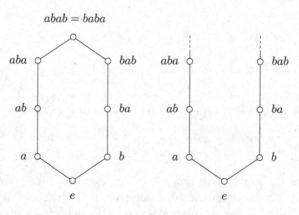

Figure 3.1. Weak order of dihedral groups.

For the symmetric group S_n, we have that $x \leq_R y$ if and only if the permutation y can be obtained from x via a sequence of adjacent transpositions that at each step increases the number of inversions. For instance,

$$263154 - 263514 - 623514 - 623541 - 632541$$

shows that $263154 <_R 632541$. The right order of S_4 is shown in Figure 3.2 and that of the group H_3 is shown in Figure 3.3. (The labels, x, y, and z in Figure 3.3 are explained in Exercise 1.)

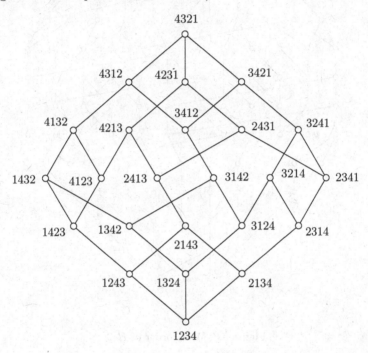

Figure 3.2. Weak order of S_4.

Let $r(w)$ denote the number of reduced decompositions of an element $w \in W$. Then,

$$r(w) = \sum_{u \lessdot_R w} r(u), \tag{3.2}$$

as is easily seen from Proposition 3.1.2(i). This recurrence can be used to count reduced decompositions in small examples. For instance, computing over the order diagram of Figure 3.2, we quickly find that $r(w_0) = 16$ for the top element $w_0 = 4321$ of S_4. In general, more sophisticated tools are needed for counting reduced decompositions; we will return to this topic in Chapter 7.

The following is a simple characterization of weak order in terms of sets of associated reflections.

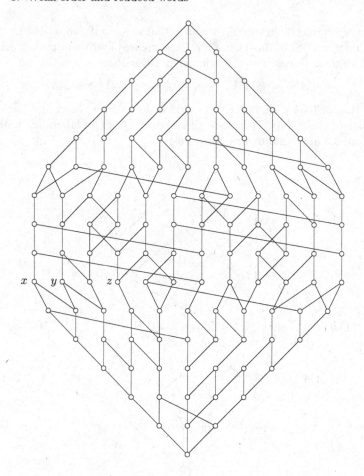

Figure 3.3. Weak order of H_3.

Proposition 3.1.3 $u \leq_R w \Leftrightarrow T_L(u) \subseteq T_L(w)$.

Proof. If $u = s_1 s_2 \ldots s_k$ and $w = s_1 s_2 \ldots s_k s_{k+1} \ldots s_q$ are reduced, then

$$T_L(u) = \{s_1 s_2 \ldots s_i \ldots s_2 s_1 : 1 \leq i \leq k\}$$
$$\subseteq \{s_1 s_2 \ldots s_i \ldots s_2 s_1 : 1 \leq i \leq q\} = T_L(w).$$

Conversely, suppose that $u = s_1 s_2 \ldots s_k$ is reduced. Let $t_i = s_1 s_2 \ldots s_i \ldots s_2 s_1$, $1 \leq i \leq k$, and assume that

$$T_L(u) = \{t_1, t_2, \ldots, t_k\} \subseteq T_L(w).$$

For $0 \leq i \leq k$, consider the following claim:

$\mathcal{C}(i)$: there exists a reduced expression $w = s_1 s_2 \ldots s_i s_1' s_2' \ldots s_{q-i}'$.

Clearly, $\mathcal{C}(0)$ is true. Now, suppose that $\mathcal{C}(i)$ is true, for some $0 \leq i < k$. The fact that $t_{i+1} \in T_L(w)$ and that $t_j \neq t_{i+1}$ for $j \leq i$ (Lemma 1.3.1) shows that

$$t_{i+1} = s_1 \ldots s_i s_1' \ldots s_m' \ldots s_1' s_i \ldots s_1$$

for some $1 \le m \le q - i$. Hence,

$$w = t_{i+1}^2 w$$
$$= (s_1 \ldots s_{i+1} \ldots s_1)(s_1 \ldots s_i s_1' \ldots \widehat{s_m'} \ldots s_{q-i}')$$
$$= s_1 \ldots s_{i+1} s_1' \ldots \widehat{s_m'} \ldots s_{q-i}',$$

which establishes $\mathcal{C}(i + 1)$. Hence, by induction, $\mathcal{C}(k)$ is true, as was to be shown. $\square$

Corollary 3.1.4 *The mapping $w \mapsto T_L(w)$ provides an order-preserving and rank-preserving embedding of W, as a poset under weak order, into the lattice of all finite subsets of T.* $\square$

Weak order on a finite Coxeter group shares with Bruhat order the symmetries induced by the top element.

Proposition 3.1.5 *For weak order on a finite W, the following hold:*

(i) $w \mapsto w_0 w$ and $w \mapsto w w_0$ are antiautomorphisms.

(ii) $w \mapsto w_0 w w_0$ is an automorphism.

Proof. This follows from the length formulas in Proposition 2.3.2 and its corollary and from the fact that if $s \in S$, then $s w_0 = w_0 s'$ for some $s' \in S$ (a consequence of $w_0 S w_0 = S$). For instance, if $w \lhd_R ws$, then $w_0 w w_0 \lhd_R w_0 w s w_0$, since $\ell(w_0 w s w_0) = \ell(ws) = \ell(w) + 1 = \ell(w_0 w w_0) + 1$ and $w_0 w s w_0 = w_0 w w_0 s'$. $\square$

The combinatorial structure of intervals is a difficult and interesting question for both Bruhat and weak order. We have already observed that an interval $[u, w]_R$ is a graded poset. For weak order, there is a simple translation principle that shows that every interval is isomorphic to a lower interval (i.e., an interval bounded below by the identity).

Proposition 3.1.6 *If $u \le_R w$, then $[u, w]_R \cong [e, u^{-1} w]_R$.*

Proof. We will show that the mapping $x \mapsto ux$ is a poset isomorphism $[e, u^{-1} w]_R \to [u, w]_R$. The basic properties of the length function give:

$$\ell(w) = \ell(u) + \ell(u^{-1} w)$$
$$\overset{(a)}{\le} \ell(u) + \ell(x) + \ell(x^{-1} u^{-1} w) \overset{(b)}{\ge} \ell(ux) + \ell(x^{-1} u^{-1} w) \overset{(c)}{\ge} \ell(w).$$

Now,

$$x \le_R u^{-1} w \Leftrightarrow \text{ equality in (a)}$$
$$\Leftrightarrow \text{ equality in (b) and (c)}$$
$$\Leftrightarrow u \le_R ux \le_R w.$$

Hence, $x \in [e, u^{-1}w]_R \Leftrightarrow ux \in [u, w]_R$, and if so, $\ell(ux) = \ell(u) + \ell(x)$. $\square$

The Boolean embeddability of weak order (Corollary 3.1.4) has the following consequences for interval structure.

Corollary 3.1.7 *Let* $u \leq_R w$ *and* $m \overset{\text{def}}{=} \ell(u, w)$. *Then, for* $0 \leq k \leq m$,

$$\#\{v \in [u, w]_R : \ell(v) = \ell(u) + k\} \leq \binom{m}{k}. \qquad \square$$

Corollary 3.1.8 *There are (up to isomorphism) only finitely many posets of each length* $m \geq 0$ *occurring as intervals in weak order of Coxeter groups.* $\square$

3.2   The lattice property

An element z in a poset is said to be the *meet* (or, greatest lower bound) of a subset A if (i) $z \leq y$ for all $y \in A$, and (ii) $u \leq y$ for all $y \in A$ implies that $u \leq z$. The meet, if it exists, is clearly unique. It is then denoted $\bigwedge A$, or if $A = \{x, y\}$, simply $x \wedge y$. A poset L for which every nonempty subset has a meet is called a *complete meet-semilattice*. Such a poset has a minimum element $\hat{0} = \bigwedge L$.

Bruhat order is not a semilattice, as is clear from a glance at Figure 2.1. It turns out, however, that weak order is. In the following, (W, S) is an arbitrary Coxeter system.

Theorem 3.2.1 *Weak order on* W *is a complete meet-semilattice.*

Proof. Given $x, y \in W$ we want to show that $x \wedge y$ exists. This is done by induction on $\ell(x)$. If $\ell(x) = 0$ or there is no $s \in S$ such that $s \leq_R x$ and $s \leq_R y$, then $x \wedge y = e$. So we may assume that $\ell(x) > 0$ and $E \overset{\text{def}}{=} [e, x]_R \cap [e, y]_R \neq \{e\}$. Pick an element $z \in E$ of maximal length. We will show that $w \leq_R z$ for all $w \in E$, implying that $z = x \wedge y$.

We first show that if $s \in E \cap S$, then $s \leq_R z$. Let $z = s_1 \ldots s_r$, $x = s_1 \ldots s_r s_1' \ldots s_p'$ and $y = s_1 \ldots s_r s_1'' \ldots s_q''$ be reduced decompositions. If $s \not\leq_R z$, then by the Exchange Property,

$$x = ss_1 \ldots s_r s_1' \ldots \widehat{s_i'} \ldots s_p' \quad \text{and} \quad y = ss_1 \ldots s_r s_1'' \ldots \widehat{s_j''} \ldots s_q''$$

are reduced. So $ss_1 \ldots s_r \in E$. However, $\ell(ss_1 \ldots s_r) > \ell(z)$, contradicting the choice of z.

Now, let $w \in E \setminus \{e\}$. We make repeated use of Proposition 3.1.2(vi), to which, for brevity, we here refer by (**). Let $s \in D_L(w)$. Then, $s \in D_L(x) \cap D_L(y)$, and, by what was shown above, also $s \in D_L(z)$. Since $\ell(sx) < \ell(x)$, by induction $sx \wedge sy$ exists, say $z' \overset{\text{def}}{=} sx \wedge sy$. By (**), we have that $sw \leq_R z'$ and $sz \leq_R z'$. Also, since $z' \leq_R sx$, (**) shows that $sz' \leq_R x$, and similarly $sz' \leq_R y$. Hence, $sz' \in E$, implying that

$\ell(sz') \leq \ell(z)$. We have shown that, on the one hand, $sz \leq_R z'$, and, on the other, $\ell(sz) = \ell(z) - 1 \geq \ell(sz') - 1 = \ell(z')$. Hence, $sz = z'$. Therefore, $sw \leq_R sz$, which (again by (**)) implies $w \leq_R z$, as was to be shown.

The existence of meets $\bigwedge A$ for arbitrary nonempty subsets $A \subseteq W$ follows from the existence of pairwise meets and the descending chain condition by a standard argument; namely let $x_0 \in A$. If $x_0 \leq_R y$ for all $y \in A$, then $\bigwedge A = x_0$. Otherwise, choose $y_0 \in A$ such that $x_0 \not\leq_R y_0$ and let $x_1 = x_0 \wedge y_0 <_R x_0$. If $x_1 \leq_R y$ for all $y \in A$, then $\bigwedge A = x_1$. Otherwise, choose $y_1 \in A$ such that $x_1 \not\leq_R y_1$ and let $x_2 = x_1 \wedge y_1 <_R x_1$, and so on. This creates a sequence $x_0, x_1, x_2, \ldots$ that either terminates for some i with $\bigwedge A = x_i$, or else is strictly descending $\ell(x_0) > \ell(x_1) > \ell(x_2) > \cdots$. Since the latter is impossible, we are done. $\square$

If W is infinite, then *joins* (least upper bounds) may fail to exist (see, e.g., the infinite dihedral group in Figure 3.1). However, if a subset $A \subseteq W$ has *some* upper bound in weak order, then it follows from Theorem 3.2.1 that the join $\bigvee A$ exists (take the meet of the set of all upper bounds). If W is finite, then there is the universal upper bound w_0, so that both joins and meets exist for arbitrary subsets. Thus, in the finite case, W is a *lattice*. In fact, it has additional structure as a lattice.

A lattice L with bottom and top elements $\widehat{0}$ and $\widehat{1}$ is called an *ortholattice* (see Birkhoff [52]) if there exists a map $x \mapsto x^\perp$ on L such that the following hold:

(α) $x \vee x^\perp = \widehat{1}$, $x \wedge x^\perp = \widehat{0}$, for all $x \in L$.

(β) $x \leq y \Rightarrow x^\perp \geq y^\perp$, for all $x, y \in L$.

(γ) $x^{\perp\perp} = x$, for all $x \in L$.

Corollary 3.2.2 *Right order on a finite Coxeter group* (W, S) *together with the translation* $x \mapsto x w_0$ *gives* W *the structure of an ortholattice.*

Proof. This follows from Propositions 2.3.2(iii) and 3.1.3. $\square$

The following lemmas show some cases when join and meet can be reasonably expressed in terms of multiplication.

Lemma 3.2.3 *Let* $J \subseteq S$. *Then,* $\bigvee J$ *exists if and only if* W_J *is finite, and if so,* $\bigvee J = w_0(J)$. *In particular, if* $|W_J| = \infty$, *then* J *has no upper bound.*

Proof. If W_J is finite, then $D_L(w_0(J)) = J$ shows, keeping Corollary 1.4.6 in mind, that $w_0(J)$ is an upper bound to the set J. We show that, conversely, if J has an upper bound w, or equivalently if $J \subseteq D_L(w)$, then W_J is finite and $w_0(J) \leq_R w$. Let $^J w$ be the minimal representative of the coset $W_J w$ (see equation (2.12)), so that

$$w = w_J \, ^J w \quad \text{and} \quad \ell(w) = \ell(w_J) + \ell(^J w), \quad w_J \in W_J. \tag{3.3}$$

By assumption, for every $s \in J$ we have

$$\ell(sw_J) + \ell({}^J w) = \ell(sw_J \, {}^J w) = \ell(sw) < \ell(w) = \ell(w_J) + \ell({}^J w).$$

Hence, $J \subseteq D_L(w_J)$, which by Proposition 2.3.1(ii) implies that W_J is finite with top element $w_0(J) = w_J$. Equation (3.3) then shows that $w_0(J) \leq_R w$. $\square$

Lemma 3.2.4 Let $w \in W$, $J \subseteq S$ with W_J finite.

(i) If $w \lhd_R ws$ for all $s \in J$, then $\bigvee \{ws : s \in J\} = ww_0(J)$.

(ii) If $ws \lhd_R w$ for all $s \in J$, then $\bigwedge \{ws : s \in J\} = ww_0(J)$.

Proof. The conditions mean that $w \in W^J$ (minimal coset representatives) and $w \in \mathcal{D}_J^S$ (maximal coset representatives), respectively. Part (i) then follows easily via Propositions 3.1.6 and 3.2.3. Part (ii) requires a few more steps that we leave to the reader. $\square$

The meet construction $x \wedge y$ in right order endows the set of group elements W with a new binary operation, different from its group multiplication. We remind the reader (see, e.g., [52]) that a semilattice $(L, \wedge)$ can be characterized either as a partial order in which pairwise meets exist or, equivalently, as an algebraic structure with a binary operation "$\wedge$" that is commutative, associative, and idempotent. We have seen that the group structure of a Coxeter group (W, S) determines the semilattice structure of $(W, \wedge)$. It is natural to ask: Does, conversely, the algebraic structure $(W, \wedge)$ determine the group structure? In other words, can Coxeter groups alternatively be thought of as algebraic systems in terms of the meet operation? For example, say that we are given (1) the group multiplication table of W (with the elements of S marked) and (2) the meet multiplication table of W. Can the information in (1) be computed from the information in (2), so that (1) and (2) provide completely equivalent information (not only up to isomorphism)? We will show that this is true in essentially all cases of interest.

Theorem 3.2.5 Let (W, S) be a Coxeter group. Then, the following two conditions are equivalent:

(i) The group multiplication is uniquely determined by the (right order) meet operation "$\wedge$."

(ii) If $\varphi : W \to W$ is an automorphism of right order such that $\varphi(s) = s$ for all $s \in S$, then $\varphi(w) = w$ for all $w \in W$.

Furthermore, both (i) and (ii) are implied by the following:

(iii) The ∞-labeled edges in W's Coxeter diagram are pairwise disjoint (i.e., if $m(s_1, s_2) = m(s_1', s_2') = \infty$, then either $\{s_1, s_2\} = \{s_1', s_2'\}$ or $\{s_1, s_2\} \cap \{s_1', s_2'\} = \emptyset$).

Proof. (i) ⇔ (ii). Begin by looking at W as an abstract set. It is clear that knowing the ∧-multiplication table on W is equivalent to knowing the right order on W. Suppose that this information is given. The set S can be recognized as the atoms of right order (the elements covering its bottom element e). The pairwise joins $s \vee s'$ ($s, s' \in S$) can also be recognized from the order structure (when they exist), and we know from Lemma 3.2.3 that $s \vee s' = w_0(\{s, s'\})$, so $m(s, s') = \ell(s \vee s')$ can be determined. Hence, the Coxeter matrix m of W is determined by right order.

Now, let (W', S) be the Coxeter group determined by m. So, we have an isomorphism $\varphi : W \to W'$, which is the identity on the common subset S. The multiplication table is known in W'. Hence, the multiplication is uniquely determined in W from the given data if and only if φ is unique.

(iii) ⇒ (ii). We will show that $\varphi(w) = w$ for all $w \in W$ by induction on $\ell(w)$, the $\ell(w) = 1$ case being assumed.

Suppose that $u \in W$ is a minimal length exception, say $\varphi(u) = v \neq u$. If u covers (at least) two elements, then since these are fixed, we would have a contradiction to the lattice property:

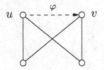

Hence, we may assume that $w \lessdot_R u$ for a unique w.

Let $a, b, c \in S$ be such that $u = wa$, $v = wb$ and $w >_R wc$:

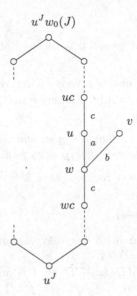

Then, $u <_R uc$ due to the uniqueness of w. We may assume that $m(a, c) \neq \infty$; if not, then by assumption (iii), we have $m(b, c) \neq \infty$ and the argument can be pursued with v replacing u. Let $J = \{a, c\}$ so that $w_0(J) = acaca\ldots$, and decompose

$$u = u^J u_J, \quad \text{with } u^J \in W^J \text{ and } u_J \in W_J.$$

We have that $u^J \lhd_R u^J a$, $u^J c$, and by induction $u^J a$ and $u^J c$ are fixed by φ. Since φ is a lattice automorphism, then also their join (by Lemma 3.2.4) $u^J w_0(J)$ is fixed by φ. Then, $u \leq_R u^J w_0(J)$ implies $v = \varphi(u) \leq_R \varphi(u^J w_0(J)) = u^J w_0(J)$. However, this would imply an equivalence of two reduced expressions $acaca\ldots = bs_1s_2\ldots$ (following two chains from w to $u^J w_0(J)$ via u and via v), which is impossible by Corollary 1.4.8(ii). $\square$

Corollary 3.2.6 *If the diagram of (W, S) has pairwise disjoint ∞-labeled edges, then the automorphism group of weak order is isomorphic to the group of diagram automorphisms.* $\square$

That some restriction on the distribution of ∞-labeled edges is needed in these results is shown by the universal groups. In the weak order of U_n, the segment above every atom is a complete $(n-1)$-ary tree; the case of $n = 3$ is shown in Figure 3.4.

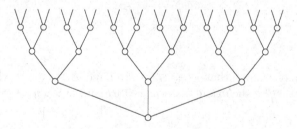

Figure 3.4. Weak order of U_3.

The lattice property makes it easy to determine the homotopy type of intervals in weak order and its Möbius function $\mu_R(u, w)$. See Appendix A2 for the relevant definitions.

Theorem 3.2.7 *Suppose that $u <_R w$ and $\ell(u, w) \geq 2$. Then, the order complex of the open interval $(u, w)_R$ is as follows:*

(i) *Homotopy equivalent to the sphere $\mathbb{S}^{|J|-2}$, if $w = uw_0(J)$ for some $J \subseteq S$.*

(ii) *Contractible, otherwise.*

Proof. By Proposition 3.1.6 we may assume that $u = e$. For each $x \in (e, w)_R$, let $f(x) = \bigvee\{s \in S : s \leq_R x\}$. Lemma 3.2.3 shows that, equivalently, $f(x) = w_0(D_L(x))$. Then, f is an order-preserving mapping of $(e, w)_R$ into itself, and $x \geq_R f(x) = f^2(x)$ for all $x \in (e, w)_R$. Hence, by Fact A2.3.2 and referring to order complexes, we have that $(e, w)_R$ is homotopy equivalent to its image $f((e, w)_R)$.

If $f(w) <_R w$, then $f((e,w)_R)$ has a top element $f(w)$; hence, its order complex is a cone and thus contractible. If $w = f(w) = w_0(D_L(w))$, then the mapping $I \mapsto w_0(I)$ gives an isomorphism from the poset of all proper subsets of $J \overset{\mathrm{def}}{=} D_L(w)$ to the poset $f((e,w)_R)$. The order complex of the latter is therefore isomorphic to the barycentric subdivision of the boundary of a $(|J| - 1)$-simplex (i.e., a triangulation of the $(|J| - 2)$-sphere). $\square$

Corollary 3.2.8

$$\mu_R(u,w) = \begin{cases} (-1)^{|J|} & , \text{ if } w = uw_0(J) \text{ for some } J \subseteq S; \\ 0 & , \text{ otherwise.} \end{cases}$$

Proof. This follows since the Möbius function is the reduced Euler characteristic of the order complex; see Fact A2.3.1. $\square$

3.3   The word property

The problem of recognizing when two words in letters from the generating set S represent the same element in W (called the "word problem") has an algorithmic solution for any Coxeter group (W, S). In this section, we describe an algorithm that solves the problem in theory but may be unwieldy in practice. It is based on an important combinatorial fact that we call the "word property." More efficient alternative procedures exist for dealing with the word problem, as will be discussed in Sections 3.4 and 4.3.

Let us go back to the definition of (W, S) in terms of its Coxeter matrix $m = (m(s,s'))_{s,s' \in S}$. For $s, s' \in S$ such that $m(s, s')$ is finite, denote by $\alpha_{s,s'}$ the alternating word $ss'ss's\ldots$ of length $m(s, s')$. Then, we have in (W, S) the two defining types of relation:

$$ss = e, \quad \text{for all } s \in S, \tag{3.4}$$

$$\alpha_{s,s'} = \alpha_{s',s}, \quad \text{for all } (s, s') \in S_{\text{fin}}^2. \tag{3.5}$$

For instance, in H_3 (the symmetry group of the icosahedron) with diagram $\overset{5}{\underset{a \quad\quad b \quad\quad c}{\circ\!\!-\!\!-\!\!-\!\!\circ\cdots\circ}}$ we have the relations

$$aa = bb = cc = e$$

of the first kind, and

$$ababa = babab, \quad ac = ca, \quad bcb = cbc$$

of the second.

Operating on words in the free monoid S^*, let us call a *nil-move* the deletion of a factor of the form ss, and a *braid-move* the replacement of a factor $\alpha_{s,s'}$ by $\alpha_{s',s}$. If a word β is changed to a word γ by either a nil-move or a braid-move, we write $\beta \sim \gamma$. For instance, the following shows a sequence of two nil-moves and two braid-moves in H_3:

$$cb\underline{c}\,\underline{a}cbabac \sim cbac\,\underline{c}babac \sim c\underline{b}\,\underline{a}\,\underline{b}\,\underline{a}\,bac \sim cabab\underline{a}\,\underline{a}c \sim cababc.$$

It is clear that the image under the surjection (1.6) $\varphi : S^* \to W$ is unaffected by these kinds of move.

Although braid-moves are symmetric (the inverse of a braid-move is a braid-move), nil-moves are not — they shorten the length of the word they operate on and are therefore irreversible. It is a nontrivial fact, which will now be shown, that if $\alpha, \beta \in S^*$ with $\varphi(\alpha) = \varphi(\beta)$ and β is reduced, then α can be transformed to β via some sequence of nil-moves and braid-moves. In particular, if α and β are both reduced expressions, then braid-moves suffice. Note that if we allowed reversible nil-moves $ss \leftrightarrow \emptyset$, then there would be nothing to prove; the statement is then equivalent to the definition of being a Coxeter group.

Theorem 3.3.1 (Word Property) *Let (W, S) be a Coxeter group and $w \in W$.*

(i) *Any expression $s_1 s_2 \ldots s_q$ for w can be transformed into a reduced expression for w by a sequence of nil-moves and braid-moves.*

(ii) *Every two reduced expressions for w can be connected via a sequence of braid-moves.*

Proof. We begin by proving (ii). This will be done by induction on $\ell(w)$, the result being clear if $\ell(w) \leq 1$. Assume that $\ell(w) > 1$ and let

$$w = s\,s_2 \ldots s_k = s' s_2' \ldots s_k'$$

be two reduced expressions. If $s = s'$, then, by induction, $s_2 \ldots s_k$ can be changed into $s_2' \ldots s_k'$ using braid-moves, and we are done.

So, assume that $s \neq s'$. We have that $s, s' \leq_R w$, so using the lattice property of weak order,

$$\varphi(\alpha_{s',s}) = \varphi(\alpha_{s,s'}) = w_0(\{s, s'\}) = s \vee s' \leq_R w.$$

This means there exists a word $\beta \in S^*$ such that $w = \varphi(\alpha_{s,s'}\beta) = \varphi(\alpha_{s',s}\beta)$, and $\alpha_{s,s'}\beta$ and $\alpha_{s',s}\beta$ are reduced. Then, we have

$$ss_2 \ldots s_k \sim \alpha_{s,s'}\beta \sim \alpha_{s',s}\beta \sim s' s_2' \ldots s_k',$$

where the first and last equivalences are given by sequences of braid-moves (by induction, since the first letter is in each case the same) and the middle equivalence is the braid-move $\alpha_{s,s'} \to \alpha_{s',s}$. The argument is easily visualized in terms of maximal chains in the weak order interval $[e, w]_R$; see Figure 3.5.

Next, (i) will be shown by induction on q, for all $q \geq \ell(w)$. Assume that $q > \ell(w)$, and let i be minimal such that $s_{i+1}s_{i+2} \ldots s_q$ is reduced. Then, by the Exchange Property, $s_{i+1}s_{i+2} \ldots s_q = s_i s_{i+1} \ldots \widehat{s_j} \ldots s_q$, for some $i + 1 \leq j \leq q$, and by part (ii), this equality of reduced expressions

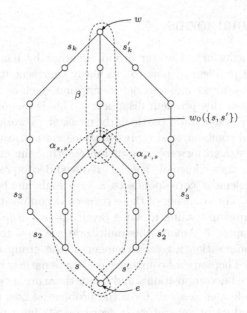

Figure 3.5. Illustration to the proof of Theorem 3.3.1.

is achieved by some sequence of braid-moves. Hence, a reduction of the length q of the given expression can be achieved by braid-moves followed by a nil-move as follows:

$$s_1 \ldots s_i s_{i+1} \ldots s_q \sim s_1 \ldots s_i s_i s_{i+1} \ldots \widehat{s}_j \ldots s_q \sim s_1 \ldots \widehat{s}_i \ldots \widehat{s}_j \ldots s_q.$$

$\square$

The result suggests the following finite algorithm for solving the word problem. Given a word $\alpha \in S^*$, let $RD(\alpha)$ be the set of reduced descendants, reachable by nil-moves and braid-moves from α. This is a finite set, computable in a finite number of steps. Then, as shown by Theorem 3.3.1, $\alpha, \beta \in S^*$ represent the same group element if and only if $RD(\alpha) \cap RD(\beta) \neq \emptyset$. Other methods for solving the word problem are discussed in Sections 3.4 (normal form) and 4.3 (the numbers game).

Let $\mathcal{R}(w)$ be the set of all reduced decompositions of an element w. Theorem 3.3.1 shows that $\mathcal{R}(w)$ can be thought of as the set of nodes of a connected graph whose edges are the braid-moves. The braid-moves can be subdivided into k-*moves*, for $k = 2, 3, 4, \ldots$, according to the order of the corresponding generating relation $(ss')^k = e$. For instance, the following are examples of a 2-move: $ab \leftrightarrow ba$, 3-move: $aba \leftrightarrow bab$, 4-move: $abab \leftrightarrow baba$, and so on.

If all edges except 2-moves are erased from the graph $\mathcal{R}(w)$, we get a (possibly disconnected) subgraph of $\mathcal{R}(w)$. The connected components of this graph are called *commutation classes*.

3.4   Normal forms

When computing in a Coxeter group, whether by hand or by machine, one faces the problem of how to efficiently represent its elements. If the group admits a nice combinatorial description, such as those discussed in Chapter 8, then this problem disappears. This is one of the reasons why those Coxeter systems are particularly pleasant to work with. In general, however, good combinatorial representations are not available.[1]

The most straightforward way of representing the elements of a Coxeter system is via a choice of "special" reduced decompositions. In general, every group element corresponds to a very high number of reduced decompositions. The counting of these reduced decompositions is in itself an interesting problem leading to some beautiful and deep combinatorial results; see Chapter 7. Among this multitude, we have to choose a specific reduced decomposition for each element of the group, usually called its *normal form*. There are, of course, several systematic ways to choose these specific reduced decompositions, and hence there are several normal forms.

Having made a choice, one faces the problem of how to algorithmically convert a product of generators (a word in S^*) into its normal form. A good solution to this problem will allow us to describe multiplication in terms of normal forms, since if we have the normal forms of two elements $u, v \in W$, then their concatenation (as words) will not, in general, be the normal form of uv. It will also allow us to solve word problems, since two words in S^* represent the same group element if and only if their normal forms are the same.

In this section, we study what is probably the most common normal form for the elements of a Coxeter group, namely the *lexicographically first reduced word*. We show how for this choice of normal form one can rather easily compute the normal form of uv, given those of u and v. As will be seen, the properties of the lexicographically first normal form are not only useful for computations but are also interesting and elegant from a combinatorial point of view.

Let (W, S) be a Coxeter group with $S = \{s_1, \ldots, s_n\}$. In this section, we identify, for simplicity, s_i with i, for $i = 1, \ldots, n$. In particular, we will identify a sequence $(s_{i_1}, \ldots, s_{i_p}) \in S^p$ with $i_1 \ldots i_p \in [n]^p$. (To simplify notation, we write $i_1 \ldots i_p$ instead of $(i_1, \ldots, i_p)$.) The *normal form* of an element $w \in W$ is $\min \mathcal{R}(w)$, where the minimum is taken with respect to

---

[1] The so-called "numbers game" (Section 4.3) offers a general method for finding combinatorial representatives of the group elements. However, there are certain shortcomings: (1) It is in general not at all clear which assignments of numbers to the nodes of the Coxeter graph correspond to actual positions in the numbers game (i.e., to actual group elements) and (2) there is in general no simple algorithmic procedure for multiplying two positions in the numbers game (i.e., to compute the product of the corresponding group elements).

lexicographic order. Given $w \in W$, we denote by $NF(w)$ its normal form (so $NF(w) \in [n]^*$). Note that $NF(w)$ depends on the choice of indexing $s_1, \ldots, s_n$ for the elements of S.

As an example, let us compute the normal form of $w = 21543 \in S_5$. With the usual indexing $s_i \stackrel{\text{def}}{=} (i, i+1)$, we have

$$\mathcal{R}(w) = \{1343, 1434, 3143, 3413, 3431, 4134, 4314, 4341\}.$$

Thus, $NF(w) = \min \mathcal{R}(w) = 1343$.

Note that if $i_1 \ldots i_p$ is a normal form, then necessarily any factor $i_j i_{j+1} \ldots i_k$, $1 \le j \le k \le p$, is also a normal form. Also note that if $J \subseteq S$, then, by Corollary 1.4.8, $\mathcal{R}(W_J) \subseteq J^*$. Hence, if $w \in W_J$, then the normal form of w as an element of W_J coincides with its normal form as an element of W.

We call an element $w \in W$ *distinguished* if $w = e$ or the normal form of w begins with "n." So, for example, $21543 \in S_5$ is not distinguished since, as we have seen, $NF(21543) = 1343$. It is easy to characterize distinguished elements, even without knowing their normal form. Recall from definition (2.11) the notation $^{[n-1]}W$ for the system of minimal right coset representatives modulo the subgroup $W_{[n-1]} = W_{\{s_1, \ldots, s_{n-1}\}}$.

Lemma 3.4.1 *Let $w \in W$. Then, w is distinguished if and only if $w \in$ $^{[n-1]}W$.*

Proof. We know from Proposition 2.4.3 that $w \in {}^{[n-1]}W$ if and only if no reduced word for w begins with one of the letters $s_1, \ldots, s_{n-1}$. However, this is precisely what it means for w to be distinguished. $\square$

The following result is a key to the combinatorics of normal forms.

Proposition 3.4.2 *Let $w \in W$. Then, w can be uniquely written as $w = x_1 \cdots x_n$, where $x_i \in {}^{[i-1]}(W_{[i]})$ for $i = 1, \ldots, n$. Furthermore, $NF(w) = NF(x_1) \cdots NF(x_n)$.*

Proof. The existence of the unique factorization $w = x_1 \cdots x_n$ is a consequence of Corollary 2.4.6. It remains only to prove the statement about normal forms.

If $w \in W_{[n-1]}$, the result holds by induction, since then $x_n = e \in {}^{[n-1]}W$. So, assume that $w \notin W_{[n-1]}$. Then, s_n appears in every reduced decomposition of w, and hence n appears in the normal form of w. So, let $i_1 \ldots i_p \stackrel{\text{def}}{=} NF(w)$ and let $r \stackrel{\text{def}}{=} \min\{j \in [p] : i_j = n\}$. Then, $s_{i_r} \ldots s_{i_p}$ is a distinguished group element, hence (by Lemma 3.4.1) $s_{i_r} \ldots s_{i_p} \in {}^{[n-1]}W$, hence $s_{i_r} \ldots s_{i_p} = x_n$ and $NF(x_n) = i_r \ldots i_p$. Let $u \stackrel{\text{def}}{=} s_{i_1} \ldots s_{i_{r-1}}$. Then, $u = x_1 \ldots x_{n-1} \in W_{[n-1]}$, $NF(u) = i_1 \ldots i_{r-1}$, and, by induction, $NF(u) = NF(x_1) \ldots NF(x_{n-1})$. $\square$

We illustrate with a small example. Let $w = 21543 \in S_5$ (so $n = 4$). Then, since $NF(w) = 1343$, we deduce from the preceding proof that $x_4 = s_4 s_3$, $x_3 = s_3$, $x_2 = e$, and $x_1 = s_1$.

Example 3.4.3 For the symmetric group S_n, the factorization $w = x_1 x_2 \ldots x_n$ of Proposition 3.4.2 amounts to a version of the sorting procedure for permutations known as "bubble sort." Namely, written as $w x_n^{-1} x_{n-1}^{-1} \ldots x_1^{-1} = e$, it can be interpreted as follows, keeping in mind that $x_i \in^{[i-1]} (W_{[i]})$ here means that $x_i = e$ or $x_i = s_i s_{i-1} \ldots s_{i-j}$ for some $0 \le j < i$. Start with the permutation w. Then, move the value "n" by adjacent transpositions to the right until it is in the last position; this gives the permutation $w s_{n-j_n} \ldots s_{n-1} s_n = w x_n^{-1}$. Next, move the value "$n-1$" to the right until it is in the next to last position; this gives $w x_n^{-1} s_{n-1-j_{n-1}} \ldots s_{n-2} s_{n-1} = w x_n^{-1} x_{n-1}^{-1}$. Then, continue with "$n-2$," and so on. $\square$

For finite groups, there is a useful way of viewing the factorization of normal forms given by Proposition 3.4.2. The *normal form forest* of a finite Coxeter group (W, S) (with a total order of the set S assumed) consists of edge-labeled rooted trees $\tau_1, \ldots, \tau_{|S|}$. The nodes of τ_i correspond to the elements of the quotient $^{[i-1]}(W_{[i]})$, and the edges are labeled by elements from S in such a way that the path from the root of τ_i to the node w generates the word $NF(w)$. Then, the set of all normal forms is obtained by concatenating such rooted paths from $\tau_1, \ldots, \tau_{|S|}$ (in this order).

Example 3.4.4 For instance, order the generators of the group F_4 as shown in Figure 3.6.

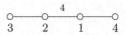

Figure 3.6. An order of the nodes of F_4.

The corresponding normal form forest is shown in Figure 3.7. We see from a glance at this forest, for example, that 13241213 is a normal form, whereas 21412141 is not. $\square$

Proposition 3.4.2 shows that it is enough to compute the normal form of distinguished elements, so we now concentrate on them.

Lemma 3.4.5 *Let $w \in W$ be distinguished and $s \in S$. Then, we have the following:*

(i) *If $s \in D_R(w)$, then ws is distinguished.*

(ii) *If $s \notin D_R(w)$, then either ws is distinguished or there is $s' \in S \setminus \{s_n\}$ such that $ws = s'w$.*

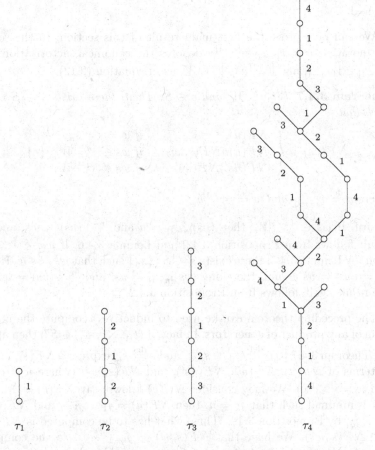

Figure 3.7. Normal form forest for F_4.

Proof. We have that $\ell(ws) = \ell(w) - 1$ and $\ell(s'w) = \ell(w) + 1$ if $s' \neq s_n$. Hence, $\ell(s'ws) = \ell(w) > \ell(ws)$ for all $s' \neq s_n$, which means that ws is distinguished.

Part (ii) follows directly from Corollary 2.5.2. $\square$

Lemma 3.4.6 *Let $w \in W$ be distinguished, and $s \in S$ (excluding the case $w = s = s_n$). Then, there exists $j \in [n-1]$ such that*

$$NF(ws) = \begin{cases} nNF(s_nws), & \text{if } ws \in {}^{[n-1]}W, \\ jNF(w), & \text{if } ws \notin {}^{[n-1]}W. \end{cases}$$

Proof. If $ws \in {}^{[n-1]}W$, then the result follows immediately from Lemma 3.4.1. If $ws \notin {}^{[n-1]}W$, then, by Lemma 3.4.5, there exists $s' \in S \setminus \{s_n\}$ such that $ws = s'w$. However, $s' \in W_{[n-1]}$ and $w \in {}^{[n-1]}W$. Hence, we conclude from Proposition 3.4.2 that $NF(ws) = NF(s'w) = NF(s')NF(w)$, as desired. $\square$

We can now prove the first main result of this section. In the sequel, the notation $v = v_{[n-1]} \cdot {}^{[n-1]}v$ denotes the canonical factorization with $v_{[n-1]} \in W_{[n-1]}$ and ${}^{[n-1]}v \in {}^{[n-1]}W$, as in equation (2.12).

Theorem 3.4.7 *Let $v \in W$ and $s \in S$. Then, there exists $s' \in S \setminus \{s_n\}$ such that*

$$NF(vs) = \begin{cases} NF(u), & \text{if } w = s = s_n, \\ NF(u)nNF(s_nws), & \text{if } ws \in {}^{[n-1]}W \setminus \{e\}, \\ NF(us')NF(w), & \text{if } ws \notin {}^{[n-1]}W, \end{cases}$$

where $u \overset{\text{def}}{=} v_{[n-1]}$ and $w \overset{\text{def}}{=} {}^{[n-1]}v$.

Proof. If $ws \in {}^{[n-1]}W$, then $(vs)_{[n-1]} = u$ and ${}^{[n-1]}(vs) = ws$, and the result follows from Proposition 3.4.2 and Lemma 3.4.6. If $ws \notin {}^{[n-1]}W$, then, by Lemma 3.4.5, there exists $s' \in S \setminus \{s_n\}$ such that $ws = s'w$. Hence, $vs = uws = us'w$ and, therefore, $(vs)_{[n-1]} = us'$ and ${}^{[n-1]}(vs) = w$, and the result again follows from Proposition 3.4.2. $\square$

The preceding theorem can be used to inductively compute the normal form of any product of generators. In fact, if $(s_{i_1}, \ldots, s_{i_p}) \in S^p$, then applying Theorem 3.4.7 to $v \overset{\text{def}}{=} s_{i_1} \ldots s_{i_{p-1}}$ and $s \overset{\text{def}}{=} s_{i_p}$ expresses $NF(s_{i_1} \ldots s_{i_p})$ in terms of $NF(u)$, $NF(w)$, $NF(us')$, and $NF(s_nws)$ (where $u = v_{[n-1]}$ and $w = {}^{[n-1]}v$). We may consider $NF(v)$ known; say $NF(v) = j_1 \ldots j_q$. If k is minimal such that $j_k = n$, then $NF(u) = j_1 \ldots j_{k-1}$ and $NF(w) = j_k \ldots j_q$, by Proposition 3.4.2. Thus, what has to be computed is $NF(us')$ and $NF(s_nws)$. We have that $NF(s_nw) = j_{k+1} \ldots j_q$, so the computation of $NF(s_nws)$ is an instance of the same problem but starting from a shorter normal form. On the other hand, us' can be expressed in terms of the first $n-1$ generators (since $s' \neq s_n$), so the computation of $NF(us')$ from $NF(u) = j_1 \ldots j_{k-1}$ takes place in the subgroup $W_{[n-1]}$ of smaller rank. Thus, by double induction on rank and length, the procedure will compute $NF(vs)$.

Note, however, that the procedure that we have just described is *as stated* not yet an *algorithm* for computing $NF(vs)$. In fact, to make the procedure entirely self-contained, one needs to be able to decide if $ws \in {}^{[n-1]}W$, and (if not) to compute s' explicitly. Both problems can be solved for the finite Coxeter groups by constructing and using appropriate tables, and the tables themselves can be computed (once and for all) with "ad hoc" algorithms. To describe those algorithms and tables would take us too far afield — we refer the interested reader to [143, 145].

We now come to the second main result of this section.

Theorem 3.4.8 *Let $v \in W$, $a_1 \ldots a_r = NF(v)$ and $s \in S$. Then, we have the following:*

(i) *If $s \in D_R(v)$, then there exists $1 \leq j \leq r$ such that $NF(vs) = a_1 \ldots \widehat{a_j} \ldots a_r$.*

(ii) *If $s \notin D_R(v)$, then there exist $0 \leq j \leq r$ and $k \in [n]$ such that $NF(vs) = a_1 \ldots a_j k a_{j+1} \ldots a_r$.*

Proof. We proceed by induction on $n + r$, the $n + r \leq 2$ cases being clear. By Proposition 3.4.2, we may write $v = uw$, where $u \in W_{[n-1]}$, w is distinguished, and $NF(v) = NF(u)NF(w)$. If $w = e$, then $v \in W_{[n-1]}$, and if $s \neq s_n$, we are done by the induction assumption, whereas if $s = s_n$, we have that $NF(vs) = a_1 \ldots a_r n$.

So, we may assume that $w \neq e$. Note that since w is distinguished, we then have that $D_L(w) = \{s_n\}$. Let, for convenience, $i_1 \ldots i_p \overset{\text{def}}{=} NF(w)$. Then, $i_1 = n$, $p = \ell(w)$, and

$$i_2 \ldots i_p = NF(s_n w). \tag{3.6}$$

We have to distinguish three cases.

Case 1: $s \in D_R(w)$. Then, by Lemma 3.4.5, ws is distinguished. The case $s = w = s_n$ is easily taken care of, so we may assume that $ws \neq e$. Hence, the normal form of ws begins with n and $s_n \in D_L(ws)$. Therefore, we have that $\ell(s_n ws) = \ell(ws) - 1 = \ell(w) - 2 = \ell(s_n w) - 1$, and hence that $s \in D_R(s_n w)$. By the induction hypothesis, applied to $s_n w$, we conclude that $NF((s_n w)s)$ is obtained by deleting one entry in $NF(s_n w)$. Hence, by equation (3.6), $NF((s_n w)s) = i_2 \ldots \widehat{i_j} \ldots i_p$ for some $j \in [2, p]$. On the other hand, if $n b_1 \ldots b_{p-2} \overset{\text{def}}{=} NF(ws)$, then $b_1 \ldots b_{p-2} = NF(s_n ws)$. Therefore, we conclude that $i_2 \ldots \widehat{i_j} \ldots i_p = b_1 \ldots b_{p-2}$ and, hence, that $NF(ws) = n i_2 \ldots \widehat{i_j} \ldots i_p$. However, since ws is distinguished, we have that $(vs)_{[n-1]} = u$ and $^{[n-1]}(vs) = ws$ and, hence, that $NF(vs) = NF(u)NF(ws)$. The result now follows.

Case 2: $s \notin D_R(w)$ and ws is distinguished. Then, $D_L(ws) = \{s_n\}$ and the normal form of ws begins with n. Therefore, $\ell(s_n ws) = \ell(ws) - 1 = \ell(w) = \ell(s_n w) + 1$ and, hence, $s \notin D_R(s_n w)$. By the induction hypothesis, we conclude that $NF((s_n w)s)$ is obtained by inserting an element in $NF(s_n w)$. Hence, by (3.6), $NF(s_n ws) = i_2 \ldots i_j k i_{j+1} \ldots i_p$ for some $k \in [n]$ and $j \in [p]$. On the other hand, if $n b_1 \ldots b_p \overset{\text{def}}{=} NF(ws)$, then $b_1 \ldots b_p = NF(s_n ws)$. Hence, $b_1 \ldots b_p = i_2 \ldots i_j k i_{j+1} \ldots i_p$ and, therefore, $n i_2 \ldots i_j k i_{j+1} \ldots i_p = NF(ws)$. Again, since ws is distinguished, we conclude, as at the end of Case 1, that $NF(vs) = NF(u)NF(ws)$, and the result follows.

Case 3: $s \notin D_R(w)$ and ws is not distinguished. Then, by Lemma 3.4.5, there exists $s' \in S \setminus \{s_n\}$ such that $ws = s'w$. Hence, $(vs)_{[n-1]} = us'$, $^{[n-1]}(vs) = w$ and, therefore, $NF(vs) = NF(us')NF(w)$. The theorem follows by induction, since $u \in W_{[n-1]}$ and $s' \in S \setminus \{s_n\}$. $\square$

The theorem shows an extremely nice property of the (lexicographically first) normal form. We already know from Chapter 1 that if $(a_1, \ldots, a_r)$ is a reduced decomposition of $v \in W$ and $s \in S$, then the following hold:

1. If $s \in D_R(v)$, there exists $j \in [r]$ such that $(a_1, \ldots, a_{j-1}, a_{j+1}, \ldots, a_r)$ is a reduced decomposition of vs (this follows from the Exchange Property).

2. If $s \notin D_R(v)$, there exists $j \in [0, r]$ and $k \in [n]$ such that $(a_1, \ldots, a_j, k, a_{j+1}, \ldots, a_r)$ is a reduced decomposition of vs (e.g., take $j = r$ and k such that $s_k = s$).

The new content of Theorem 3.4.8 is therefore the assertion that if $(a_1, \ldots, a_r)$ is the *lexicographically first* reduced expression for v, then j and k can be chosen so that the above reduced expressions are again *lexicographically first*.

Exercises

1. Consider the three elements x, y, and z appearing in Figure 3.3. Exactly one of them is the product of the other two. Determine the correct identity (among the six possible ones).

2. Let $(S_n)^J$ with $|J| = n - 2$ be a quotient modulo a maximal parabolic subgroup of S_n. Show that for $x, y \in (S_n)^J$:

$$x \leq y \ \text{(Bruhat order)} \quad \Leftrightarrow \quad x \leq_L y \ \text{(left weak order)}.$$

3. Suppose that W is finite. Let $w \in W$ and $J = S \setminus D_L(w)$. Show that

$$x \vee w = w_0 \quad \Leftrightarrow \quad x \geq w_0(J).$$

4. Suppose that W is finite and let $w \in W$. Show that $\{x \in W : x \wedge w = e \text{ and } x \vee w = w_0\}$ (i.e., the set of lattice-theoretic complements of w) forms an interval in weak order.

5. Let $I, J \subseteq S$ and assume that W_I and W_J are finite. Show that the following are equivalent:

    (a) $I \subseteq J$,
    (b) $w_0(I) \leq_R w_0(J)$,
    (c) $w_0(I) \leq w_0(J)$ in Bruhat order.

6.* For $m = 1, 2, \ldots$, let a_m be the number of posets of length m that occur as intervals in weak order of Coxeter groups (see Corollary 3.1.8). What can be said about the sequence $a_1, a_2, \ldots$?

7. For the definition of order dimension, see Exercise 2.24. Consider here the finite irreducible Coxeter groups (W, S) under weak order.

(a) Show that $|S| \leq \operatorname{odim}(W) \leq \ell(w_0)$.

(b) Show that $\operatorname{odim}(W) = |S|$ in the following cases: type A_n, B_n, D_n, H_3, and $I_2(m)$.

(c)* Determine the order dimension of weak order in the remaining open cases. Is it always true that $\operatorname{odim}(W) = |S|$?

8. Define *two-sided weak order* "$\leq_{LR}$" as the transitive closure of the union of left and right weak order. So, $u \leq_{LR} w$ means that $u <_{X_1} u_1 <_{X_2} u_2 <_{X_3} \cdots <_{X_k} w$ for some elements u_i and some choices $X_i \in \{L, R\}$. Show the following:

(a) Two-sided weak order is not a semilattice.

(b) Let $u <_{LR} w$, and assume that w has a unique reduced decomposition. Then, the order complex of the open interval $(u, w)_{LR}$ is either contractible or homotopy equivalent to a sphere. In particular, $\mu_{LR}(u, w) \in \{0, +1, -1\}$ for all $u \leq_{LR} w$.

(c)* What can be said about general intervals $(u, w)_{LR}$?

9. Let S be a finite set and $m = (m(x, y))_{x,y \in S}$ a Coxeter matrix. Let P be a finite poset with bottom element $\hat{0}$, having exactly $|S|$ atoms. Suppose that there exists a labeling $\lambda : \operatorname{Cov}(P) \to S$ (where $\operatorname{Cov}(P) \overset{\text{def}}{=} \{(x, y) \in P^2 : x \lessdot y\}$) such that the following hold:

(a) If $e, f \in \operatorname{Cov}(P)$, $e \neq f$, are incident, then $\lambda(e) \neq \lambda(f)$.

(b) If $a, b, x \in P$ are such that $x \lessdot a$, $x \lessdot b$, $a \neq b$, then there exist two unrefinable chains $a_0 \lessdot a_1 \lessdot \cdots \lessdot a_m$ and $b_0 \lessdot b_1 \lessdot \cdots \lessdot b_m$ such that $a_0 = b_0 = x$, $a_1 = a$, $b_1 = b$, $a_m = b_m$, $a_i \neq b_i$ if $i = 1, \ldots, m - 1$, where $m = m(\lambda(x, a), \lambda(x, b))$, and

$$(\lambda(a_0, a_1), \ldots, \lambda(a_{m-1}, a_m)) = (\lambda(x, a), \lambda(x, b), \lambda(x, a), \ldots),$$
$$(\lambda(b_0, b_1), \ldots, \lambda(b_{m-1}, b_m)) = (\lambda(x, b), \lambda(x, a), \lambda(x, b), \ldots).$$

Show that then P is isomorphic to $(W, \leq_R)$ (weak order), where (W, S) is the Coxeter system having Coxeter matrix m.

10. Show that the (undirected) order diagram of weak order (equivalently, the Cayley graph) of a finite Coxeter group is Hamiltonian; that is, it contains a cycle which visits each vertex exactly once.

11. Consider the property of an infinite Coxeter group that every antichain in weak order is finite.

(a) Give examples of one group that has this property and one (of rank 3) that does not.

(b)* Characterize the infinite groups with this property.

12. (a) Describe an algorithm, based on the numbers game (see Section 4.3), for generating the normal form forest of a finite Coxeter group (W, S) with ordered S.

(b) Say you want to work with elements w of bounded length $\ell(w) \leq m$ in an infinite Coxeter group. How would you modify the concept of a normal form forest, and the algorithm from part (a) for constructing it, to suit this situation?

13. Suppose that (W, S) is an infinite Coxeter group. Its *nerve* is the simplicial complex

$$\mathcal{N}(W, S) \stackrel{\text{def}}{=} \{J \subseteq S : |W_J| < \infty\}.$$

(a) Show that the order complex of $W \setminus \{e\}$ under weak order is homotopy equivalent to $\mathcal{N}(W, S)$.
(b) Show that the order complex of $W \setminus \{e\}$ under Bruhat order is contractible.

14. Call a chain $u = x_0 < x_1 < \cdots < x_k = w$ in right order of a Coxeter group (W, S) *essential* if $k \geq 2$ and for each $i = 1, \ldots, k$ there exists $J \in \mathcal{N}(W, S)$ such that $x_i = x_{i-1}w_0(J)$. For instance, if $\ell(u, w) \geq 2$, then every saturated chain from u to w is essential (with $|J| = 1$ at each step).

For a finite Coxeter group of rank > 1, let $\mathrm{ess}(W, S)$ be the minimal length of an essential chain from the identity e to the top element w_0.

(a) Show that $\mathrm{ess}(A_n) = n$ if n is even, and $\mathrm{ess}(A_n) = n + 1$ if n is odd.
(b) Show that $\mathrm{ess}(W, S) = \frac{2|T|}{|S|}$, for all other finite irreducible Coxeter groups[2].
(c) Assume that $\ell(u, w) \geq 2$. Show that the order complex of the open interval $(u, w)_R$ has the same homotopy type as the poset of essential chains from u to w ordered by inclusion.

15. Let $\alpha = s_1 \ldots s_m \in S^*$ be a word. For subsets $J = \{i_1, \ldots, i_k\}_< \subseteq [m]$, let α_J denote the subword $s_{i_1} \ldots s_{i_k}$. Let w be an element in a Coxeter group (W, S). The *subword complex* $\Delta(\alpha, w)$ is the complex on the vertex set $[m]$ determined by

$$[m] \setminus J \text{ is a facet of } \Delta(\alpha, w) \Leftrightarrow \alpha_J \text{ is a reduced decomposition of } w.$$

(a) Show that $\Delta(\alpha, w)$ is shellable.
(b) Show that $\Delta(\alpha, w)$ is homeomorphic to either a sphere or a ball.
(c) Characterize when $\Delta(\alpha, w)$ is homeomorphic to a sphere.

16. Let (W, S) be a Coxeter group, $|S| < \infty$. To simplify notation, we write $(s) \stackrel{\text{def}}{=} S \setminus \{s\}$, for $s \in S$, and $V \stackrel{\text{def}}{=} \bigcup_{s \in S} W/W_{(s)}$ for the collection of all left cosets of all maximal parabolic subgroups.

---

[2]The quantity $2|T|/|S|$ is known as the *Coxeter number* of the group.

The *Coxeter complex* $\Delta(W, S)$ is by definition the simplicial complex on the vertex set V with facets $C_w \overset{\text{def}}{=} \{wW_{(s)} : s \in S\}$, for $w \in W$. Show the following for $\Delta = \Delta(W, S)$:

(a) Δ is a pure $(|S| - 1)$-dimensional complex, and it is naturally colored by $V = \biguplus V_s$, with $V_s = W/W_{(s)}$.

(b) The mapping $w \mapsto C_w$ gives a bijection $W \leftrightarrow \mathcal{F}(\Delta)$.

(c) Two facets C_w and $C_{w'}$ are adjacent if and only if $w' = ws$ for some $s \in S$.

(d) Δ is thin.

(e) Any linear extension of the weak ordering of W assigns a shelling order to the facets C_w of Δ.

(f) The restriction map $\mathcal{R}$ of such a shelling sends each facet C_w to its face of type $D_R(w)$.

(g) If W is finite, the h-vector of Δ satisfies

$$h_0 + h_1 t + \cdots + h_{|S|} t^{|S|} = \sum_{w \in W} t^{d(w)},$$

where $d(w) = |D_R(w)|$ is the *descent number* of w.

(h) If W is finite, then Δ is homeomorphic to the sphere $\mathbb{S}^{|S|-1}$.

(i) If W is infinite, then Δ is contractible. Furthermore, W is homeomorphic to Euclidean space $\mathbb{R}^{|S|-1}$ if and only if all maximal parabolic subgroups are finite.

(j) For $J \subseteq S$, the type-selected subcomplex Δ_J has the homotopy type of a wedge of $|\mathcal{D}_J|$ spheres.

Notes

For the case of symmetric groups, the weak order seems to have been considered first by statisticians in the 1960s; see Lehmann [359], Savage[452], and Yanagimoto and Okamoto [557]. For S_n, see also the book by Berge [25].

Sections 3.1 and 3.2 are drawn from Björner [53]; these results were published without proofs in [54]. The word property is due to Tits [537]. Section 3.4 is based on the work of du Cloux [143, 145].

Exercise 4. For the case of symmetric groups, see Markowsky [388].

Exercise 7. See Reading [431].

Exercise 8(b). See Björner [59].

Exercise 9. See Eriksson [230].

Exercise 10. See Conway, Sloane, and Wilks [157].

Exercise 14. See Ungar [541] for part (a), Eriksson [223] for (b), and Björner [58] for (c).

Exercise 15. See Knutson and Miller [329].

Exercise 16. See Bourbaki [79, pp. 40–44], Tits [538] and Björner [56].

Coxeter complexes (as defined in Exercise 16) come in two other guises. On the one hand, there exists a canonical geometric realization in the space $V^* \cong \mathbb{R}^S$ (induced by the decomposition of the Tits cone into Weyl chambers, briefly described in Section 4.9), and on the other hand, there is an axiomatic characterization (due to Tits [538]). The theory of Coxeter complexes provide an alternative approach to Coxeter groups, since there is a one-to-one correspondence between Coxeter groups and (axiomatically defined) Coxeter complexes. They also provide a crucial stepping stone to the theory of buildings; see the books by Brown [106], Ronan [446], and Tits [538].

4

Roots, games, and automata

The main goal of this chapter is to obtain deeper algorithmic and structural properties of the system of reduced words of a Coxeter group. In particular, we prove the existence of a finite state automaton that recognizes the language of reduced words, showing that this language is regular.

The construction of automata uses the so-called "root system" of a Coxeter group and a partial order structure on the set of roots. These concepts are of fundamental importance also in their own right. We begin by developing the properties of root systems that are needed. As a by-product, we obtain the so-called "numbers game," a handy computational device for working with reduced decompositions.

We adhere to the following convention: *Throughout this chapter, (W, S) is a Coxeter system with S finite.*

4.1 A linear representation

By a *linear representation* of W we understand a homomorphism $\varphi : W \to GL(U)$, where $GL(U)$ denotes the group of invertible linear transformations of some vector space U into itself. For the representation that we construct in this section, U is the real vector space $\mathbb{R}^S$.

We begin with a simple geometric lemma. Let $m \geq 3$ be an integer, let $\gamma = \frac{\pi}{m}$, and let $k, k' > 0$ be real numbers such that

$$kk' = 4\cos^2 \gamma.$$

Choose basis vectors β and β' in the Euclidean plane $\mathbb{E}^2$ such that
(i) the angle between β and β' equals γ, and
(ii) their lengths are related by

$$|\beta'| = \frac{2\cos\gamma}{k}|\beta| \quad \text{and} \quad |\beta| = \frac{2\cos\gamma}{k'}|\beta'|.$$

See Figure 4.1 for an illustration.

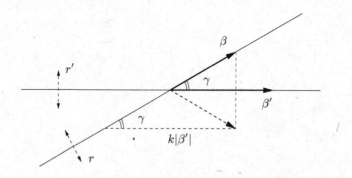

Figure 4.1. Illustration for Lemma 4.1.1.

Let r (resp. r') denote orthogonal reflection of $\mathbb{E}^2$ in the line spanned by β (resp. β').

Lemma 4.1.1 *The coordinates (q, q') of a point $q\beta + q'\beta'$ are transformed as follows by the orthogonal reflections:*

$$r' : (q, q') \mapsto (-q, q' + kq),$$
$$r : (q, q') \mapsto (q + k'q', -q').$$

Proof. We have that $r(1,0) = (1,0)$ and $r'(0,1) = (0,1)$, since points on the reflecting lines remain fixed. The geometry indicated in Figure 4.1 shows that $r'(1,0) = (-1, \frac{2|\beta|\cos\gamma}{|\beta'|}) = (-1, k)$. By symmetry, $r(0,1) = (\frac{2|\beta'|\cos\gamma}{|\beta|}, -1) = (k', -1)$. Thus, the stated formulas are correct for the basis vectors and, therefore, by linearity for all vectors in $\mathbb{E}^2$. $\square$

Let (W, S) be a Coxeter system given by a Coxeter matrix m. Define a function $k : S \times S \to \mathbb{R}$ as follows. First, put

$$k_{s,s} = -2 \text{ for all } s \in S, \tag{4.1}$$

and

$$k_{s,s'} = 0 \text{ if } m(s, s') = 2. \tag{4.2}$$

Then, for each ordered pair (s, s') such that $m(s, s') \geq 3$, choose a real number $k_{s,s'}$ subject to the constraints

$$\begin{cases} k_{s,s'} > 0 \\ k_{s,s'} k_{s',s} = 4 \cos^2 \frac{\pi}{m(s,s')}, & \text{if } m(s, s') \neq \infty \\ k_{s,s'} k_{s',s} \geq 4, & \text{if } m(s, s') = \infty. \end{cases} \qquad (4.3)$$

Thus, in relations (4.3), we have infinitely many choices available for $k_{s,s'}$; the condition concerns only the product $k_{s,s'} k_{s',s}$. (The two factors in this product can be viewed as corresponding to the two directions of the (undirected) Coxeter graph edge s—s'.)

Let[1] $\{\alpha_s^*\}_{s \in S}$ be the canonical basis for $\mathbb{R}^S$, meaning that $\alpha_s^*(s') \stackrel{\text{def}}{=} \delta_{s,s'}$. For each $s \in S$, define a linear transformation $\sigma_s^* : \mathbb{R}^S \to \mathbb{R}^S$ by

$$\sigma_s^*(p) = p + p_s \sum_{s' \in S} k_{s,s'} \alpha_{s'}^*, \qquad (4.4)$$

where $p = \sum p_{s'} \alpha_{s'}^* \in \mathbb{R}^S$. Note that $\sigma_s^*(p)$ differs from p only in coordinates corresponding to nodes s' that are neighbors of s in the Coxeter graph. More precisely, the s'-coordinate of $\sigma_s^*(p)$ is $p_{s'} + k_{s,s'} p_s$, which can be rewritten

$$(\sigma_s^*(p))_{s'} = \begin{cases} -p_s, & \text{if } s' = s \\ p_{s'} + k_{s,s'} p_s, & \text{if } m(s, s') \geq 3 \\ p_{s'}, & \text{if } m(s, s') = 2. \end{cases} \qquad (4.5)$$

Proposition 4.1.2 *For all $s, s' \in S$, the following hold:*

(i) $(\sigma_s^*)^2 = \text{id}.$

(ii) *The order of $\sigma_s^* \sigma_{s'}^*$ is $m(s, s')$.*

Proof. Part (i) and the $m(s, s') = 2$ case of part (ii) are easy to see directly from equation (4.5). So, we may assume that $m(s, s') \geq 3$.

Suppose first that $m(s, s') \neq \infty$. Let $p^0 \in \mathbb{R}^S$ and consider the successive images $p^i = \sigma_s^* \sigma_{s'}^*(p^{i-1})$, $i \geq 1$. The coordinate pair $(p_s^i, p_{s'}^i)$ determines a vector $\nu^i = p_s^i \beta + p_{s'}^i \beta'$ in the Euclidean plane $\mathbb{E}^2$ considered in Figure 4.1, where we now put $m = m(s, s')$, $k = k_{s,s'}$, and $k' = k_{s',s}$.

The product of two orthogonal reflections through lines at an angle γ in $\mathbb{E}^2$ equals a rotation through the angle 2γ. Hence, Lemma 4.1.1 shows, in view of equation (4.5), that ν^i is obtained from ν^{i-1} by a rotation of $\mathbb{E}^2$ through the angle $\frac{2\pi}{m(s,s')}$. It follows that

$$\nu^0, \nu^1, \ldots, \nu^{m(s,s')-1} \text{ are distinct, and } \nu^{m(s,s')} = \nu^0, \qquad (4.6)$$

---

[1]The reason for using asterisks in the notation for basis vectors and other objects in this section will become clear in Section 4.2.

and that

$$\nu^0 + \nu^1 + \cdots + \nu^{m(s,s')-1} = 0. \tag{4.7}$$

From statement (4.6) follows that $p_s^{m(s,s')} = p_s^0$ and $p_{s'}^{m(s,s')} = p_{s'}^0$, and that no exponent smaller than $m(s,s')$ has this property. Furthermore, equation (4.7) shows that $p_{s''}^{m(s,s')} = p_{s''}^0$ for all $s'' \neq s, s'$, since by equation (4.4),

$$p_{s''}^{m(s,s')} = p_{s''}^0 + k_{s,s''}\left(p_s^0 + p_s^1 + \cdots + p_s^{m(s,s')-1}\right)$$
$$+ k_{s',s''}\left(p_{s'}^0 + p_{s'}^1 + \cdots + p_{s'}^{m(s,s')-1}\right).$$

Hence, the order of $\sigma_s^* \sigma_{s'}^*$ is indeed $m(s,s')$.

Suppose now that $m(s,s') = \infty$. Define p^i as the iterated images of a point p^0 under the mapping $\sigma_s^* \sigma_{s'}^*$, as earlier. A small computation shows that the coordinate pairs $(p_s^i, p_{s'}^i)$, $i \geq 0$, satisfy

$$\begin{cases} p_s^{i+1} = -p_s^i - k_{s',s}p_{s'}^i, \\ p_{s'}^{i+1} = k_{s,s'}p_s^i + (k_{s,s'}k_{s',s} - 1)\,p_{s'}^i. \end{cases}$$

Hence, for $i \geq 1$,

$$p_{s'}^{i+1} + p_{s'}^i = k_{s,s'}\left(p_s^i + p_s^{i-1}\right) + (k_{s,s'}k_{s',s} - 1)\left(p_{s'}^i + p_{s'}^{i-1}\right)$$

and

$$p_s^i + p_s^{i-1} = -k_{s',s}p_{s'}^{i-1},$$

which together yield

$$p_{s'}^{i+1} = (k_{s,s'}k_{s',s} - 2)p_{s'}^i - p_{s'}^{i-1}. \tag{4.8}$$

Choosing $p^0 = \alpha_{s'}^*$ we have that $p_{s'}^0 = 1$ and $p_{s'}^1 = k_{s,s'}k_{s',s} - 1$. Since $k_{s,s'}k_{s',s} \geq 4$, we can from this and the recurrence relation (4.8) deduce that

$$p_{s'}^0 < p_{s'}^1 < p_{s'}^2 < p_{s'}^3 < \cdots.$$

Hence, $\sigma_s^* \sigma_{s'}^*$ is of infinite order. $\square$

The following is an immediate consequence.

Theorem 4.1.3 *Let (W, S) be a Coxeter system. Then, the mapping $s \mapsto \sigma_s^*$ $(s \in S)$ extends uniquely to a homomorphism $\sigma^* : W \to GL(\mathbb{R}^S)$.* $\square$

We end this section by filling a gap left open since Section 1.1.

Proof of Proposition 1.1.1. The homomorphism σ^* maps s and s' to distinct elements in $GL(\mathbb{R}^S)$, whose product has order $m(s,s')$. This proves (i) and that $m(s,s')$, if finite, divides the order of ss'. However, since $(ss')^{m(s,s')} = e$, the order of ss' then divides $m(s,s')$. The $m(s,s') = \infty$ case is clear. $\square$

4.2   The geometric representation

In this section, we continue the study of the linear representation σ^* from Section 4.1. Its contragredient σ, better known as "the geometric representation," is introduced. The material provides a foundation for the rest of the chapter.

Let V be a real vector space of dimension $|S|$, and let V^* be the space dual to V. Choose a basis $\alpha = \{\alpha_s\}_{s \in S}$ for V, indexed by the elements of S, and let $\alpha^* = \{\alpha_s^*\}_{s \in S}$ be the dual basis for V^*. Via this choice we have specific identifications $V \cong \mathbb{R}^S$ and $V^* \cong \mathbb{R}^S$, and elements of V and V^* can, whenever convenient, be expressed as column vectors. There is a natural pairing

$$\langle p \mid \beta \rangle \overset{\text{def}}{=} p(\beta) \tag{4.9}$$

for $p \in V^*$ and $\beta \in V$. By definition, $\langle \alpha_s^* \mid \alpha_{s'} \rangle = \delta_{s,s'}$, so if p and β are expressed as column vectors, then $\langle p \mid \beta \rangle = p^{tr} \beta$.

Choose real numbers $k_{s,s'}$ for all pairs $s, s' \in S$ subject to the constraints (4.1), (4.2), and (4.3). Define a bilinear form $(\cdot \mid \cdot)$ on V by prescribing its value on basis vectors as follows:

$$(\alpha_s \mid \alpha_{s'}) = -\frac{k_{s,s'}}{2}, \tag{4.10}$$

for all $s, s' \in S$. This implies, among other things, that

$$(\alpha_s \mid \alpha_s) = 1, \tag{4.11}$$
$$(\alpha_s \mid \alpha_{s'}) = 0, \quad \text{if } m(s, s') = 2 \tag{4.12}$$
$$(\alpha_s \mid \alpha_{s'}) < 0, \quad \text{if } m(s, s') \geq 3. \tag{4.13}$$

Note that, in general, this bilinear form is *not* symmetric.

For each $s \in S$ define a linear mapping $\sigma_s \colon V \to V$ by

$$\sigma_s(\beta) = \beta - 2(\alpha_s \mid \beta)\alpha_s. \tag{4.14}$$

Clearly, $\sigma_s(\alpha_s) = -\alpha_s$.

Proposition 4.2.1 *For all $s, s' \in S$, the following hold:*

(i) $\sigma_s^2 = \mathrm{id}$.

(ii) The order of $\sigma_s \sigma_{s'}$ is $m(s, s')$.

Proof. Let Σ_s be the matrix representing σ_s in the α basis; that is, for all $\beta \in V$,

$$\sigma_s(\beta) = \Sigma_s \beta. \tag{4.15}$$

Definitions (4.10) and (4.14) show that $\Sigma_s = I + K_s$, where I is the identity matrix and K_s is the zero matrix with its s-th row replaced by the row vector $(k_{s,s'})_{s' \in S}$.

Now, let $\sigma_s^* : V^* \to V^*$ be the linear mapping defined in Section 4.1. Definition (4.5) shows that the matrix representing σ_s^* in the α^* basis is $I + K_s^{tr}$; thus,

$$\sigma_s^*(p) = \Sigma_s^{tr}\, p. \tag{4.16}$$

The result therefore follows from Proposition 4.1.2. $\square$

We draw the following immediate conclusion.

Theorem 4.2.2 *The mapping $s \mapsto \sigma_s$ has a unique extension to a homomorphism $\sigma : w \mapsto \sigma_w$ from W to $GL(V)$.* $\square$

The linear mapping $\sigma_w : V \to V$ is defined by $\sigma_w = \sigma_{s_1}\sigma_{s_2}\ldots\sigma_{s_k}$ for any expression $w = s_1 s_2 \ldots s_k$, $s_i \in S$. Its action on a vector $\beta \in V$ will be notationally simplified to

$$w(\beta) \stackrel{\text{def}}{=} \sigma_w(\beta). \tag{4.17}$$

Similarly, the action of the linear mapping $\sigma_w^* : V^* \to V^*$ of Theorem 4.1.3 on a vector $p \in V^*$ will be denoted

$$w(p) \stackrel{\text{def}}{=} \sigma_w^*(p). \tag{4.18}$$

The similarity of notation will not cause ambiguity, since we have the convention that elements of V are named by Greek letters $\alpha, \beta, \ldots$ and elements of V^* by Roman letters $p, q, \ldots$ (except for the basis vectors α_s^*).

The representations $\sigma : W \to GL(V)$ and $\sigma^* : W \to GL(V^*)$ are *contragredient* in the following sense.

Proposition 4.2.3 *For all $w \in W$, $\beta \in V$ and $p \in V^*$,*

$$\langle w(p) \mid \beta \rangle = \langle p \mid w^{-1}(\beta) \rangle.$$

Proof. Equations (4.15) and (4.16) give that

$$\langle s(p) \mid \beta \rangle = \left(\Sigma_s^{tr} p\right)^{tr} \beta = p^{tr}\Sigma_s\beta = \langle p \mid s(\beta) \rangle.$$

The formula for $\ell(w) > 1$ follows by repeated application. $\square$

For $s, s' \in S$ and $j \geq 0$, define $(ss's\ldots)_j$ to be the alternating word of length j beginning with s. Similarly, let $(\ldots ss's)_j$ denote the alternating word of length j ending with s. As usual, the same symbols also denote the corresponding group element.

Lemma 4.2.4 *If $j < m(s, s')$, then $(\ldots s'ss')_j(\alpha_s) = c\alpha_s + d\alpha_{s'}$, with $c, d \geq 0$.*

Proof. Since acting on a vector by s adds a multiple of α_s (equation (4.14)), and similarly for s', the element in question is clearly a linear combination of α_s and $\alpha_{s'}$. Thus, all that has to be shown is that c and d are non-negative.

We have that

$$c = \langle \alpha_s^* \,|\, (\ldots s'ss')_j(\alpha_s) \rangle = \langle (s'ss'\ldots)_j(\alpha_s^*) \,|\, \alpha_s \rangle \qquad (4.19)$$

and

$$d = \langle \alpha_{s'}^* \,|\, (\ldots s'ss')_j(\alpha_s) \rangle = \langle (s'ss'\ldots)_j(\alpha_{s'}^*) \,|\, \alpha_s \rangle. \qquad (4.20)$$

Equations (4.19) and (4.20) express c and d as the α_s^*-coordinates of the images of α_s^* and $\alpha_{s'}^*$ under the mapping $\sigma_{s'}^* \sigma_s^* \sigma_{s'}^* \ldots$ (j factors). The situation can, therefore, be transferred to that studied in Section 4.1.

Assume first that $m(s,s') \neq \infty$. Choose basis vectors β and β' in the Euclidean plane $\mathbb{E}^2$ as in Lemma 4.1.1 and the proof of Proposition 4.1.2. Define a linear mapping $\psi : V^* \to \mathbb{E}^2$ by letting $\psi(\alpha_s^*) = \beta$, $\psi(\alpha_{s'}^*) = \beta'$ and $\psi(\alpha_{s''}^*) = 0$ for all $s'' \notin \{s, s'\}$, and extending linearly. Then,

$$r \circ \psi(p) = \psi \circ \sigma_{s'}^*(p) \quad \text{and} \quad r' \circ \psi(p) = \psi \circ \sigma_s^*(p)$$

for all $p \in V^*$. Namely, for $p = p_s \alpha_s^* + p_{s'} \alpha_{s'}^* + \cdots$,

$$\begin{aligned}
r(\psi(p)) &= r(p_s \beta + p_{s'} \beta') \\
&= (p_s + k_{s',s} p_{s'})\beta - p_{s'}\beta' \\
&= \psi((p_s + k_{s',s} p_{s'})\alpha_s^* - p_{s'} \alpha_{s'}^*) \\
&= \psi(\sigma_{s'}^*(p)),
\end{aligned}$$

using Lemma 4.1.1 for the second equality and equation (4.5) for the fourth, and similarly for $r'(\psi(p))$. Hence, we get

$$\psi \circ (\sigma_{s'}^* \sigma_s^* \sigma_{s'}^* \ldots)_j(\alpha_s^*) = (rr'r\ldots)_j(\beta)$$

and

$$\psi \circ (\sigma_{s'}^* \sigma_s^* \sigma_{s'}^* \ldots)_j(\alpha_{s'}^*) = (rr'r\ldots)_j(\beta').$$

It follows that c and d are the β-coordinates in $\mathbb{E}^2$ of the images of β and β' under the mapping $R = (rr'r\ldots)_j$. Now, look at the geometric situation of Figure 4.1 and keep in mind that the mapping rr' is a counterclockwise rotation of the $\mathbb{E}^2$ plane through the angle $2\gamma = \frac{2\pi}{m(s,s')}$. Since $R = (rr')^{\frac{j}{2}}$ if j is even, $R = (rr')^{\frac{j-1}{2}} r$ if j is odd, and $j < m(s,s')$, one sees from simple geometric considerations that $R(\beta)$ and $R(\beta')$ must lie in the upper half-plane; that is, their β-coordinates c and d must be non-negative.

Suppose next that $m(s,s') = \infty$. For any point $p^0 \in V^*$ define the iterated images $p^i = \sigma_{s'}^* \sigma_s^*(p^{i-1})$, $i \geq 1$. Then, we derive as in the proof of Proposition 4.1.2 (switching the roles of s and s'), via the recurrence relation (4.8), that $p_s^i = \langle p^i \,|\, \alpha_s \rangle$ satisfies

$$0 \leq p_s^0 \leq p_s^1 \quad \Rightarrow \quad p_s^i \leq p_s^{i+1}, \quad \text{for all } i \geq 0.$$

We must check four cases:

$$p^0 = \alpha_s^* \quad\Rightarrow\quad p_s^1 = k_{s,s'}k_{s',s} - 1 \geq 3 > 1 = p_s^0,$$
$$p^0 = \sigma_{s'}^*(\alpha_s^*) \quad\quad \text{same, since } \sigma_{s'}^*(\alpha_s^*) = \alpha_s^*,$$
$$p^0 = \alpha_{s'}^* \quad\Rightarrow\quad p_s^1 = k_{s',s} > 0 = p_s^0,$$
$$p^0 = \sigma_{s'}^*(\alpha_{s'}^*) \quad\Rightarrow\quad p_s^1 = k_{s',s}(k_{s,s'}k_{s',s} - 2) > k_{s',s} = p_s^0.$$

It follows via equations (4.19) and (4.20) that c and d are non-negative. $\square$

Call a vector $\gamma = \sum_{s \in S} c_s \alpha_s \in V$ *positive* (and write $\gamma > 0$) if $c_s \geq 0$ for all $s \in S$, and *negative* (denoted $\gamma < 0$) if $c_s \leq 0$ for all $s \in S$.

Proposition 4.2.5 *For all $w \in W$ and $s \in S$, the following hold:*

(i) *$\ell(ws) > \ell(w)$ implies $w(\alpha_s) > 0$.*

(ii) *$\ell(ws) < \ell(w)$ implies $w(\alpha_s) < 0$.*

Proof. We prove part (i) by induction on $\ell(w)$, the $\ell(w) = 0$ case being clearly true.

So, assume that $\ell(ws) > \ell(w) > 0$, and let $s' \in D_R(w)$. Put $J = \{s, s'\}$, and let $w = w^J w_J$ be the canonical decomposition (see Proposition 2.4.4). From $\ell(ws) = \ell(w)+1 = \ell(w^J)+\ell(w_J)+1$, we see that $\ell(w_J s) = \ell(w_J)+1$, which implies that $w_J = (\ldots s'ss')_j$ with $j < m(s, s')$. Hence, Lemma 4.2.4 shows that

$$w(\alpha_s) = w^J(\ldots s'ss')_j(\alpha_s) = w^J(c\alpha_s + d\alpha_{s'}),$$

with $c, d \geq 0$.

By definition, w^J satisfies $\ell(w^J s) > \ell(w^J)$ and $\ell(w^J s') > \ell(w^J)$, and $\ell(w^J) < \ell(w)$ since $s' \in D_R(w)$. Hence, by induction we have that $w^J(\alpha_s) > 0$ and $w^J(\alpha_{s'}) > 0$ and, therefore,

$$w(\alpha_s) = c\,w^J(\alpha_s) + d\,w^J(\alpha_{s'}) > 0.$$

The second part follows easily from the first. If $\ell(ws) < \ell(w)$, then $\ell(wss) > \ell(ws)$ and, hence, $0 < ws(\alpha_s) = w(-\alpha_s) = -w(\alpha_s)$, so $w(\alpha_s) < 0$. $\square$

Corollary 4.2.6 *Let $p \in \mathbb{R}_+^S \subseteq V^*$, and $u, v \in W$. Then, the following hold:*

(i) *$D_L(u) = \{s \in S : \langle u(p) \,|\, \alpha_s \rangle \text{ is negative}\}$.*

(ii) *If $u \neq v$, then $u(p) \neq v(p)$.*

Proof. Since all coefficients of p are positive, Proposition 4.2.5 implies that $\langle u(p) \,|\, \alpha_s \rangle = \langle p \,|\, u^{-1}(\alpha_s) \rangle$ is negative if and only if $\ell(u^{-1}s) < \ell(u^{-1})$.

Suppose that $u(p) = p$. Then, by part (i), $D_L(u) = \varnothing$ and, hence, $u = e$. This implies part (ii). $\square$

As an immediate consequence of Corollary 4.2.6(ii), we obtain the following.

Theorem 4.2.7 *The homomorphisms* $\sigma : W \to GL(V)$ *and* $\sigma^* : W \to GL(V^*)$ *are injective.*

The mapping $\sigma : W \to GL(V)$ is called the *geometric representation* of W. Note that it depends on the choice of bilinear form $(\cdot \mid \cdot)$, or, equivalently, via equation (4.10), on the choice of edge weights $k_{s,s'}$. In the special case when the bilinear form is for all $s, s' \in S$ given by

$$(\alpha_s \mid \alpha_{s'}) = -\cos \frac{\pi}{m(s,s')}, \tag{4.21}$$

σ is called the *standard geometric representation*. In this case, the bilinear form is symmetric, and equation (4.21) essentially (except for the value assumed at $m(s, s') = \infty$ edges) characterizes the symmetric case.

4.3   The numbers game

In this section, we present a combinatorial rendition of the contragredient geometric representation $\sigma^* : W \to GL(V^*)$. Vectors in V^* are assignments of real numbers p_s to the nodes $s \in S$ of the Coxeter graph, and each such assignment is here thought of as a position in a certain "game." The "moves" in the game are local rearrangements of the assigned values at a chosen node s and its neighbors, governed by the labels of the edges surrounding s in the Coxeter graph. The point of this game is that it gives a combinatorial model of the Coxeter group (W, S), where group elements correspond to positions and reduced decompositions correspond to play sequences.

Before proceeding we have to choose for each *ordered* pair (s, s') such that $m(s, s') \geq 3$ a real number $k_{s,s'} > 0$ such that

$$\begin{cases} k_{s,s'} k_{s',s} = 4\cos^2 \frac{\pi}{m(s,s')}, & \text{if } m(s, s') \neq \infty, \\ k_{s,s'} k_{s',s} \geq 4, & \text{if } m(s, s') = \infty. \end{cases}$$

These numbers, which we refer to as *weights*, remain fixed once chosen. The edge weights can, of course, always be chosen symmetrically:

$$k_{s,s'} = k_{s',s} = 2\cos \frac{\pi}{m(s,s')}. \tag{4.22}$$

However, if $m(s, s') \in \{3, 4, 6, \infty\}$ for all edges $s{-}s'$, the following choices, of which the middle three are asymmetric, have the advantage that all edge weights are integers:

$m(s, s')$	$k_{s,s'}$	$k_{s',s}$
3	1	1
4	2	1
6	3	1
∞	4	1
∞	2	2

The starting position for our game can be any distribution $s \mapsto p_s$ of real numbers p_s to the nodes $s \in S$ of the Coxeter graph. A position is called *positive* if $p_s > 0$ for all $s \in S$. The special position with $p_s = 1$ for all $s \in S$ is called the *unit position* and denoted **1**.

Moves are defined as follows. A *firing of node s* changes a position $p \in \mathbb{R}^S$ in the following way:

- Switch sign of the value at s.

- Add $k_{s,s'}p_s$ to the value at each neighbor s' of s.

- Leave all other values unchanged.

Such a move will be called *positive* if $p_s > 0$, and *negative* if $p_s < 0$. A *positive game* is one that is played with positive moves from a given starting position, and similarly for a negative game. A *play sequence* is a word $s_1 s_2 \ldots s_k$ ($s_i \in S$) recording a game in which s_1 was fired first, then s_2, then s_3, and so on. Similarly, a *positive play sequence* records a positive game and a *negative play sequence* records a negative game.

For example, let us consider the infinite group with Coxeter graph

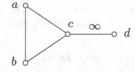

Choose edge weights as in the above table (with $k_{c,d} = k_{d,c} = 2$). The following is an example of a positive game (from the unit position):

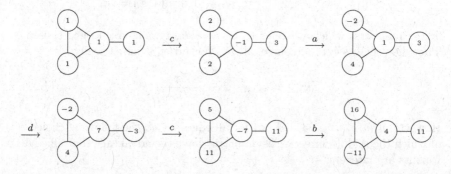

Thus, *cadcb* is a positive play sequence. Note (from the illustration) that the play sequences *cada* and *cadd* are not positive.

Suppose a starting position $p \in \mathbb{R}^S$ is given. Then, every play sequence $s_1 s_2 \ldots s_k$ will by composition of mappings lead to some other position, which we denote by $p^{s_1 s_2 \cdots s_k}$. For instance, in the previous example (with $p = 1$), we computed $p^{cadcb} = (16, -11, 4, 11)$. Let $\mathcal{P}_p \subseteq \mathbb{R}^S$ denote the set of all positions that can be reached this way. The relevance of this is the following.

Theorem 4.3.1 *Consider play sequences starting from some positive position $p \in \mathbb{R}_+^S$.*

(i) *Two play sequences $s_1 s_2 \ldots s_k$ and $s_1' s_2' \ldots s_q'$ lead to the same position (i.e., $p^{s_1 s_2 \cdots s_k} = p^{s_1' s_2' \cdots s_q'}$) if and only if $s_1 s_2 \ldots s_k = s_1' s_2' \ldots s_q'$ as elements of W.*

(ii) *The induced mapping $w \mapsto p^w$ is a bijection $W \to \mathcal{P}_p$.*

(iii) *$D_R(w) = \{s \in S : \text{the } s\text{-entry of position } p^w \text{ is negative}\}$.*

(iv) *The play sequence $s_1 s_2 \ldots s_k$ is positive if and only if $s_1 s_2 \ldots s_k$ is a reduced decomposition.*

Proof. The rule for changing the "position" p to p^s by firing node s coincides with the mapping $\sigma_s^* : p \mapsto s(p)$ considered in Section 4.1 (cf. equation (4.5)). Hence, the point denoted p^w here is the same as the point denoted by $w^{-1}(p)$ there. This implies the "if" direction of part (i). The rest of the theorem follows from Corollary 4.2.6.   □

Let us summarize the theorem in less technical language. It states that we should think about the positions in the numbers game (starting from a positive position) as representing the group elements. The negative components of a position indicate the descent set of the corresponding group element. The set of reduced decompositions is realized as the set of positive play sequences.

The numbers game offers an interesting alternative for computations with Coxeter groups, since it so easily solves word problems and determines reducedness. For instance, if the game is played from the unit position $\mathbf{1}$ (or any other positive starting position) according to two words α and β in the alphabet S, then α and β represent the same group element if and only if the final positions are the same. Or, say that one wants to know the set $\mathcal{R}(w)$ of all reduced decompositions of an element w. One can then proceed as follows. First, compute the position $p^{w^{-1}}$, using any expression $w = s_1 s_2 \ldots s_k$ and playing from $\mathbf{1}$ according to the play sequence $s_k, s_{k-1}, \ldots, s_1$. Then, choosing among the negative moves available at each step, find all negative games leading from $p^{w^{-1}}$ back to $\mathbf{1}$. These games constitute the set $\mathcal{R}(w)$. If, instead, one wants to know only the lexicographically minimal element in $\mathcal{R}(w)$ (i.e., the normal form), one can play from $p^{w^{-1}}$ to $\mathbf{1}$ by firing at each step the minimal negative node.

From a computational point of view, it is an advantage if all edge weights $k_{s,s'}$ used in the numbers game are integers. This is possible if and only if $m(s, s') \in \{3, 4, 6, \infty\}$ for all edges $s\text{---}s'$, as was already discussed.

The numbers game is related to the question of finding combinatorial models for Coxeter groups. For instance, for a permutation $x = x_1 x_2 \ldots x_n \in S_n$ let

$$d(x) = (x_2 - x_1, x_3 - x_2, \ldots, x_n - x_{n-1}).$$

A small computation, left to the reader, shows that $d(x)$ is the position corresponding to x in the numbers game played from the unit position $\mathbf{1}$ on the type A_{n-1} Coxeter graph $\circ\text{---}\circ\text{---} \ldots \text{---}\circ\text{---}\circ$. The details of the correspondence $x \mapsto p^x = d(x)$ lead to the realization of A_{n-1} as the full permutation group S_n; see Exercise 2. For another example, see Exercise 3.

We end with a small computational example. Say we want to compute the normal form of the permutation $x = 362415 \in S_6$. Then, since $x^{-1} = 531462$ and $p^{x^{-1}} = d(x^{-1}) = (-2, -2, 3, 2, -4)$, we are led to play the following game (at each step firing the minimal negative node):

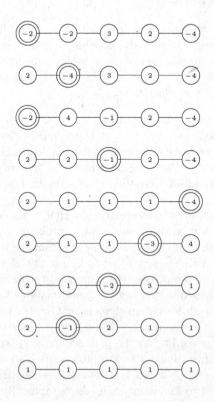

Hence, $NF(362415) = 12135432$.

4.4 Roots

We now return to the general setup of Section 4.2, but from this point on, we specialize the discussion to the *standard* geometric representation σ and its contragredient σ^*. Let us, for emphasis, restate this:

> *From here and to the end of this chapter the bilinear form $(\cdot \mid \cdot)$ is defined by equation (4.21) and the corresponding edge weights by* $k_{s,s'} = 2\cos\frac{\pi}{m(s,s')}$.

One immediate consequence of this assumption is that the W-action on V preserves the bilinear form:

$$(w(\beta) \mid w(\gamma)) = (\beta \mid \gamma) \tag{4.23}$$

for all $w \in W$ and $\beta, \gamma \in V$. It suffices to check this for generators $s \in S$:

$$(s(\beta) \mid s(\gamma)) = (\beta - 2(\alpha_s \mid \beta)\alpha_s \mid \gamma - 2(\alpha_s \mid \gamma)\alpha_s) = (\beta \mid \gamma),$$

using that $(\alpha_s \mid \beta) = (\beta \mid \alpha_s)$ and $(\alpha_s \mid \alpha_s) = 1$.

We now make the following important definition of a certain distinguished subset of $V \cong \mathbb{R}^S$.

Definition 4.4.1 *The* root system *of* (W, S) *is* $\Phi \stackrel{\text{def}}{=} \{w(\alpha_s)\}_{w \in W, s \in S}$. *Its elements are called* roots. *The elements of* $\Pi \stackrel{\text{def}}{=} \{\alpha_s\}_{s \in S}$ *are the* simple roots.

Let Φ^+ (respectively, Φ^-) consist of those roots that have non-negative (respectively, nonpositive) coefficients when expressed in the basis Π of simple roots. They are called *positive roots* and *negative roots*, respectively. Proposition 4.2.5 shows that Φ decomposes into a disjoint union

$$\Phi = \Phi^+ \biguplus \Phi^-, \tag{4.24}$$

and implies the following characterization of positive roots:

$$w(\alpha_s) \in \Phi^+ \quad \Leftrightarrow \quad \ell(ws) > \ell(w). \tag{4.25}$$

Furthermore, since $s(\alpha_s) = -\alpha_s$, we obtain

$$\Phi^- = -\Phi^+, \tag{4.26}$$

and since all roots are unit vectors (due to equation (4.23)), we have that

if $\beta, \gamma \in \Phi$ and $\gamma = r\beta$ for some $r \in \mathbb{R}$, then $r \in \{+1, -1\}$. $\tag{4.27}$

Example 4.4.2 It is instructive to visualize the W-action on V and its induced root system, and to compare it to the geometry of the W-action on V^*, in the accessible case of rank 2.

Figure 4.2 shows the case of the Weyl group $G_2 = I_2(6) =$ ⊶⁶ , for which $(\alpha_a \mid \alpha_b) = -\cos\frac{\pi}{6}$. The picture of the space V indicates the six positive roots and how they are generated by simple reflections from the two simple roots α_a and α_b. The view of V^* shows the arrangement of six reflecting lines, dual to the positive roots via $H_\alpha \overset{\text{def}}{=} \{p \in V^* : \langle p \mid \alpha \rangle = 0\}$, $\alpha \in \Phi^+$.

Figure 4.3 shows the the affine group $\widetilde{A}_1 = I_2(\infty) =$ ⊶^∞ , for which $(\alpha_a \mid \alpha_b) = -1$. The picture of V here shows the positive roots and how they are generated by simple reflections from the two simple roots α_a and α_b, whereas that of V^* shows the arrangement of reflecting lines H_α, dual to the positive roots.

Notice the differences between the two cases. In the finite case, one can, without much danger, blur the distinction between V and its dual. However, this is not so in the infinite case, where the geometry of the W-action is quite different for the two spaces. For instance, observe in Figure 4.3 how both simple reflections σ_a and σ_b share *the same* fixed line in V, namely the dashed diagonal. In V^*, on the other hand, σ_a^* and σ_b^* have different fixed lines H_{α_a} and H_{α_b}.

In the finite case, the W-action on V^* preserves every circle of fixed radius centered at the origin, whereas in the infinite case, the W-action on V^* preserves any line L of the form $x + y = \text{const.} > 0$. The analysis of the numbers game in the preceding section shows how to interpret the game geometrically in V^*-space. Namely, the starting position $\mathbf{1}$ (indicated in Figures 4.2 and 4.3) is moved along the circle in the finite case, and along the line L in the infinite case, according to the W-action on V^*, and the positions in the game are its images. $\square$

Lemma 4.4.3 *The mapping s permutes the set $\Phi^+ \smallsetminus \{\alpha_s\}$.*

Proof. Let $\gamma \in \Phi^+ \smallsetminus \{\alpha_s\}$. From statement (4.27) follows that γ is not a scalar multiple of α_s. Thus, $\langle \alpha_{s'}^* \mid \gamma \rangle$ is positive for some $s' \neq s$. Since $\langle \alpha_{s'}^* \mid s(\gamma) \rangle = \langle s(\alpha_{s'}^*) \mid \gamma \rangle = \langle \alpha_{s'}^* \mid \gamma \rangle$ is positive, we conclude that $s(\gamma) \in \Phi^+$. Also, $s(\gamma) \neq \alpha_s$ since $s(-\alpha_s) = \alpha_s$. $\square$

To each group element w we associate a set of positive roots as follows:

$$N(w) \overset{\text{def}}{=} \{\beta \in \Phi^+ : w(\beta) \in \Phi^-\}. \tag{4.28}$$

Proposition 4.4.4 *For all $w \in W$,*

$$\ell(w) = \operatorname{card} N(w).$$

Proof. The formula holds for $\ell(w) \leq 1$ by Lemma 4.4.3. We continue by induction on $\ell(w)$.

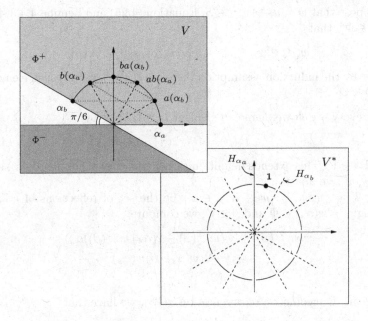

Figure 4.2. Finite case: $I_2(6)$.

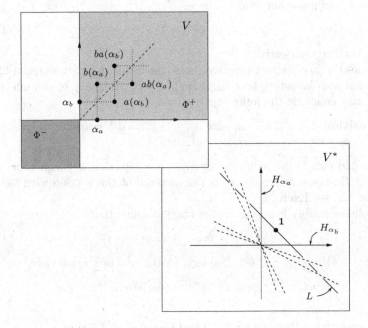

Figure 4.3. Infinite case: $I_2(\infty)$.

Suppose that $w = us > u$, $s \in S$. Equation (4.25) and Lemma 4.4.3 show for $\beta \in \Phi^+$ that

$$w(\beta) \in \Phi^- \quad \Leftrightarrow \quad \beta = \alpha_s \text{ or } s(\beta) \in N(u).$$

Hence, by the induction assumption there are $\ell(u) + 1 = \ell(w)$ such elements $\beta \in \Phi^+$. $\square$

For every $\gamma \in \Phi$ we define an element $t_\gamma \in GL(V)$ by

$$t_\gamma(\beta) = \beta - 2(\gamma \,|\, \beta)\gamma, \tag{4.29}$$

for all $\beta \in V$. This extends definition (4.14) (the case of simple roots), and, clearly, $t_\gamma^2 = e$ and $t_{-\gamma} = t_\gamma$.

Let $T = \{wsw^{-1} : w \in W, s \in S\}$ be the set of reflections of (W, S). Letting $\gamma = w(\alpha_s) \in \Phi$ and $\beta \in V$, we compute

$$\begin{aligned}
wsw^{-1}(\beta) &= w(w^{-1}(\beta) - 2(\alpha_s \,|\, w^{-1}(\beta))\alpha_s) \\
&= \beta - 2(w(\alpha_s) \,|\, \beta)w(\alpha_s) \\
&= t_\gamma(\beta).
\end{aligned}$$

Hence, in the geometric representation $\sigma(W)$, we have that

$$t_\gamma = wsw^{-1}, \tag{4.30}$$

for any $w \in W$ and $s \in S$ such that $\gamma = w(\alpha_s)$. Since σ is injective, $\rho : \gamma \mapsto t_\gamma$ defines a mapping

$$\rho : \Phi^+ \to T, \tag{4.31}$$

which is clearly surjective.

If γ and γ' are distinct positive roots, then (since, by statement (4.27), one is not a scalar multiple of the other) $t_\gamma \neq t_{\gamma'}$. Hence, ρ is also injective, so we may conclude the following.

Proposition 4.4.5 *The mapping* $\rho : \gamma \mapsto t_\gamma$ *is a bijective correspondence between positive roots and reflections.* $\square$

The mapping ρ immediately extends to a bijection between roots and signed reflections, which connects the material of this section with that of Section 1.3; see Exercise 7.

It follows readily from our earlier computations that

$$w(\gamma) = \beta \quad \Longrightarrow \quad wt_\gamma w^{-1} = t_\beta, \tag{4.32}$$

for all $w \in W$ and $\beta, \gamma \in \Phi^+$. Namely, letting $\gamma = u(\alpha_s)$, we have

$$wt_\gamma w^{-1} = wusu^{-1}w^{-1} = (wu)s(wu)^{-1} = t_\beta,$$

since $wu(\alpha_s) = w(\gamma) = \beta$.

We know from Proposition 4.4.4 and Corollary 1.4.5 that

$$\mathrm{card}\{\gamma \in \Phi^+ : w(\gamma) \in \Phi^-\} = \ell(w) = \mathrm{card}\{t \in T : \ell(wt) < \ell(w)\}.$$

Thus, it is natural to expect that the two equicardinal sets are related via the bijection ρ. The following proposition establishes this link, thus generalizing equation (4.25).

Proposition 4.4.6 *For all $w \in W$ and $\gamma \in \Phi^+$,*

$$w(\gamma) \in \Phi^- \quad \Leftrightarrow \quad \ell(w\,t_\gamma) < \ell(w).$$

Proof. Suppose that $\ell(w\,t_\gamma) < \ell(w)$, and let $w = s_1 s_2 \ldots s_k$ be a reduced expression. Then, by Corollary 1.4.4, $t_\gamma = s_k s_{k-1} \ldots s_i \ldots s_{k-1} s_k$ for some $i \in [k]$ and, hence, $\gamma = s_k s_{k-1} \ldots s_{i+1}(\alpha_{s_i})$. We deduce that

$$w(\gamma) = (s_1 \ldots s_k)(s_k \ldots s_{i+1})(\alpha_{s_i}) = s_1 \ldots s_{i-1}(-\alpha_{s_i}) < 0,$$

since, by Proposition 4.2.5, $s_1 \ldots s_{i-1}(\alpha_{s_i}) > 0$. Hence,

$$\{\gamma \in \Phi^+ : \ell(w\,t_\gamma) < \ell(w)\} \subseteq \{\gamma \in \Phi^+ : w(\gamma) \in \Phi^-\},$$

and since the two sets are of equal finite cardinality, they are in fact identical. $\square$

4.5 Roots and subgroups

This section is devoted to proving some algebraic and geometric results about subgroups in a Coxeter group that will later be needed for combinatorial purposes. We begin with a couple of lemmas.

For $p \in V^*$, let

$$M(p) \overset{\text{def}}{=} \{\beta \in \Phi^+ : \langle p \mid \beta \rangle < 0\}$$

and

$$C \overset{\text{def}}{=} \{p \in V^* : M(p) = \emptyset\}.$$

Note that if $p \in C$, then $\langle p \mid \Phi^+ \rangle \subseteq \mathbb{R}_{\geq 0}$ and $\langle p \mid \Phi^- \rangle \subseteq \mathbb{R}_{\leq 0}$.

Recall that we have an action of W on V^* such that

$$\langle w^{-1}(p) \mid \beta \rangle = \langle p \mid w(\beta) \rangle$$

for all $w \in W$, $\beta \in V$, and $p \in V^*$.

Lemma 4.5.1 *Let $p \in C$. Then,*

$$\text{Stab}\,(p) = W_J,$$

where $J \overset{\text{def}}{=} \{s \in S : \langle p \mid \alpha_s \rangle = 0\}$.

Proof. If $s \in J$, then

$$\langle s(p) \mid \beta \rangle = \langle p \mid s(\beta) \rangle = \langle p \mid \beta - 2(\alpha_s \mid \beta)\alpha_s \rangle = \langle p \mid \beta \rangle$$

for all $\beta \in V$ and, hence, $s \in \text{Stab}(p)$. This shows that $W_J \subseteq \text{Stab}(p)$.

Conversely, let $w \in \mathrm{Stab}(p)$. We will show, by induction on $\ell(w)$, that $w \in W_J$. This is clear if $\ell(w) = 0$. So, suppose that $\ell(w) > 0$ and let $s \in D_R(w)$. Then, by relation (4.25), $w(\alpha_s) < 0$ and, hence,

$$0 \leq \langle p \mid \alpha_s \rangle = \langle w^{-1}(p) \mid \alpha_s \rangle = \langle p \mid w(\alpha_s) \rangle \leq 0,$$

which implies that $s \in J$. Since we have already shown that $W_J \subseteq \mathrm{Stab}(p)$, we conclude that $ws \in \mathrm{Stab}(p)$. Since $\ell(ws) < \ell(w)$, we have from our induction hypothesis that $ws \in W_J$, and this concludes the proof. $\square$

Lemma 4.5.2 *Let $p \in V^*$ be such that $M(p)$ is finite. Then, there exists a $w \in W$ such that $w(p) \in C$.*

Proof. Let $p \in V^* \setminus C$ be such that $0 < |M(p)| < \infty$. Then, there exists $s \in S$ such that $\alpha_s \in M(p)$. However, it is clear from our definitions that

$$s(M(s(p))) \subseteq M(p) \setminus \{\alpha_s\}$$

(note that $\alpha_s \notin M(s(p))$). Hence, $|M(s(p))| < |M(p)|$. Continuing in this way, we conclude that there exists a $w \in W$ such that $|M(w(p))| = 0$. $\square$

The next theorem is the main result of this section. It has the interesting consequence that up to conjugacy there are only finitely many finite subgroups of an infinite Coxeter group.

Theorem 4.5.3 *Let H be a finite subgroup of W. Then, there exists $w \in W$ and $J \subseteq S$ such that W_J is finite and $wHw^{-1} \subseteq W_J$.*

Proof. We proceed by induction on $|S|$, the result being clear if $|S| = 1$. We may assume that $|W| = \infty$, otherwise there is nothing to prove.

Let

$$p \overset{\mathrm{def}}{=} \sum_{s \in S} \alpha_s^* \qquad \text{and} \qquad q \overset{\mathrm{def}}{=} \sum_{h \in H} h(p).$$

Now, if $\beta \in \Phi^+ \setminus \bigcup_{h \in H} N(h)$, then $h(\beta) > 0$ for all $h \in H$ and, hence,

$$\langle q \mid \beta \rangle = \sum_{h \in H} \langle h^{-1}(p) \mid \beta \rangle = \sum_{h \in H} \langle p \mid h(\beta) \rangle > 0. \qquad (4.33)$$

This shows that $M(q) \subseteq \bigcup_{h \in H} N(h)$ and, hence, that $|M(q)| < \infty$. By Lemma 4.5.2, we conclude that there exists $w \in W$ such that $w(q) \in C$. However,

$$wHw^{-1} \subseteq \mathrm{Stab}(w(q))$$

(since $h(q) = q$ for all $h \in H$). Hence, by Lemma 4.5.1,

$$wHw^{-1} \subseteq W_J,$$

where $J = \{s \in S : \langle w(q) \mid \alpha_s \rangle = 0\}$. However, $J \neq S$ (or else $w(q) = 0$ and therefore $q = 0$, which contradicts equation (4.33)). Therefore, wHw^{-1} is a finite subgroup of a parabolic subgroup of rank $< |S|$, and this, by the induction hypothesis, concludes the proof. $\square$

The next two results are of a technical nature. They serve to shorten several proofs later.

Proposition 4.5.4 *Let* $\alpha, \beta \in \Phi^+$.

(i) *If* $|(\alpha \mid \beta)| < 1$, *then the subgroup generated by* t_α *and* t_β *is a finite dihedral group.*

(ii) *If* $(\alpha \mid \beta) \leq -1$, *then the subgroup generated by* t_α *and* t_β *is an infinite dihedral group. Furthermore, the roots* $(t_\alpha t_\beta)^n(\alpha)$, *for* $n = 0, 1, 2, \ldots$, *are all positive linear combinations of* α *and* β, *and are all distinct.*

Proof. Assume that $|(\alpha \mid \beta)| < 1$. Let $w \in W$ be such that $w(\beta) \in \Pi$. Since $(\alpha \mid \beta) = (w(\alpha) \mid w(\beta))$, $t_{w(\alpha)} = wt_\alpha w^{-1}$, $t_{w(\beta)} = wt_\beta w^{-1}$, and $t_{-w(\alpha)} = t_{w(\alpha)}$, we may assume that $\beta = \alpha_s$ for some $s \in S$, and $\alpha \in \Phi^+$.

Let V_0 be the subspace of V spanned by α and α_s. Write $\alpha = \sum_{r \in S} a_r \alpha_r$, and let $\gamma \overset{\text{def}}{=} \alpha - a_s \alpha_s$. Then, $\gamma \neq 0$, and

$$\{\lambda\gamma + \mu\alpha_s : \lambda, \mu \in \mathbb{R}, \ \lambda\mu < 0\} \cap \Phi = \emptyset, \tag{4.34}$$

because of the decomposition (4.24). Since $|(\alpha \mid \alpha_s)| < 1$, a simple computation shows that the restriction of the bilinear form (4.21) to V_0 is positive definite. Hence, V_0 is a Euclidean plane, and we are in the situation discussed in Example 1.2.7. Namely, $t_\alpha s$ acts on V_0 as a rotation through the origin of $2x$ radians, where $0 < x < \pi$ is such that $\cos(x) = (\alpha \mid \alpha_s)$. If x is not a rational multiple of π, then the roots $(t_\alpha s)^n(\alpha)$ for $n = 0, 1, 2, \ldots$ are dense in the unit circle, and this contradicts equation (4.34). Hence, $x = q\pi$ for some $q \in \mathbb{Q}$, and the subgroup D generated by t_α and s acts on V_0 in the same way as a finite dihedral group.

Since V_0 is Euclidean, we have that $V = V_0^\perp \oplus V_0$. Furthermore, D fixes $V_0^\perp$ pointwise. By Theorem 4.2.7 the geometric representation of W on V is faithful (in other words, the only element of W that acts on V as the identity is e). It follows that also the action of D on V_0 is faithful and, therefore, D is isomorphic to a finite dihedral group. This proves part (i).

Next, put $x \overset{\text{def}}{=} -(\alpha \mid \beta)$ and assume that $x \geq 1$. Direct computation shows that for any $\lambda, \mu \in \mathbb{R}$,

$$
\begin{aligned}
t_\alpha t_\beta(\lambda\alpha + \mu\beta) &= t_\alpha(\lambda\alpha + \mu\beta - 2(\lambda\alpha + \mu\beta \mid \beta)\beta) \\
&= t_\alpha(\lambda\alpha - (2x\lambda + \mu)\beta) \\
&\quad \vdots \\
&= ((4x^2 - 1)\lambda - 2x\mu)\alpha + (2x\lambda - \mu)\beta.
\end{aligned}
$$

Thus, putting $(t_\alpha t_\beta)^n(\alpha) \overset{\text{def}}{=} \lambda_n \alpha + \mu_n \beta$, we have that

$$
\begin{cases}
\lambda_{n+1} = (4x^2 - 1)\lambda_n - 2x\mu_n, \\
\mu_{n+1} = 2x\lambda_n - \mu_n.
\end{cases}
$$

We now prove for all n that

$$\lambda_n > \mu_n \geq 0. \tag{4.35}$$

This is true for $n = 0$, since $\lambda_0 = 1$ and $\mu_0 = 0$. We continue by induction on n. Let $A \stackrel{\text{def}}{=} 4x^2 - 2x - 1$ and $B \stackrel{\text{def}}{=} 2x - 1$. Then,

$$\lambda_{n+1} - \mu_{n+1} = A\lambda_n - B\mu_n, \tag{4.36}$$

and

$$A \geq B \geq 1, \text{ since } x \geq 1 \text{ and } A - B = (2x - 1)^2 - 1 \geq 0. \tag{4.37}$$

The induction assumption (4.35) together with relations (4.36) and (4.37) imply that $\lambda_{n+1} > \mu_{n+1}$. That $\mu_{n+1} \geq 0$ is seen directly from relation (4.35) and $\mu_{n+1} = 2x\lambda_n - \mu_n$. This finishes the induction step.

Equation (4.35) also implies that $\mu_{n+1} - \mu_n = 2(x\lambda_n - \mu_n) > 0$. Hence,

$$\cdots > \mu_{n+1} > \mu_n > \cdots > \mu_1 > \mu_0 = 0,$$

and the proof is complete. $\square$

Proposition 4.5.5 *The set*

$$\{(\alpha \mid \alpha_s) : \alpha \in \Phi^+, \ s \in S, \ |(\alpha \mid \alpha_s)| < 1\} \tag{4.38}$$

is finite.

Proof. Let

$$\mathcal{P}_{\text{fin}} \stackrel{\text{def}}{=} \{J \subseteq S : |W_J| < \infty\},$$

$$\mathcal{A}_{\text{fin}} \stackrel{\text{def}}{=} \left\{(\alpha \mid \beta) : \alpha, \beta \in \Phi, \ t_\alpha, t_\beta \in \bigcup_{J \in \mathcal{P}_{\text{fin}}} W_J\right\}.$$

Clearly, $\mathcal{A}_{\text{fin}}$ is a finite set. We claim that

$$\{(\alpha \mid \alpha_s) : \alpha \in \Phi^+, \ s \in S, \ |(\alpha \mid \alpha_s)| < 1\} \subseteq \mathcal{A}_{\text{fin}}.$$

Indeed, let $\alpha \in \Phi^+$ and $s \in S$ be such that $|(\alpha \mid \alpha_s)| < 1$. By Proposition 4.5.4, the subgroup D generated by t_α and s is finite. Therefore, by Theorem 4.5.3, there exists a $w \in W$ and a $J \in \mathcal{P}_{\text{fin}}$ such that $wDw^{-1} \subseteq W_J$. However, $(\alpha \mid \alpha_s) = (w(\alpha) \mid w(\alpha_s))$ and $t_{w(\alpha)} = wt_\alpha w^{-1}$, $t_{w(\alpha_s)} = wsw^{-1}$. Hence, $(\alpha \mid \alpha_s) \in \mathcal{A}_{\text{fin}}$. $\square$

4.6   The root poset

A certain interesting partial order on the set Φ^+ of positive roots plays an important role in the sequel. In this section, we study some of its basic properties and show that it is related to a "numbers game" dual to that of Section 4.3.

Definition 4.6.1 *The* depth *of* $\beta \in \Phi^+$ *is*

$$\mathrm{dp}\,(\beta) = \min\{k:\ w(\beta) \in \Phi^-\ \textit{for some } w \in W \textit{ with } \ell(w) = k\}.$$

Since $t_\beta(\beta) \in \Phi^-$, the concept of depth is well defined, and by Lemma 4.4.3, the roots of depth 1 are precisely the simple roots.

It is important to know how the depth changes when acting on a positive root by a simple reflection. The answer is very elegant.

Lemma 4.6.2 *Let* $s \in S$ *and* $\beta \in \Phi^+ - \{\alpha_s\}$. *Then,*

$$\mathrm{dp}\,(s(\beta)) = \begin{cases} \mathrm{dp}\,(\beta) - 1, & \textit{if } (\beta|\alpha_s) > 0, \\ \mathrm{dp}\,(\beta), & \textit{if } (\beta|\alpha_s) = 0, \\ \mathrm{dp}\,(\beta) + 1, & \textit{if } (\beta|\alpha_s) < 0. \end{cases}$$

Proof. If $(\beta|\alpha_s) = 0$, then $s(\beta) = \beta$, so trivially $\mathrm{dp}\,(s(\beta)) = \mathrm{dp}\,(\beta)$.

Suppose that $(\beta|\alpha_s) > 0$. Clearly, $\mathrm{dp}\,(s(\beta)) \geq \mathrm{dp}\,(\beta) - 1$. Hence, it will suffice to show that $\mathrm{dp}\,(s(\beta)) < \mathrm{dp}\,(\beta)$. For this, choose $w \in W$ such that $w(\beta) \in \Phi^-$ and $\ell(w) = \mathrm{dp}\,(\beta)$. Now, consider the two possibilities: $ws < w$ and $ws > w$.

If $ws < w$, we are done, since $ws(s(\beta)) = w(\beta) \in \Phi^-$ shows that $\mathrm{dp}\,(s(\beta)) \leq \ell(ws) < \mathrm{dp}\,(\beta)$.

Assume that $ws > w$. Consider the root

$$\gamma = ws(\beta) = w(\beta - 2(\beta|\alpha_s)\alpha_s) = w(\beta) - 2(\beta|\alpha_s)w(\alpha_s).$$

By assumption, $w(\beta) \in \Phi^-$ and $(\beta|\alpha_s) > 0$, and (by Proposition 4.2.5) $w(\alpha_s) \in \Phi^+$. Hence, $\gamma \in \Phi^-$. Furthermore, $\gamma \neq -\alpha_{s'}$ for all $s' \in S$, since $-\alpha_{s'}$ can never be the sum of two negative roots. Now, choose $s' \in S$ such that $s'w < w$. Then, $s'w(s(\beta)) = s'(\gamma)$, and $s'(\gamma) \in \Phi^-$ by Lemma 4.4.3, since $\gamma \in \Phi^- \setminus \{-\alpha_{s'}\}$. Therefore, $\mathrm{dp}\,(s(\beta)) \leq \ell(s'w) < \ell(w) = \mathrm{dp}\,(\beta)$, as desired.

Finally, suppose that $(\beta|\alpha_s) < 0$. Then, $(s(\beta)|\alpha_s) = (\beta - 2(\beta|\alpha_s)\alpha_s|\alpha_s) = -(\beta|\alpha_s) > 0$, so by the previous case $\mathrm{dp}\,(\beta) = \mathrm{dp}\,(s(s(\beta))) = \mathrm{dp}\,(s(\beta)) - 1$. $\square$

Using the concept of depth we can now define the root poset.

Definition 4.6.3 *For* $\beta, \gamma \in \Phi^+$, *let* $\beta \leq \gamma$ *if there exist* $s_1, s_2, \ldots, s_k \in S$ *such that*

(i) $\gamma = s_k s_{k-1} \ldots s_1(\beta)$,

(ii) $\mathrm{dp}\,(s_i s_{i-1} \ldots s_1(\beta)) = \mathrm{dp}\,(\beta) + i$, *for all* $1 \leq i \leq k$.

What we have proved so far shows that the root poset $(\Phi^+, \leq)$ has the following structure. The minimal elements are the simple roots. All maximal chains in an interval $[\beta, \gamma]$ have the same length $\mathrm{dp}\,(\gamma) - \mathrm{dp}\,(\beta)$, and all maximal chains in $\{\beta|\beta \leq \gamma\}$ have the same length $\mathrm{dp}\,(\gamma) - 1$. Hence, depth is a rank function. See Figures 4.4 and 4.5 for two examples of root posets.

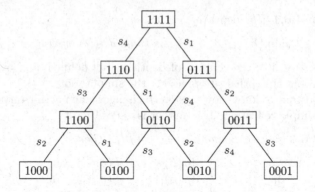

Figure 4.4. Root poset of A_4.

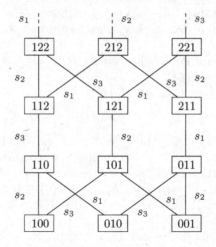

Figure 4.5. Root poset of $\widetilde{A}_2$.

Root posets have a natural edge labeling by elements of S. Namely, for every covering $\beta \lessdot \gamma$, there is a unique $s \in S$ such that $s(\beta) = \gamma$, which provides the label $\lambda(\beta, \gamma) = s$. The labels are indicated in the figures.

Let $s \in S$ and suppose that $\beta = \sum_{s' \in S} b_{s'} \alpha_{s'}$. We have from equations (4.10) and (4.14) that

$$s(\beta) = \beta + \left(\sum_{s' \in S} k_{s,s'} b_{s'} \right) \alpha_s,$$

which if we define

$$B_s \stackrel{\text{def}}{=} -b_s + \sum_{s' : s' \longrightarrow s} k_{s,s'} b_{s'} \tag{4.39}$$

can be written

$$s(\beta) = \beta + (B_s - b_s)\alpha_s. \tag{4.40}$$

The sum in the definition of B_s is over all neighbors s' of s in the Coxeter diagram. We then get the following criterion for moving up or down along an s-labeled edge in the root poset.

Lemma 4.6.4 $s(\beta) > \beta \Leftrightarrow B_s > b_s$.

Proof. It is clear from our definitions that $B_s - b_s = -2(\alpha_s|\beta)$. Now use Lemma 4.6.2. $\square$

This lemma has the following useful consequence.

Corollary 4.6.5 Let $\beta = \sum_{s \in S} b_s \alpha_s$ and $\gamma = \sum_{s \in S} c_s \alpha_s$, $\beta, \gamma \in \Phi^+$. If $\beta \leq \gamma$, then $b_s \leq c_s$ for all $s \in S$. $\square$

Note, for example, from Figure 4.5, that the converse of this corollary is not true.

Lemma 4.6.4 gives a simple algorithmic procedure for generating the edge-labeled root poset. Namely, start with the roots of depth 1, (i.e., the unit basis vectors $\{\alpha_s | s \in S\}$). Recursively, assume that we have constructed the edge-labeled root poset up to (and including) the roots of depth j. Then, for each root β of depth j and each $s \in S$ such that no s-labeled edge leads down from β, compute the quantity B_s defined by (4.39); that is, compute the $k_{s,s'}$-weighted sum of β's coordinates at all neighbors of s in the Coxeter diagram minus its coordinate b_s at s. If $B_s > b_s$, let γ be the vector that you get by replacing b_s by B_s as the s-coordinate of β. Then, γ is a root of depth $j + 1$ and (β, γ) is an s-labeled edge. If $B_s = b_s$, then do nothing ($B_s < b_s$ cannot occur, since then an s-labeled edge would lead down from β to a root of depth $j - 1$). After performing this for all pairs β and s (of the specified kinds), the root poset up to depth $j + 1$ will be constructed.

The algorithm described in the preceding paragraph can be thought of as a "dual numbers game." Namely, positive roots $\beta = \sum_{s \in S} b_s \alpha_s$ are assignments of numbers b_s to the nodes s of the Coxeter diagram, or "dual positions." It is allowed to move from one such position to another by "firing" the node s, which replaces the number b_s attached to s by the number B_s. By only allowing firings that increase the number at the fired node we create the "legal" games. These correspond to unrefinable ascending chains in the root poset. Playing legal games from the starting positions with "1" on one node and "0" on all the others, all positive roots will be generated.

Thus, an appealing picture emerges. The dual numbers game — modeling the root poset ($\Phi^+, \leq$) — takes place in the space V, whereas the numbers game — a combinatorial model of the Coxeter group W and its right weak order — lives in the dual space V^*. See Figures 4.2 and 4.3 for an illustration. The Coxeter graph itself is the game board for both games, the positions are assignments of numbers to its vertices, and the moves in both cases are certain firings of the vertices involving only numbers at the fired node and its neighbors in the diagram. The two kinds of firing, both given by simple combinatorial rules, are dual in the sense of Proposition 4.2.3.

Every position p^w of the numbers game (i.e., each group element $w \in W$) assigns a number $\langle p^w | \beta \rangle$ to each root $\beta \in \Phi$. (Note that if $p^w = \sum p_s \alpha_s^*$ and $\beta = \sum b_s \alpha_s$, then $\langle p^w | \beta \rangle = \sum p_s b_s$.) We know from Proposition 4.4.6 that the negative such numbers occurring at positive roots indicate the reflection descent set $T_R(w)$. What happens with the distribution of such numbers when we pass from w to ws? Since

$$\langle p^{ws} | \beta \rangle = \langle s(p^w) | \beta \rangle = \langle p^w | s(\beta) \rangle, \tag{4.41}$$

we see that ws assigns to β the same number that w assigns to $s(\beta)$. Recall that s permutes the set $\Phi^+ \setminus \{\alpha_s\}$. More precisely, it switches the positive roots connected by s-labeled edges in the root poset, and the remaining ones are fixed points. In summary, we have proved the following.

Proposition 4.6.6 *Define $\varphi_w : \Phi^+ \to \mathbb{R}$ by $\varphi_w(\beta) = \langle p^w | \beta \rangle$ and let $s \in S$. Then, the distribution φ_{ws} of values to the positive roots differs from the distribution φ_w as follows:*

(i) *The value at α_s changes sign.*

(ii) *The values are interchanged between pairs β —— $s(\beta)$ connected by s-labeled edges in the root poset.*

(iii) *The values of all remaining roots are unchanged.*

Example 4.6.7 Let $w = 52314$ and $ws = 53214$. The corresponding positions are (by the rule of successive differences explained at the end of Section 4.3): $p^w = (-3, 1, -2, 3)$ and $p^{ws} = (-2, -1, -1, 3)$. The values assigned to the positive roots (see Figure 4.4) are shown in Figure 4.6, where the s-labeled edges are solid and all other root poset edges are dashed. $\square$

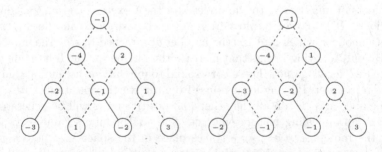

Figure 4.6. Illustration for Example 4.6.7.

4.7 Small roots

The crucial tool for the construction of finite state automata in the next section is a remarkable subset of the positive roots of W, which includes the simple roots. In this section, we define and study this subset.

Let Σ be the smallest subset of Φ^+ such that the following hold:

(i) $\Pi = \{\alpha_s : s \in S\} \subseteq \Sigma$.

(ii) If $\alpha \in \Sigma$, $s \in S$ and $-1 < (\alpha|\alpha_s) < 0$, then $s(\alpha) \in \Sigma$.

We call Σ the set of *small roots* of (W, S).

Note that α and $s(\alpha)$ are equal except in the s-th coordinate where they differ by $-2(\alpha|\alpha_s)$. Thus, the definition of small roots can be rephrased:

> *All simple roots are small, and if α is small and the difference in the s-th coordinate between α and $s(\alpha)$ is positive and < 2, then $s(\alpha)$ is also small.*

This is the reason for the terminology "small." In other words, the small roots are the ones that can be reached from the simple roots by successive "small changes," moving up along saturated chains in the root poset.

Let us call a covering edge $\beta \lhd \gamma$ in the root poset *short* if $|(\beta|\alpha_s)| < 1$, where $s \in S$ is such that $s(\beta) = \gamma$, and *long* otherwise. Since $(\gamma|\alpha_s) = -(\beta|\alpha_s)$, the characterization of a short edge takes the same form in terms of the pair (γ, s). Small roots are then characterized by the property:

> *A root is small if and only if it is reachable from a simple root along an up-directed path of short edges.*

Referring to Figures 4.4 and 4.5, we see that all positive roots of the finite group A_4 are small and that if (W, S) is a Coxeter system of type $\widetilde{A}_2$, then $\Sigma = \{\alpha_1, \alpha_2, \alpha_3, \alpha_1 + \alpha_2, \alpha_2 + \alpha_3, \alpha_1 + \alpha_3\}$. It is a general property of finite groups that all positive roots are small; see Exercise 19.

Figure 4.7 shows the root poset up to depth 3 of the infinite group $\overset{\infty}{\underset{a \quad b \quad c}{\circ - \circ - \circ}}$, previously discussed in Example 1.2.9. Long edges are dashed. The separating set of long edges above the bottom four roots shows that $\Sigma = \{\alpha_1, \alpha_2, \alpha_3, \alpha_1 + \alpha_2\}$.

We have seen two examples of infinite groups for which the set of small roots is finite. It is a surprising fact that this is always the case, as we now proceed to prove.

The following lemma tells us that all edges *down* from a small root are necessarily short.

Lemma 4.7.1 Let $\alpha \in \Sigma$ and $s \in S$, $\alpha \neq \alpha_s$. Then, $(\alpha|\alpha_s) < 1$.

Proof. We proceed by induction on $\mathrm{dp}\,(\alpha)$, the result being clear if $\mathrm{dp}\,(\alpha) = 1$.

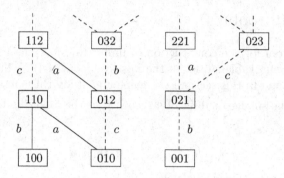

Figure 4.7. Bottom part of root poset of the group $\underset{a}{\circ}\!\!-\!\!\underset{b}{\circ}\overset{\infty}{-}\underset{c}{\circ}$.

Suppose $\mathrm{dp}\,(\alpha) \geq 2$. Since $\alpha \in \Sigma$, there is $\beta \in \Sigma$ and $r \in S$ such that $\beta \lhd \alpha$, $\alpha = r(\beta)$ and $-1 < (\beta \mid \alpha_r) < 0$. If $s = r$, we are done, so suppose that $s \neq r$. If $\beta = \alpha_s$ then the result is easy to check. If $\beta \neq \alpha_s$ then, using our induction hypothesis, we have that

$$
\begin{aligned}
(\alpha \mid \alpha_s) &= (r(\beta) \mid \alpha_s) \\
&= (\beta - 2(\beta \mid \alpha_r)\alpha_r \mid \alpha_s) \\
&= (\beta \mid \alpha_s) - 2(\beta \mid \alpha_r)(\alpha_r \mid \alpha_s) \\
&\leq (\beta \mid \alpha_s) < 1,
\end{aligned}
$$

as desired. $\square$

Given $\alpha \in \Sigma$, we let

$$
\mathcal{N}(\alpha) \overset{\mathrm{def}}{=} \{s \in S : |(\alpha|\alpha_s)| < 1\}.
$$

Lemma 4.7.2 *Let $\beta, \gamma \in \Sigma$ be such that $\beta \lhd \gamma$ in the root poset, $\mathrm{dp}\,(\beta) \geq 2$. Then, $\mathcal{N}(\beta) \supseteq \mathcal{N}(\gamma)$.*

Proof. Let $s \in S \setminus \mathcal{N}(\beta)$. Then, we have from Lemma 4.7.1 that $(\beta|\alpha_s) \leq -1$. On the other hand, since $\gamma = r(\beta)$ for some $r \in S$ and $\gamma > \beta$, we conclude from Lemma 4.7.1 (applied to γ) and Lemma 4.6.2 that $-1 < (\beta|\alpha_r) < 0$. Hence, $r \neq s$, and

$$
\begin{aligned}
(\gamma|\alpha_s) &= (r(\beta)|\alpha_s) \\
&= (\beta - 2(\beta|\alpha_r)\alpha_r|\alpha_s) \\
&= (\beta|\alpha_s) - 2(\beta|\alpha_r)(\alpha_r|\alpha_s) \\
&\leq (\beta|\alpha_s).
\end{aligned}
$$

Therefore, $(\gamma|\alpha_s) \leq -1$ and, hence, $s \notin \mathcal{N}(\gamma)$. $\square$

We can now prove the main result of this section. It is a fundamental fact.

Theorem 4.7.3 $|\Sigma| < \infty$.

Proof. Suppose that $|\Sigma| = \infty$. Since there are only finitely many elements in the root poset of any given depth, we conclude that there are small roots of arbitrarily large depth. For each small root α, we have (by the definition of Σ) a saturated chain in the root poset, entirely contained in Σ, from some simple root to α. Hence, there are saturated chains (in the root poset) consisting entirely of small roots, of arbitrarily great length.

By Proposition 4.5.5, there are only finitely many pairs (J, v), with $J \subseteq S$ and $v \in \mathbb{R}^J$, such that there exists a $\gamma \in \Sigma$ with $\mathcal{N}(\gamma) = J$ and $((\gamma \mid \alpha_s))_{s \in J} = v$. Let C be a saturated chain in $\Sigma \setminus \Pi$ of length greater than the number of such pairs. Taking a suitable segment of C, we conclude that there exists a saturated chain $\gamma_j \vartriangleleft \gamma_{j+1} \vartriangleleft \cdots \vartriangleleft \gamma_k$, such that $\mathcal{N}(\gamma_j) = \mathcal{N}(\gamma_k)$ and

$$(\gamma_j | \alpha_s) = (\gamma_k | \alpha_s) \tag{4.42}$$

for all $s \in \mathcal{N}(\gamma_k)$, with $\mathrm{dp}\,(\gamma_i) = i$ for $i = j, \ldots, k$.

Let, for brevity, $\mathcal{N} \overset{\text{def}}{=} \mathcal{N}(\gamma_k)$. Let $s_i \in S$ be such that $s_i(\gamma_i) = \gamma_{i+1}$ for $i = j, \ldots, k - 1$. It follows from Lemma 4.7.2 that $s_j, s_{j+1}, \ldots, s_{k-1} \in \mathcal{N}$. Let $\gamma_{j-1} \overset{\text{def}}{=} s_{k-1}(\gamma_j)$. By equation (4.42), we have that

$$
\begin{aligned}
(\gamma_{j-1} \mid \alpha_s) &= (s_{k-1}(\gamma_j) \mid \alpha_s) \\
&= (\gamma_j \mid s_{k-1}(\alpha_s)) \\
&= (\gamma_j \mid \alpha_s - 2(\alpha_{s_{k-1}} | \alpha_s) \alpha_{s_{k-1}}) \\
&= (\gamma_k \mid \alpha_s - 2(\alpha_{s_{k-1}} | \alpha_s) \alpha_{s_{k-1}}) \\
&= (\gamma_k \mid s_{k-1}(\alpha_s)) \\
&= (s_{k-1}(\gamma_k) \mid \alpha_s) \\
&= (\gamma_{k-1} \mid \alpha_s) \tag{4.43}
\end{aligned}
$$

for all $s \in \mathcal{N}$. In particular, since $\gamma_{k-1} < \gamma_k$, equation (4.43) implies by Lemma 4.6.2 that $\gamma_{j-1} \vartriangleleft \gamma_j$.

Hence, we have obtained a saturated chain of roots $\gamma_{j-1} \vartriangleleft \gamma_j \vartriangleleft \cdots \vartriangleleft \gamma_{k-1}$ such that

$$(\gamma_{j-1} \mid \alpha_s) = (\gamma_{k-1} \mid \alpha_s)$$

for all $s \in \mathcal{N}$, and $s_{k-1}(\gamma_{j-1}) = \gamma_j$, $s_i(\gamma_i) = \gamma_{i+1}$ for $i = j, \ldots, k - 2$. Continuing in this way, we construct a saturated chain

$$\gamma_k \vartriangleright \gamma_{k-1} \vartriangleright \cdots \vartriangleright \gamma_j \vartriangleright \gamma_{j-1} \vartriangleright \cdots \vartriangleright \gamma_2 \vartriangleright \gamma_1$$

in the root poset, with γ_1 a simple root, and

$$(\gamma_1 \mid \alpha_s) = (\gamma_{k-j+1} \mid \alpha_s) \tag{4.44}$$

for all $s \in \mathcal{N}$.

Let $r \in S$ be such that $\gamma_1 = \alpha_r$. If $r \in \mathcal{N}$, then from equation (4.44) we have that

$$(\gamma_{k-j+1} \mid \alpha_r) = (\gamma_1 \mid \alpha_r) = 1,$$

whereas, on the other hand,

$$|(\gamma_{k-j+1} \mid \alpha_r)| < 1$$

since $\mathcal{N}(\gamma_{k-j+1}) = \mathcal{N}$. A contradiction!

Thus, assume that $r \notin \mathcal{N}$. Let $j \le i \le k - 1$ be such that $s_i(\gamma_{k-j}) = \gamma_{k-j+1}$. Then, $r \ne s_i$ and we conclude from equation (4.44) that

$$0 \ge (\gamma_1 \mid \alpha_{s_i}) = (\gamma_{k-j+1} \mid \alpha_{s_i}) > 0,$$

thus again reaching a contradiction. $\square$

We close this section by giving a useful and interesting characterization of small roots. Let $\beta, \gamma \in \Phi^+$. We say that β *dominates* γ, denoted β dom γ, if $w(\beta) < 0$ implies $w(\gamma) < 0$ for all $w \in W$. This relation is clearly transitive. Note that if β dom γ, then $\mathrm{dp}\,(\beta) \ge \mathrm{dp}\,(\gamma)$. Also, if β dom γ, then $w(\beta)$ dom $w(\gamma)$ for all $w \in W$ such that $w(\gamma) \in \Phi^+$.

Lemma 4.7.4 *Let $\beta \in \Phi^+$ and $s \in S$. Then, β dominates α_s if and only if $(\beta \mid \alpha_s) \ge 1$.*

Proof. We may clearly assume that $\beta \ne \alpha_s$ and, hence, that $s(\beta) \in \Phi^+$. Suppose that β dominates α_s and that $(\beta \mid \alpha_s) < 1$. Since β dominates α_s and $t_\beta(\beta) = -\beta < 0$, there follows that $t_\beta(\alpha_s) = \alpha_s - 2(\alpha_s \mid \beta)\beta \in \Phi^-$ and, therefore,

$$(\alpha_s \mid \beta) > 0. \tag{4.45}$$

So we conclude that $|(\alpha_s \mid \beta)| < 1$. By Proposition 4.5.4, this implies that the subgroup D generated by s and t_β is a finite dihedral group and that $t_\beta s$ acts as a rotation of finite order on the subspace of V spanned by β and α_s. However, it is not hard to see that this implies that there exists $w \in D$ such that $w(\beta) < 0$ but $w(\alpha_s) > 0$, and this contradicts our hypothesis.

Conversely, suppose that $(\beta \mid \alpha_s) \ge 1$. Then,

$$(s(\beta) \mid \beta) = (\beta - 2(\alpha_s \mid \beta)\alpha_s \mid \beta) = 1 - 2(\alpha_s \mid \beta)^2 \le -1.$$

By Proposition 4.5.4, this implies that there are infinitely many positive roots of the form

$$\lambda\beta + \mu s(\beta) \tag{4.46}$$

with $\lambda, \mu > 0$. Now, let $w \in W$ be such that $w(\beta) < 0$. If $w(\alpha_s) > 0$, then we conclude that

$$w(s(\beta)) = w(\beta - 2(\alpha_s \mid \beta)\alpha_s) = w(\beta) - 2(\alpha_s \mid \beta)w(\alpha_s) < 0,$$

and, hence, that all the positive roots of the form (4.46) are also in $N(w)$, which contradicts Proposition 4.4.4. Hence, $w(\alpha_s) < 0$, showing that β dominates α_s, as desired. $\square$

A positive root $\alpha \in \Phi^+$ is said to be *humble* if α dominates no positive root except itself. All simple roots are clearly humble. The preceding lemma has the following useful consequence.

Lemma 4.7.5 *Let $\beta, \alpha \in \Phi^+$ be such that $\beta \lhd \alpha$ in the root poset. Then, we have that the following hold:*

(i) If $\beta \lhd \alpha$ is long, then α is not humble.

(ii) If $\beta \lhd \alpha$ is short, then α is humble if and only if β is humble.

Proof. Let $s \in S$ be such that $\alpha = s(\beta)$.

Assume that (i) holds. Then, $(\alpha \mid \alpha_s) \geq 1$ and, hence, α dominates α_s by Lemma 4.7.4. So α is not humble.

Assume now that (ii) holds. Then, $0 > (\beta \mid \alpha_s) > -1$. Let $\gamma \in \Phi^+ \backslash \{\beta\}$ be such that $\beta \text{ dom } \gamma$. Then, by Lemma 4.7.4 , $\gamma \neq \alpha_s$. Hence, $s(\beta) \text{ dom } s(\gamma)$ and $s(\gamma) \in \Phi^+ \backslash \{s(\beta)\}$. So β is humble if α is humble. For the converse statement a similar argument holds. $\square$

We can now prove the promised characterization of small roots.

Theorem 4.7.6 *Let $\alpha \in \Phi^+$. Then, $\alpha \in \Sigma$ if and only if α is humble.*

Proof. Assume that $\alpha \in \Sigma$. Then, by definition, there is a saturated chain in the root poset, consisting entirely of short edges, from some simple root to α. However, simple roots are humble, so α is humble by part (ii) of Lemma 4.7.5.

Conversely, suppose that α is humble and let $\alpha_1 \lhd \alpha_2 \lhd \cdots \lhd \alpha_p = \alpha$ be a saturated chain in the root poset from some simple root α_1 to α. By part (i) of Lemma 4.7.5, the edge $\alpha_{p-1} \lhd \alpha_p$ is short and, hence, by part (ii), α_{p-1} is humble. Continuing in this way, we conclude that all edges $\alpha_1 \lhd \alpha_2$, $\alpha_2 \lhd \alpha_3, \ldots, \alpha_{p-1} \lhd \alpha_p$ are short and, hence, that $\alpha_p = \alpha \in \Sigma$, as desired. $\square$

As an immediate consequence, we obtain the following fact.

Corollary 4.7.7 *Σ is an order ideal in the root poset.*

Proof. Let $\alpha \in \Sigma$ and $\beta \in \Phi^+$ be such that $\beta \lhd \alpha$. By Theorem 4.7.6, α is humble, and, hence, by Lemma 4.7.5, $\beta \lhd \alpha$ is short and β is humble. Hence, by Theorem 4.7.6, $\beta \in \Sigma$. $\square$

4.8   The language of reduced words is regular

Consider the set $\mathcal{R}(\widetilde{A}_1)$ of all reduced words of the infinite group $\widetilde{A}_1$. If the two Coxeter generators are a and b, then

$$\mathcal{R}(\widetilde{A}_1) = \{abab\ldots, babab\ldots\}$$

(i.e., the set of alternating words of all lengths, including the empty word). A compact way of describing this set is via the labeled directed graph in Figure 4.8.

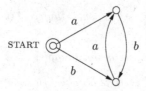

Figure 4.8. An automaton for $\mathcal{R}(\tilde{A}_1)$.

Namely, the words in $\mathcal{R}(\tilde{A}_1)$ correspond to the finite directed paths emanating from the start node, for which the edge labels are read in sequence as we travel along the path (see Figure 4.9).

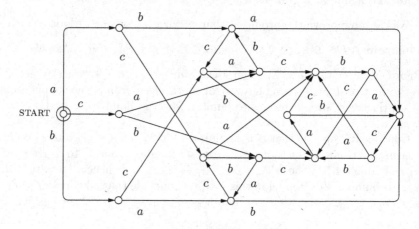

Figure 4.9. An automaton for $\mathcal{R}(\tilde{A}_2)$.

Is there such a finite representation for the set $\mathcal{R}(W, S)$ of reduced words for *every* infinite Coxeter group? The surprising answer is *yes*, as will be shown in this section. (Remember our standing assumption in this chapter that S is finite.) The general idea should be clear from this example. Here is a quick review of some definitions.

A *formal language* (or just *language*) $\mathcal{L}$ is a subset of the set A^* of words in a given finite *alphabet* A. A *finite state automaton* (or just *automaton*) is a finite directed graph, with one distinguished node labeled "start" and every edge labeled by an element of A. We say that a word $w \in \mathcal{L}$ is *accepted* by the automaton $\mathcal{A}$ if the sequence of edge labels along some directed path from the start node equals w, otherwise w is *rejected*.

The language $\mathcal{L}$ is *recognized* by the automaton $\mathcal{A}$ if the words of $\mathcal{L}$ are precisely those that are accepted by $\mathcal{A}$. A language is *regular* if it is recognized by some finite state automaton. An example of a language that is not regular is $\{a^n b^n : n \geq 0\}$.

In this section, we reach one of the main goals of this chapter. Namely, we construct, for each Coxeter system (W, S), a finite state automaton that recognizes the language $\mathcal{R}(W, S)$ of reduced words, thus showing that this language is regular.

Notice that if the group W is finite, then such an automaton is given by the order diagram of the right weak order of W. In fact, for any Coxeter group (W, S) (finite or infinite), we may take W as the set of nodes of an automaton, with e as the start node. For each $w \in W$ and $s \in S$, we then put a directed edge labeled by s from w to ws if $s \notin D_R(w)$. It is clear that this automaton (the right weak order diagram) recognizes the reduced expressions of W — this bijection between reduced words and chains in weak order was already used several times in Chapter 3. If the group is infinite, however, this automaton is also infinite.

We begin the general construction with a technical lemma.

Lemma 4.8.1 *Let $\alpha \in \Sigma$, $s \in S$ and $w \in W$ be such that $s(\alpha) \in \Phi^+ \setminus \Sigma$ and $s \notin D_R(w)$. Then, $ws(\alpha) > 0$.*

Proof. Since $ws > w$, we have that $w(\alpha_s) > 0$ and, hence, $ws(\alpha_s) < 0$. Suppose that $ws(\alpha) < 0$. Since $\alpha \in \Sigma$ but $s(\alpha) \notin \Sigma$, and Σ is an order ideal in the root poset, we have that $(\alpha|\alpha_s) \leq -1$. Hence, by Proposition 4.5.4, the roots $(t_\alpha s)^n(\alpha)$, for $n = 0, 1, 2, \ldots$, are all non-negative linear combinations of α and α_s and are all distinct. Hence, $ws((t_\alpha s)^n(\alpha)) < 0$ for all $n = 0, 1, 2, \ldots$, which contradicts Proposition 4.4.4. $\square$

Define the *small descent set* of $w \in W$ by

$$D_\Sigma(w) \overset{\text{def}}{=} \{\alpha \in \Sigma : w(\alpha) < 0\}.$$

These sets have two crucial properties: (1) There are finitely many of them (by Theorem 4.7.3) and (2) if $w < ws$ and we know $D_\Sigma(w)$, then with no further information we can compute $D_\Sigma(ws)$, as we now show.

Proposition 4.8.2 *Let $w \in W$ and $s \notin D_R(w)$. Then,*

$$D_\Sigma(ws) = \{\alpha_s\} \cup (\{s(\beta) : \beta \in D_\Sigma(w)\} \cap \Sigma).$$

Proof. Let $\alpha \in D_\Sigma(ws)$. Then, $ws(\alpha) < 0$ and, hence, by Lemma 4.8.1, $s(\alpha) \notin \Phi^+ \setminus \Sigma$. Therefore, either $s(\alpha) \in \Phi^-$, which implies $\alpha = \alpha_s$, or $s(\alpha) \in \Sigma$, which implies that $s(\alpha) \in D_\Sigma(w)$. This shows that $D_\Sigma(ws) \subseteq \{\alpha_s\} \cup (s(D_\Sigma(w)) \cap \Sigma)$. The opposite inclusion is obvious. $\square$

It is now easy to deduce the main result of this section.

Theorem 4.8.3 *The language of reduced expressions is regular.*

Proof. We construct a finite state automaton for the language of reduced expressions as follows. Take

$$\mathcal{S} \overset{\text{def}}{=} \{D_\Sigma(w) : w \in W\}$$

as the set of nodes of the automaton, with $D_\Sigma(e)$ $(= \emptyset)$ as the start node. The set S is finite by Theorem 4.7.3.

For each $D \in S$ and $s \in S$ such that $\alpha_s \notin D$, we put a labeled directed edge

$$D \xrightarrow{s} \{\alpha_s\} \cup \big(s(D) \cap \Sigma\big).$$

Note that $\{\alpha_s\} \cup (s(D) \cap \Sigma) \in S$ if $D \in S$ and $\alpha_s \notin D$, by Proposition 4.8.2. It is clear from Proposition 4.8.2 that this automaton recognizes the language of reduced expressions of (W, S). $\square$

The finite state automaton for the language of reduced words of (W, S) constructed in the proof of Theorem 4.8.3 will henceforth be referred to as the *canonical automaton*. It is by no means necessarily optimal in terms of size, although it turns out to have minimal size, for example, for the $\widetilde{A}_n$ groups.

Note that using $T_R(w)$ in place of $D_\Sigma(w)$ in the proof of Theorem 4.8.3 also yields an automaton recognizing the language of reduced expressions, since $T_R(ws) = \{s\} \cup s(T_R(w))s$ if $w \in W$ and $s \notin D_R(w)$. However, this automaton is finite if and only if W is finite. In fact, this is precisely the "weak order automaton" already considered earlier in this section.

On the other hand, using $D_R(w)$ in place of $D_\Sigma(w)$ does give a finite number of nodes, but the same construction would have been impossible since $D_R(ws)$ does *not* depend only on $D_R(w)$ and s (if $w \in W$ and $s \notin D_R(w)$).

Example 4.8.4 To illustrate the general construction of a finite state automaton in the proof, we end this section with an example. Namely, we construct the canonical automaton for the group ∘—∘ $\overset{\infty}{—}$∘ .
                                                              a b c

The first step is to determine the set of small roots and the edge-labeled order ideal of the root poset that they determine. This task was achieved in Figure 4.7, from which we extract the simplified picture (Figure 4.10) of the edge-labeled order ideal.

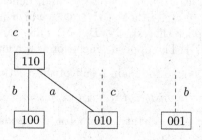

Figure 4.10. Edge-labeled order ideal of small roots in ∘—∘ $\overset{\infty}{—}$∘ .
                                                        a b c

Figure 4.10 contains all the information we need for the construction of the automaton. The nodes of the automaton will be denoted by symbols such as

$$\boxed{\begin{array}{ccc} - & \bullet & \\ + & - & + \end{array}}$$

Referring to the relative positions of the four small roots in Figure 4.10, this symbolizes the node $D_\Sigma(w) = \{110, 010\}$, or, equivalently, the node $D_\Sigma(w)$ for any $w \in W$ such that

$$w(110) < 0, \quad w(100) > 0, \quad w(010) < 0, \quad w(001) > 0.$$

Of the 16 possible symbols, the ones that actually occur as nodes in the automaton (there are 8 of them) can be generated from the start node (the one with four +), in parallel with generating the labeled edges of the automaton. Namely, from a given node $D_\Sigma(w)$, there is one outgoing edge for each $s \in S$ such that $w(\alpha_s) > 0$ (i.e., for each position marked with + in the bottom row of our symbol). Such an s-marked edge leads to the node $D_\Sigma(ws)$, the distribution of + and - signs in whose symbol can be determined by Proposition 4.8.2.

At this point, it is convenient to keep the rule given in Proposition 4.6.6 in mind. It says that the distribution of + and - signs in the node symbol will change only (1) by switching from + to - at the simple root α_s and (2) by exchanging the values at the two ends of each s-labeled edge in the root poset. There is only one catch: What if such an edge is a long edge leading from an element $u \in \Sigma$ to some $v \notin \Sigma$? Then, from knowing $D_\Sigma(w)$ we do not know the sign at v that will be traded to become the sign at u. However, actually we do! Lemma 4.8.1 comes to our rescue. It says that in this situation, the value at v is necessarily a +. The following edge illustrates this combinatorial rule:

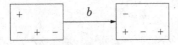

The automaton is constructed from Figure 4.10 by repeated application of this edge- and node-forming rule. Figure 4.11 shows the result. □

4.9 Complement: Counting reduced words and small roots

Theorem 4.8.3 has a purely enumerative corollary. For a Coxeter system (W, S), let r_k denote the number of reduced words of length k, for $k \geq 0$. Equivalently, r_k is the number of weak order saturated chains of length k starting from e. So, for example, $r_0 = 1$, $r_1 = |S|$, and $r_2 = |S|(|S| - 1)$.

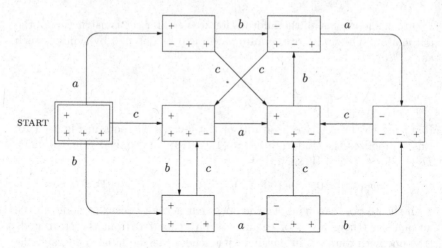

Figure 4.11, Canonical automaton for $\underset{a}{\circ}\!-\!\underset{b}{\circ}\overset{\infty}{-}\underset{c}{\circ}$.

Let

$$\mathcal{R}_{(W,S)}(q) \overset{\text{def}}{=} \sum_{k \geq 0} r_k q^k.$$

Theorem 4.9.1 *The formal power series* $\mathcal{R}_{(W,S)}(q)$ *is rational.*

Proof. We have that r_k equals the number of directed paths of length k in the automaton from the start node to any node in $\mathcal{S}$. Since $|\mathcal{S}| < \infty$ the result follows via the well-known "Transfer Matrix Method"; see Corollary A4.1.3. $\square$

The following are some examples of initial segments of such formal power series (where $(3, \infty) = \underset{a}{\circ}\!-\!\underset{b}{\circ}\overset{\infty}{-}\underset{c}{\circ}$):

$$\mathcal{R}_{\tilde{A}_1}(q) = 1 + 2q + 2q^2 + 2q^3 + 2q^4 + 2q^5 + 2q^6 + 2q^7 + 2q^8 + \cdots,$$
$$\mathcal{R}_{\tilde{A}_2}(q) = 1 + 3q + 6q^2 + 12q^3 + 18q^4 + 30q^5 + 42q^6 + 66q^7 + 90q^8 + \cdots,$$
$$\mathcal{R}_{(3,\,\infty)}(q) = 1 + 3q + 6q^2 + 10q^3 + 16q^4 + 24q^5 + 38q^6 + 60q^7 + 92q^8 + \cdots,$$

and here are the rational expressions:

$$\mathcal{R}_{\tilde{A}_1}(q) = \frac{1+q}{1-q},$$
$$\mathcal{R}_{\tilde{A}_2}(q) = \frac{(1+2q)(1+q^2)}{(1-q)(1-2q^2)},$$
$$\mathcal{R}_{(3,\,\infty)}(q) = \frac{(1+q)(1+2q+3q^2+2q^3+2q^4+q^6)}{1-q^2-2q^3-q^6}.$$

Not much is known about the enumeration of reduced words. In the case of finite Coxeter groups, this topic has been studied mainly for the symmetric groups; see Section 7.4. For infinite groups, we know of no general results beyond Theorem 4.9.1. Rational expressions for the series $\mathcal{R}_{(W,S)}(q)$, in the case of infinite groups of rank 3, have been computed by Avasjö [15]. Unfortunately, the results do not seem to suggest any general pattern.

As has been shown, the counting of reduced words in (W, S) has close ties to the structure of the recognizing finite state automaton. Thus, it becomes of interest to adress the question: *How large is the canonical automaton?* This means to try to determine or estimate the number $|\Sigma|$ of small roots and then to try to determine or estimate how many of the $2^{|\Sigma|}$ potential nodes actually occur in the canonical automaton.

Surprisingly, these questions have exact answers for the class of affine Coxeter groups. The explanation goes via details of the geometric representation that lie just a bit beyond the scope of this book. We will sketch the connection and state the results. For more details about the geometric picture, see, for example, [306, Section 5.13].

Let (W, S) be a Coxeter group. To each positive root $\beta \in \Phi^+$ is associated a hyperplane

$$H_\beta \stackrel{\text{def}}{=} \{p \in V^* : \langle p \mid \beta \rangle = 0\}$$

in V^*, with its two half-spaces $H_\beta^+ = \{p \in V^* : \langle p \mid \beta \rangle > 0\}$ and $H_\beta^- = -H_\beta^+$. The closure $\overline{C}$ of the open simplicial cone $C = \cap_{s \in S} H_{\alpha_s}^+$ is a fundamental domain for the action of W on the *Tits cone* $U = \cup_{w \in W} w(\overline{C})$. If all the hyperplanes H_β ($\beta \in \Phi^+$) are removed from U, what remains is a collection of pairwise disjoint open simplicial cones $w(C)$, called *chambers* (or *Weyl chambers*), in bijection with W via $w \mapsto w(C)$.

We say that a chamber $w(C)$ is *beneath* the hyperplane H_β if it is contained in H_β^+ (i.e., it lies on the same side of H_β as the fundamental chamber C), otherwise it is *beyond* H_β. A crucial observation is that $w(C)$ is beneath H_β if and only if $w(\beta) > 0$.

Interpreting positive roots β as hyperplanes H_β, we may consider the *arrangement of small hyperplanes*

$$\mathcal{A}_\Sigma \stackrel{\text{def}}{=} \{H_\beta : \beta \in \Sigma\}.$$

In terms of this arrangement, the geometric meaning of the small descent set of w is this: $D_\Sigma(w)$ is the set of small hyperplanes separating $w(C)$ from C. Hence, $D_\Sigma(w) = D_\Sigma(w')$ if and only if $w(C)$ and $w'(C)$ are contained in the same connected component of $U \setminus \mathcal{A}_\Sigma$ (the complement of the arrangement $\mathcal{A}_\Sigma$ in the Tits cone U). From this, we get the following.

Lemma 4.9.2 *The number of nodes of the canonical automaton equals the number of connected regions of* $U \setminus \mathcal{A}_\Sigma$.

What is the geometric meaning of saying that a hyperplane H_β is small? By Theorem 4.7.6, we know that this means that β dominates no positive root α other than itself; that is, for all $\alpha \in \Phi^+ \setminus \{\beta\}$, there is some $w \in W$ such that $w(\alpha) > 0$ and $w(\beta) < 0$. In other words, we have the following.

Lemma 4.9.3 *A hyperplane H_β is small if and only if for every other hyperplane H_α, there is some chamber $w(C)$ that is beneath H_α and beyond H_β.*

Informally speaking, the lemma says that a hyperplane is small if it lies "close" to the fundamental chamber, so that it has sufficiently many chambers on the "beyond" side.

Now, we specialize the discussion to the case of an affine group (W, S), where it is possible to be much more precise. In this case, we pass from the geometric description used so far to the realization as an affine reflection group of a crystallographic root system in a space of dimension $|S| - 1$, obtained by intersecting with an affine hyperplane E in V^*. See [306, Chapter 4] for details about such affine reflection groups and [306, Section 6.5] for the transition.

After the transition, the arrangement $\mathcal{A} = \{H_\beta \cap E : \beta \in \Sigma\}$ of affine hyperplanes decomposes into parallelism classes, one for each positive root of the associated finite Weyl group $\overline{W}$. For each such class, consider the two adjacent hyperplanes between which the fundamental chamber $C \cap E$ is sandwiched. It is geometrically clear from Lemma 4.9.3 that these two hyperplanes are small and that no other hyperplane in the class is small. Hence, we have proved the first part of the following theorem.

Theorem 4.9.4 *Let (W, S) be an affine Weyl group, with corresponding finite Weyl group $\overline{W}$. Let t be the number of reflections of $\overline{W}$. Then, we have the following:*

(i) $|\Sigma| = 2t$.

(ii) *The number of connected regions of* $U \setminus \mathcal{A}_\Sigma$ *is* $\left(\frac{2t}{|S|-1} + 1 \right)^{|S|-1}$.

The second part of the theorem is due to Shi [456], who introduced and studied arrangements of small hyperplanes in the affine case. In fact, in that case, these arrangements are usually known as *Shi arrangements*.

Since the number t of reflections is known for all finite Weyl groups (see Appendix A1) we can deduce from Theorem 4.9.4 and Lemma 4.9.2 the following explicit information in the case of affine irreducible groups. Here, N stands for the number of nodes of the canonical automaton.

Group	$|\Sigma|$	N
$\widetilde{A}_n$	$n(n+1)$	$(n+2)^n$
$\widetilde{B}_n, \widetilde{C}_n$	$2n^2$	$(2n+1)^n$
$\widetilde{D}_n$	$2n(n-1)$	$(2n-1)^n$
$\widetilde{E}_6$	72	13^6
$\widetilde{E}_7$	126	19^7
$\widetilde{E}_8$	240	31^8
$\widetilde{F}_4$	48	13^4
$\widetilde{G}_2$	12	7^2

As was mentioned, the canonical automaton is not necessarily minimal. Eriksson [223] shows that for type $\widetilde{A}_n$, it is minimal, but for type $\widetilde{C}_n$, it is not. In any case, the gathered information gives the following upper bound in the affine case, for fixed $d = |S|$:

$$N \leq \frac{|\Sigma|^{d-1}}{(d-1)^{d-1}} + O(|\Sigma|^{d-2}). \tag{4.47}$$

What about the general case? We do not know of any estimates of the number of small roots, but in terms of this number $|\Sigma|$, there is the following bound:

$$N \leq 2\frac{|\Sigma|^{d-1}}{(d-1)!} + O(|\Sigma|^{d-2}). \tag{4.48}$$

This follows from the formula $2\sum_{i=0}^{d-1} \binom{h-1}{i}$ for the number of connected components in the complement of a generic arrangement of h hyperplanes in $\mathbb{R}^d$, together with the fact that the number of components is maximized by the generic case.

Exercises

1. The following problem was given at the International Olympiad of Mathematics in 1986:

   Five integers with positive sum are arranged on a circle. The following game is played. If there is at least one negative number, the player may pick one of them, add it to its neighbors, and reverse its sign. The game terminates when all the numbers are nonnegative. Prove that this game must always terminate.

   (a) Find an elementary proof.

(b) Analyze the situation in terms of the numbers game and prove the following stronger conclusion: The game will terminate in the same number of steps and in the same final position no matter how it is played.

2. Prove that the group A_{n-1} is finite and isomorphic to S_n using the numbers game.

3. Let $x = [x_1, \ldots, x_n]$, $x_i \in [\pm n]$, be an element of the group S_n^B in window notation; see Section 8.1 for the definitions. Show that

$$\overline{d}(x) = (x_1, x_2 - x_1, x_3 - x_2, \ldots, x_n - x_{n-1})$$

is the position corresponding to x in the numbers game played from the unit position $\mathbf{1}$ on the type B_n Coxeter graph (with a suitable choice of edge weights).

4. Let $p \in \mathcal{P}$ be a position in the numbers game (played from a positive starting position). Show that all entries of p are nonzero.

5. Show that for an affine Coxeter group there exists a vector $\gamma = \sum c_s \alpha_s$, with $c_s > 0$ for all $s \in S$, such that $\langle p^w | \gamma \rangle$ takes the same value for all positions p^w of the numbers game.
   [*Hint:* See [306, Section 6.5].]

6. In the situation of Lemma 4.2.4, what are the precise conditions for which $c > 0$? Same question for $d > 0$.

7. Define $R = T \times \{+1, -1\}$ and $\pi_w : R \to R$ as in Section 1.3, and let Φ be the root system of (W, S). Define a map $\phi : R \to \Phi$ by

$$
\begin{aligned}
(t, +1) &\mapsto \gamma \in \Phi^+ \text{ such that } t_\gamma = t, \\
(t, -1) &\mapsto \gamma \in \Phi^- \text{ such that } t_\gamma = t.
\end{aligned}
$$

(a) Show that ϕ is bijective.
(b) Show that

$$\phi(\pi_w(t, \varepsilon)) = w(\phi(t, \varepsilon))$$

for all $(t, \varepsilon) \in R$ and $w \in W$.
(c) Using this, show that Propositions 4.4.4 and 4.4.6 can be deduced from the results of Section 1.4 (assuming that these propositions were not used for part (b)).

8. Let $(\,\cdot\,|\,\cdot\,)$ be the standard symmetric bilinear form (4.21).

(a) Show that for rank 2 Coxeter groups $(\,\cdot\,|\,\cdot\,)$ is either positive definite or positive semidefinite.

For a rank 3 Coxeter system having $S = \{s_1, s_2, s_3\}$ and

$$(p, q, r) = (m(s_1, s_2), m(s_1, s_3), m(s_2, s_3)),$$

let d be the quantity

$$d \overset{\text{def}}{=} \frac{1}{p} + \frac{1}{q} + \frac{1}{r}.$$

(b) Show that $(\cdot \mid \cdot)$
   - is positive definite if and only if $d > 1$,
   - is positive semidefinite but degenerate if and only if $d = 1$,
   - has signature $(2, 1)$ (i.e., its associated symmetric matrix has one negative eigenvalue and two positive eigenvalues) if $d < 1$.

   [*Hint:* It helps to break up the analysis into two cases: the case where one of p, q, and r is 2, so that the Coxeter diagram contains no cycle, and the other case where the Coxeter diagram is a triangle.]

(c) Explictly write down all the possible values (p, q, r) in the positive definite and semidefinite $(d \geq 1)$ cases.

(d) Show that in the cases with $d > 1$, one has $|W| = \frac{4}{d-1}$.

9. The reason for specializing to the case of a symmetric bilinear form in Section 4.4 is that important properties of root systems would otherwise not be true. For example, consider the Coxeter system A_2 on generators s and s' and with edge weights $k_{s,s'} = 2$ and $k_{s',s} = \frac{1}{2}$. Extending the definitions of Section 4.4, show that this example has *six* "positive roots," namely $\alpha_s = (1, 0)$, $s'(\alpha_s) = (1, \frac{1}{2})$, $ss'(\alpha_s) = (0, \frac{1}{2})$, $\alpha_{s'} = (0, 1)$, $s(\alpha_{s'}) = (2, 1)$, and $s's(\alpha_{s'}) = (2, 0)$. Then, find counterexamples showing that the generalizations of equation (4.27), Lemma 4.4.3, and Proposition 4.4.4 are false.

10. Let (W, S) be finite and irreducible. Consider the standard geometric representation $\sigma : W \to \mathbb{R}^S$ and let $\alpha(y) = -y$ be the antipodal map on $\mathbb{R}^S$. Show that the following are equivalent:

   (a) $w_0 x w_0 = x$ for all $x \in W$.
   (b) $\sigma_{w_0} = \alpha$.
   (c) $\alpha \in \sigma(W)$.
   (d) All exponents of W are odd.

   Conclude that $x \mapsto w_0 x w_0$ is *not* the identity mapping if and only if W is one of A_n ($n \geq 2$), D_n (n odd), E_6, or $I_2(m)$ (m odd).

   [*Hint:* The implications (b) $\Leftrightarrow$ (c) and (b) $\Rightarrow$ (a) are easy to see, but (a) $\Rightarrow$ (b) requires a small geometric argument. For (c) $\Leftrightarrow$ (d), see "Corollaire 3" on p. 123 of [79]. The argument is concluded by referring to a table of exponents; see, for example, Appendix A1.]

11. (a) Prove the statements made•about rank 2 groups in Example 4.4.2.

(b) Show that in the standard geometric representation of $\tilde{A}_2$ in $V \cong \mathbb{R}^3$ there are *only three* planes left fixed by its reflections, with infinitely many roots (pointing in different directions) associated with each such plane.

12. Let (W, S) be a finite Coxeter system and $\Phi \subset V$ its root system. Let $\Gamma_2(V)$ be the set of 2-dimensional subspaces of V and define

$$g(\Phi) \overset{\text{def}}{=} |\{U \in \Gamma_2(V) : |U \cap \Phi| \geq 5\}|.$$

Show the following:

(a) If W is of type A_n, then $g(\Phi) = \binom{n+1}{3}$.

(b) If W is of type B_n $(n \geq 2)$, then $g(\Phi) = \frac{n(n-1)(4n-5)}{6}$.

(c) If W is of type D_n $(n \geq 4)$, then $g(\Phi) = 4\binom{n}{3}$.

13. Let (W, S) be a finite Coxeter system, and consider the geometric representation of (W, S). For $\alpha \in \Phi$, let $H_\alpha \overset{\text{def}}{=} \{p \in V^* : \langle p \mid \alpha \rangle = 0\}$ be the hyperplane orthogonal to α. Clearly, the complement in V^* of the hyperplane arrangement $\bigcup_{\alpha \in \Pi} H_\alpha$ has $2^{|\Pi|}$ connected components. Each such connected component C is a union of a finite number (call it $s(C)$) of Weyl chambers (i.e., connected components of $V^* \setminus (\bigcup_{\alpha \in \Phi} H_\alpha)$). The *Springer number* of (W, S) is

$$\mathrm{Spr}(W, S) \overset{\text{def}}{=} \max_C \{s(C)\},$$

where C runs over all the connected components of $V^* \setminus (\bigcup_{\alpha \in \Pi} H_\alpha)$. Show that

$$\mathrm{Spr}(W, S) = \max_{J \subseteq S} \{|\mathcal{D}_J|\}.$$

In particular, show the following:

(a) If (W, S) is a dihedral group, then $\mathrm{Spr}(W, S) = \frac{|W|}{2} - 1$.

(b) If (W, S) has rank 3, then $\mathrm{Spr}(W, S) = \frac{|W|}{4} - 1$.

(c) If (W, S) is of type A_n, then $\mathrm{Spr}(W, S)$ equals the number of *alternating* permutations of S_{n+1} (i.e., $|\{\sigma \in S_{n+1} : \sigma(1) < \sigma(2) > \sigma(3) < \cdots\}|$).

(d) If (W, S) is of type B_n, then

$$\mathrm{Spr}(W, S) = |\{\sigma \in S_n^B : 0 < \sigma(1) > \sigma(2) < \sigma(3) > \cdots\}|$$

(notation as in Section 8.1).

(e) If (W, S) is of type D_n, then

$$\mathrm{Spr}(W, S) = |\{\sigma \in S_n^D : -\sigma(2) < \sigma(1) < \sigma(2) > \sigma(3) < \cdots\}|$$

(notation as in Section 8.2).

14. Construct the root posets of B_2 and B_3.

15. Is the root poset $(\Phi^+, \leq)$, with a bottom element appended, a meet-semilattice?

16. Show that the set

$$\{(\alpha \mid \beta) : \alpha, \beta \in \Phi^+, \ |(\alpha \mid \beta)| < 1\}$$

is finite.

17. Let $u \in W$, $u^2 = e$. Show that there exists $J \subseteq S$ and $w \in W$ such that $u = ww_0(J)w^{-1}$.
[*Hint:* Use Theorem 4.5.3 together with [79, Exc. 17b, p. 225].]

18. Can Lemma 4.7.2 be extended to $\mathrm{dp}\,(\beta) = 1$?

19. Let (W, S) be a finite Coxeter system. Show that all positive roots are small.

20. Let (W, S) be a universal Coxeter system (see Example 1.2.2).
    (a) Show that $\Sigma = \{\alpha_s : s \in S\}$.
    (b) Construct a finite state automaton for reduced words in this group.

21. Given $(i_1, \ldots, i_p) \in \mathbb{N}^p$, let

$$\{z_1, \ldots, z_t\}_< \overset{\mathrm{def}}{=} \{j \in [2, p-1] : \ i_{j-1} = i_{j+1}\}.$$

    (a) Show that $(i_1, \ldots, i_p) \in [3]^p$, with $i_k \neq i_{k+1}$, is a reduced decomposition of some element of $\widetilde{A}_2$ if and only if $z_j - z_{j-1}$ is even for $j = 2, \ldots, t$.
    (b) Show that the statement in part (a) is false for $\widetilde{A}_n$, $n \geq 3$.
    (c) Use part (a) to construct a finite state automaton that recognizes the language of reduced expressions of $\widetilde{A}_2$, and compare it with the canonical automaton.
    (d) Deduce the rational expression for $\mathcal{R}_{\widetilde{A}_2}(q)$ given on page 122.

22. Deduce the rational expression for $\mathcal{R}_{(3, \infty)}(q)$ given on page 122.
    [*Hint:* Use Fact A4.1.2 together with Figure 4.11.]

23. (a) Show the following:

$$\mathcal{R}_{\widetilde{A}_3}(q) = 1 + 4q + 12q^2 + 32q^3 + 80q^4 + 184q^5$$
$$+ 416q^6 + 864q^7 + \cdots,$$

$$\mathcal{R}_{\widetilde{A}_4}(q) = 1 + 5q + 20q^2 + 70q^3 + 230q^4 + 700q^5$$
$$+ 2080q^6 + 5910q^7 + \cdots.$$

    (b)* Determine the rational expressions for $\mathcal{R}_{\widetilde{A}_n}(q)$, $n \geq 3$.

24. Let (W, S) be a Coxeter group with S finite.

(a) Construct a finite state automaton that recognizes the language of lexicographically minimal reduced words with respect to some total ordering of S (or, equivalently, the set of normal forms).

(b) Deduce that the formal power series $W(q) = \sum_{w \in W} q^{\ell(w)}$ is rational.

Notes

Basic references for the geometric representation are the books by Bourbaki [79, Ch. 5] and Humphreys [306, Ch. 5]; for root systems, see also Deodhar's paper [179].

As is shown in the proof of Theorem 4.3.1, the numbers game is mathematically just a particular choice of a generic orbit for the contragredient of the geometric representation of a Coxeter group. From this point of view, the content of Section 4.3 is quite special. The point is, however, the recognition that "playing" in this simple way with numbers on the nodes of a Coxeter graph can be a useful way to perform computations. The numbers game first appeared, in a somewhat restricted version related to Kac-Moody Lie algebras, in the work of Mozes [400]. The general version given here is due to Eriksson [227].

Theorem 4.5.3 appears as an exercise in Bourbaki [79], and Proposition 4.5.4 appears in Dyer's paper [200] (see also [104]).

The root poset was introduced independently by Brink and Howlett [104] and by H. and K. Eriksson [223, 227]. Most of the results in Section 4.6 appear in these sources. Small roots were introduced in [223] and humble roots (not by this name) in [104]. Our treatment in Section 4.7 borrows ideas from both [104] and [223] in about equal measure.

The existence of a finite state automaton that recognizes the language of normal forms is an important problem in the computational theory of finitely generated groups. In the case of Coxeter groups, the first complete proof was given by Brink and Howlett [104]. Via an earlier paper of Davis and Shapiro [171], the result of [104] implies regularity also for the language of reduced words. Other sources for this result include the theses of Eriksson [223] and Headley [290]. Various aspects of automata for Coxeter group computations are discussed in [103, 117, 118, 119].

Sections 4.8 and 4.9 are based mainly on [223].

Exercises 8 and 17. See Bourbaki [79].
Exercise 12. See Shi [463].
Exercise 13. See Arnol'd [11].
Exercises 16 and 24(a). See Brink and Howlett [104].

5

Kazhdan-Lusztig and R-polynomials

In their fundamental paper [322], Kazhdan and Lusztig defined, for every Coxeter group W, a family of polynomials with integer coefficients, indexed by pairs of elements of W. These polynomials, which have become known as the Kazhdan-Lusztig polynomials of W, are intimately related to the Bruhat order of W and to the algebraic geometry and topology of Schubert varieties. They have also proven to be of fundamental importance in representation theory. In order to prove the existence of these polynomials, Kazhdan and Lusztig used another family of polynomials that arises from the multiplicative structure of the Hecke algebra associated to W. These auxiliary polynomials are known as the R-polynomials of W. Their importance stems mainly from the fact that knowing them is equivalent to knowing the Kazhdan-Lusztig polynomials.

In this chapter, we examine, from a combinatorial point of view, the R-polynomials and Kazhdan-Lusztig polynomials of a Coxeter group. We describe two combinatorial interpretations of the R-polynomials (one in terms of subexpressions of reduced decompositions and one in terms of certain paths in the Bruhat graph) and a nonrecursive combinatorial formula for the Kazhdan-Lusztig polynomials.

5.1 Introduction and review

In this section, we review a few basic facts about Kazhdan-Lusztig and R-polynomials that we need as a starting point. Their proofs can all be found in Chapter 7 of [306], and more precise references are given after each one of them. Throughout this chapter, let (W, S) be a Coxeter system.

Kazhdan-Lusztig and R-polynomials can be defined in several equivalent ways. We choose here the one that is best suited for our purposes. This involves defining first the R-polynomials and then using these to define the Kazhdan-Lusztig polynomials. In both cases, we have a "Theorem-Definition."

Theorem 5.1.1 *There is a unique family of polynomials* $\{R_{u,v}(q)\}_{u,v \in W} \subseteq \mathbb{Z}[q]$ *satisfying the following conditions:*

(i) $R_{u,v}(q) = 0$, *if* $u \not\leq v$.

(ii) $R_{u,v}(q) = 1$, *if* $u = v$.

(iii) If $s \in D_R(v)$, *then*

$$R_{u,v}(q) = \begin{cases} R_{us,vs}(q), & \text{if } s \in D_R(u), \\ qR_{us,vs}(q) + (q-1)R_{u,vs}(q), & \text{if } s \notin D_R(u). \end{cases} \tag{5.1}$$

The uniqueness part of this theorem is trivial. What is not obvious is the existence. This follows from the invertibility of certain basis elements of the Hecke algebra $\mathcal{H}$ of W and is proved in §§7.4 and 7.5 of [306]. The polynomials whose existence and uniqueness are guaranteed by Theorem 5.1.1 are called the R-*polynomials* of (W, S).

Theorem 5.1.1 can be used to compute the polynomials $\{R_{u,v}(q)\}_{u,v \in W}$ by induction on $\ell(v)$. In fact, if $v \neq e$, it is possible to find some $s \in D_R(v)$, and then since $\ell(vs) < \ell(v)$, we may assume by induction that we have already computed all of the R-polynomials appearing on the right-hand side of equation (5.1). Thus, we may use part (iii) of Theorem 5.1.1 as a recurrence relation for the computation of the R-polynomials, using parts (i) and (ii) as "initial conditions."

Example 5.1.2 Suppose that we want to compute $R_{123,321}(q)$ in S_3. Choosing $s = (1, 2) \in D_R(321)$, we have from part (iii) of Theorem 5.1.1 that

$$R_{123,321}(q) = q\,R_{213,231}(q) + (q-1)\,R_{123,231}(q).$$

Now, choosing $s = (2, 3) \in D_R(231)$, we obtain that

$$R_{213,231}(q) = q\,R_{231,213}(q) + (q-1)\,R_{213,213}(q) = q - 1,$$

and

$$\begin{aligned} R_{123,231}(q) &= q\,R_{132,213}(q) + (q-1)\,R_{123,213}(q) \\ &= (q-1)\,R_{123,213}(q), \end{aligned}$$

by Theorem 5.1.1. Finally, choosing $s = (1, 2) \in D_R(213)$, we get that

$$R_{123,213}(q) = q\,R_{213,123}(q) + (q-1)\,R_{123,123}(q) = q - 1,$$

again by Theorem 5.1.1. Therefore, we conclude that

$$R_{123,321}(q) = q\,(q-1) + (q-1)^3 = q^3 - 2q^2 + 2q - 1.$$

$\square$

In the same way, the reader is encouraged to compute as an exercise that

$$R_{123,132}(q) = R_{123,213}(q) = q - 1$$

and

$$R_{123,312}(q) = R_{123,231}(q) = q^2 - 2q + 1.$$

It may then come as a relief to learn that there are simpler and faster ways to compute the R-polynomials of a Coxeter group.

Theorem 5.1.1 can also be used to prove, by induction on $\ell(v)$, some simple basic properties of the R-polynomials.

Proposition 5.1.3 Let $u, v \in W$, $u \leq v$. Then, $R_{u,v}(q)$ is a monic polynomial of degree $\ell(u,v)$ and with constant term $(-1)^{\ell(u,v)}$.

Proof. All three statements are true if $\ell(v) = 0$ (i.e., if $v = e$) by part (ii) of Theorem 5.1.1. So let $\ell(v) > 0$ and assume by induction that they hold whenever the second indexing element has length $< \ell(v)$. Let $s \in D_R(v)$. If $s \in D_R(u)$, then by part (iii) of Theorem 5.1.1, we have that

$$R_{u,v}(q) = R_{us,vs}(q),$$

and the result holds by induction since $\ell(us, vs) = \ell(u, v)$ in this case. If $s \notin D_R(u)$, then we have that

$$R_{u,v}(q) = qR_{us,vs}(q) + (q-1)R_{u,vs}(q),$$

and the result again holds since $\ell(us, vs) = \ell(u, v) - 2$ and $\ell(u, vs) = \ell(u, v) - 1$ (note that it could happen that $us \not\leq vs$ and, hence, that $R_{us,vs}(q) = 0$, but this does not affect our conclusion). Thus, the result holds for $R_{u,v}(q)$ in either case, and this concludes the induction step. $\square$

In a similar way, other basic properties of the R-polynomials can be proved; see Exercises 1, 2, and 3.

We now come to the definition of the Kazhdan-Lusztig polynomials. As mentioned above, this is again a "Theorem-Definition."

Theorem 5.1.4 There is a unique family of polynomials $\{P_{u,v}(q)\}_{u,v \in W} \subseteq \mathbb{Z}[q]$ satisfying the following conditions:

(i) $P_{u,v}(q) = 0$, if $u \not\leq v$.

(ii) $P_{u,v}(q) = 1$, if $u = v$.

(iii) $\deg(P_{u,v}(q)) \leq \frac{1}{2}(\ell(u,v) - 1)$, if $u < v$.

(iv)

$$q^{\ell(u,v)} P_{u,v}\left(\frac{1}{q}\right) = \sum_{a\in[u,v]} R_{u,a}(q) P_{a,v}(q), \quad \text{if } u \leq v.$$

Once again, the uniqueness part is easy. A proof of Theorem 5.1.4 appears in §§7.9, 7.10, and 7.11 of [306]. The polynomials $\{P_{u,v}(q)\}_{u,v\in W}$ whose existence and uniqueness are guaranteed by the previous theorem are called the *Kazhdan-Lusztig polynomials* of (W, S).

Theorem 5.1.4 can be used to prove some simple basic properties of the Kazhdan-Lusztig polynomials.

Proposition 5.1.5 *Let $u, v \in W$, $u \leq v$. Then, $P_{u,v}(0) = 1$.*

Proof. We proceed by induction on $\ell(u, v)$, the result being true by part (ii) of Theorem 5.1.4 if $\ell(u, v) = 0$. So let $\ell(u, v) > 0$. Then, we conclude from parts (iii) and (iv) of Theorem 5.1.4 that

$$0 = \sum_{a\in[u,v]} R_{u,a}(0) P_{a,v}(0).$$

Using Proposition 5.1.3 and the induction hypothesis, we may rewrite this as

$$P_{u,v}(0) = -\sum_{u<a\leq v} (-1)^{\ell(u,a)}.$$

Now, using Corollary 2.7.10, we conclude from this that

$$P_{u,v}(0) = -\sum_{u<a\leq v} \mu(u, a) = \mu(u, u) = 1,$$

as desired. □

The preceding proof already shows that the Kazhdan-Lusztig polynomials are considerably more subtle than the R-polynomials. In fact, whereas for the latter ones we have been able to deduce in a fairly straightforward way their degree, leading term, and constant term, for the Kazhdan-Lusztig polynomials we have had to use a more substantial result (the computation of the Möbius function of Bruhat order) just to compute their constant term. In fact, at present, there are no simple formulas known for computing the leading term and degree of Kazhdan-Lusztig polynomials.

Once the R-polynomials have been computed, then Theorem 5.1.4 can be used to recursively compute the polynomials $\{P_{u,v}(q)\}_{u,v\in W}$, by induction on $\ell(u, v)$. In fact, by induction we may assume that we have already computed the polynomials $P_{a,v}(q)$ for all $a \in [u, v]$, $a \neq u$. This, by part (iv) of Theorem 5.1.4, means that we can compute

$$q^{\ell(u,v)} P_{u,v}\left(\frac{1}{q}\right) - P_{u,v}(q) \tag{5.2}$$

(recall that $R_{u,u}(q) = 1$ by part (ii) of Theorem 5.1.1). However, by part (iii) of Theorem 5.1.4, the coefficient of q^i in the polynomial (5.2) is the same as the coefficient of q^i in $-P_{u,v}(q)$ for all $i = 0, \ldots, \lfloor \frac{1}{2}(\ell(u,v) - 1) \rfloor$ (we assume that $u < v$ for otherwise we already know $P_{u,v}(q)$ by parts (i) and (ii) of Theorem 5.1.4) and thus we can compute $P_{u,v}(q)$ from the polynomial (5.2). (The reader will notice that the reasoning that we have just explained proves the uniqueness part of Theorem 5.1.4).

Example 5.1.6 Let us compute $P_{123,321}(q)$ in S_3. From part (iv), we deduce that

$$
\begin{aligned}
q^3 P_{123,321}(q^{-1}) - P_{123,321}(q) &= R_{123,213}(q) P_{213,321}(q) \\
&\quad + R_{123,132}(q) P_{132,321}(q) \\
&\quad + R_{123,231}(q) P_{231,321}(q) \\
&\quad + R_{123,312}(q) P_{312,321}(q) \\
&\quad + R_{123,321}(q) P_{321,321}(q).
\end{aligned}
$$

However, by parts (ii) and (iii) of Theorem 5.1.4 and Proposition 5.1.5, we know that $P_{u,321}(q) = 1$ for all $u \in S_3 \setminus \{123\}$. Hence, we obtain that

$$
\begin{aligned}
q^3 P_{123,321}(q^{-1}) - P_{123,321}(q) &= R_{123,213}(q) + R_{123,132}(q) + R_{123,231}(q) \\
&\quad + R_{123,312}(q) + R_{123,321}(q).
\end{aligned}
$$

Assuming, as we are, that we have already computed the R-polynomials, we then get

$$
\begin{aligned}
q^3 P_{123,321}(q^{-1}) - P_{123,321}(q) &= (q-1) + (q-1) + (q-1)^2 \\
&\quad + (q-1)^2 + (q^3 - 2q^2 + 2q - 1) \\
&= q^3 - 1.
\end{aligned}
$$

Now, since $\frac{1}{2}(\ell(321) - \ell(123) - 1) = 1$, we deduce from this, by part (iii) of Theorem 5.1.4, that

$$
P_{123,321}(q) = 1.
$$

□

As a further (although quite a bit longer) example, the reader may want to verify that

$$
P_{2134,4231}(q) = P_{1324,3412}(q) = 1 + q.
$$

We conclude this review section by recalling a few more fundamental properties of the Kazhdan-Lusztig polynomials. Although they are not needed in the text of this chapter, they are used in some of the exercises, as well as in Chapter 6.

For $u, w \in W$, $u \leq w$, we let

$$
\bar{\mu}(u, w) \overset{\text{def}}{=} \begin{cases} [q^{\frac{1}{2}(\ell(u,w)-1)}](P_{u,w}(q)), & \text{if } \ell(u,w) \text{ is odd}, \\ 0, & \text{otherwise}. \end{cases} \tag{5.3}
$$

We use here "$\overline{\mu}$" to distinguish this function from the Möbius function "μ" encountered in Section 2.7 and elsewhere.

Theorem 5.1.7 *Let $u, v \in W$, $u \leq v$, and $s \in D_R(v)$. Then,*

$$P_{u,v}(q) = q^{1-c}P_{us,vs}(q) + q^c P_{u,vs}(q) - \sum_{\{z:\, s \in D_R(z)\}} q^{\frac{\ell(z,v)}{2}} \overline{\mu}(z,vs)P_{u,z}(q)$$

$$(5.4)$$

where $c = 1$ if $s \in D_R(u)$, and $c = 0$ otherwise.

A proof of this result can be found in [306, §7.11]. Two simple but important consequences of Theorem 5.1.7 are the following (see [306, Theorem 7.9, part (b), and Corollary 7.14]).

Proposition 5.1.8 *Let $u, v \in W$, $u \leq v$. If $s \in D_R(v)$, then*

$$P_{u,v}(q) = P_{us,v}(q).$$

Proof. Applying Theorem 5.1.7 first to the pair (u, v), and then to (us, v), one sees that the right-hand side of equation (5.4) does not change, except that $P_{u,z}(q)$ gets replaced by $P_{us,z}(q)$. However, these two polynomials are by induction equal, since $s \in D_R(z)$ and $\ell(z) \leq \ell(vs) < \ell(v)$. The result follows. □

So, for example, $P_{2147563,6157243}(q) = P_{1245736,6157243}(q)$. It also follows that

$$P_{u,w_0}(q) = 1,$$

for all $u \in W$, if W is finite and w_0 is its longest element.

Proposition 5.1.9 *Let $z, w \in W$, $z \leq w$, be such that $\overline{\mu}(z,w) \neq 0$ and $\ell(z,w) > 1$. Then, $D_R(z) \supseteq D_R(w)$.*

Proof. Let $s \in D_R(w)$ and suppose that $s \notin D_R(z)$. Then, from Proposition 5.1.8 we conclude that $P_{z,w}(q) = P_{zs,w}(q)$. Hence,

$$\overline{\mu}(z,w) = [q^{\frac{1}{2}(\ell(z,w)-1)}](P_{z,w}) = [q^{\frac{1}{2}(\ell(z,w)-1)}](P_{zs,w}) = 0$$

by part (iii) of Theorem 5.1.4, since $\deg(P_{zs,w}) \leq \frac{1}{2}(\ell(zs,w) - 1) = \frac{1}{2}(\ell(z,w) - 2)$. This contradiction shows that $s \in D_R(z)$. □

5.2   Reflection orderings

Reflection orderings play an important role in the combinatorics of the Kazhdan-Lusztig and R-polynomials, as well as in other parts of the theory of Coxeter groups. Although many of the results in this chapter depend on reflection orderings, not all do (for example, Theorems 5.3.7 and 5.5.2 do not). In this section, we define reflection orderings and derive those properties that we need. We refer the reader to Sections 4.2 and 4.4 for the notation that we use for root systems and the standard geometric representation.

A total ordering $<$ on Φ^+ is a *reflection ordering* if for all $\alpha, \beta \in \Phi^+$ and $\lambda, \mu \in \mathbb{R}_{>0}$, such that $\lambda\alpha + \mu\beta \in \Phi^+$, we have that either

$$\alpha < \lambda\alpha + \mu\beta < \beta$$

or

$$\beta < \lambda\alpha + \mu\beta < \alpha.$$

For example, if $W = S_4$ with its standard root system, then it is easy to check that

$$\alpha_1 < \alpha_1 + \alpha_2 < \alpha_1 + \alpha_2 + \alpha_3 < \alpha_2 < \alpha_2 + \alpha_3 < \alpha_3 \qquad (5.5)$$

is a reflection ordering.

It is not obvious that reflection orderings always exist.

Proposition 5.2.1 *There exists a reflection ordering on Φ^+.*

Proof. Fix an indexing (i.e., a total ordering) of the elements of S, say $S = \{s_1, \ldots, s_n\}$. Let

$$\mathcal{U} \overset{\mathrm{def}}{=} \left\{ \sum_{i=1}^n c_{s_i} \alpha_{s_i} : \sum_{i=1}^n c_{s_i} = 1 \right\}.$$

Note that if $\alpha \in \Phi^+$, then there is a unique $\lambda_\alpha \in \mathbb{R}_{>0}$ such that $\lambda_\alpha \alpha \in \mathcal{U}$. For $\alpha, \beta \in \Phi^+$, define $\alpha < \beta$ to mean that $\lambda_\alpha \alpha <_{lex} \lambda_\beta \beta$. Here, $<_{lex}$ denotes the total ordering of V obtained by comparing coordinates lexicographically; that is, $\sum_{i=1}^n c_{s_i} \alpha_{s_i} < \sum_{i=1}^n b_{s_i} \alpha_{s_i}$ if and only if $(c_{s_1}, \ldots, c_{s_n}) < (b_{s_1}, \ldots, b_{s_n})$ lexicographically.

Now, if $\alpha, \beta \in \Phi^+$ and $a, b \in \mathbb{R}_{>0}$ are such that $\alpha < \beta$ and $a\alpha + b\beta \in \Phi^+$, then $\lambda_{a\alpha+b\beta}(a\alpha + b\beta) = c(\lambda_\alpha \alpha) + (1-c)(\lambda_\beta \beta)$ for some $0 < c < 1$. However, this, since $\lambda_\alpha \alpha <_{lex} \lambda_\beta \beta$, implies that $\lambda_\alpha \alpha <_{lex} \lambda_{a\alpha+b\beta}(a\alpha + b\beta) <_{\mathrm{lex}} \lambda_\beta \beta$, as desired. $\square$

There is one property of reflection orderings that is crucial for our purposes. To state and prove it, we need first the following preliminary observation.

Proposition 5.2.2 *Let $<$ be a reflection ordering, $s \in S$, and $\beta \in \Phi^+ \setminus \{\alpha_s\}$. Then, $\beta < \alpha_s$ if and only if $s(\beta) < \alpha_s$.*

Proof. We may assume that $s(\beta) \neq \beta$ and, hence, that $(\beta \mid \alpha_s) \neq 0$. If $(\alpha_s \mid \beta) < 0$, then since $s(\beta) = \beta - 2(\alpha_s \mid \beta)\alpha_s$, there follows from the definition of reflection ordering that $s(\beta)$ lies between α_s and β, and the result follows in this case.

Similarly, if $(\alpha_s \mid \beta) > 0$, then β lies between $s(\beta)$ and α_s, and the result again follows. $\square$

Let $<$ be a reflection ordering, and $s \in S$. We define a total ordering $<^s$ on Φ^+ as follows. For $\beta, \gamma \in \Phi^+$, we set $\beta <^s \gamma$ if and only if either one of the following (mutually exclusive) conditions apply:

(i) $\beta, \gamma < \alpha_s$ and $\beta < \gamma$,

(ii) $\beta, \gamma > \alpha_s$ and $s(\beta) < s(\gamma)$,

(iii) $\beta < \alpha_s < \gamma$,

(iv) $\gamma = \alpha_s$.

We call $<^s$ the *upper s-conjugate* of $<$. Note that the definition of $<^s$ is consistent by Proposition 5.2.2.

Similarly, we define the *lower s-conjugate* of $<$, denoted $<_s$, by letting $\beta <_s \gamma$ if and only if either one of the following conditions is satisfied:

(i') $\beta, \gamma < \alpha_s$ and $s(\beta) < s(\gamma)$,

(ii') $\beta, \gamma > \alpha_s$ and $\beta < \gamma$,

(iii') $\beta < \alpha_s < \gamma$,

(iv') $\beta = \alpha_s$.

For example, if $<$ is the reflection ordering of S_4 given in (5.5), and $s = (2,3)$, then

$$\alpha_1 <^s \alpha_1 + \alpha_2 <^s \alpha_1 + \alpha_2 + \alpha_3 <^s \alpha_3 <^s \alpha_2 + \alpha_3 <^s \alpha_2.$$

Note that α_s is the maximum (respectively, minimum) element of $<^s$ (respectively, $<_s$). If we think of the elements of Φ^+ as arranged on an infinite line so that $\alpha < \beta$ if and only if α is to the left of β, then $<^s$ is obtained from $<$ by applying s to the roots to the right of α_s, and then moving α_s to "$+\infty$." Similarly, $<_s$ is obtained from $<$ by applying s to the roots to the left of α_s and then moving α_s to "$-\infty$." Note that

$$(<_s)^s = <^s \tag{5.6}$$

for any $s \in S$ and reflection ordering $<$.

We can now state and prove the main result of this section.

Proposition 5.2.3 *Let $<$ be a reflection ordering, and $s \in S$. Then, $<_s$ and $<^s$ are also reflection orderings.*

Proof. We show that $<^s$ is a reflection ordering, the proof for $<_s$ being entirely similar.

Let $\alpha, \beta \in \Phi^+$ and $a, b \in \mathbb{R}_{>0}$ be such that $a\alpha + b\beta \in \Phi^+$. Suppose that $\alpha < \beta$. Then,

$$\alpha < a\alpha + b\beta < \beta. \tag{5.7}$$

We will show that either

$$\alpha <^s a\alpha + b\beta <^s \beta \tag{5.8}$$

or

$$\beta <^s a\alpha + b\beta <^s \alpha. \tag{5.9}$$

Note first that if $\beta = \alpha_s$ then, by relation (5.7) and the definition of $<^s$, relation (5.8) follows immediately. If $\alpha = \alpha_s$, then we have that

$$s(\beta) = \frac{1}{b}(-a\alpha_s + bs(\beta)) + \frac{a}{b}\alpha_s$$

and $-a\alpha_s + bs(\beta) = s(a\alpha_s + b\beta) \in \Phi^+$. Since $<$ is a reflection ordering, this implies, by relation (5.7) and Proposition 5.2.2, that $\alpha_s < s(\beta) < s(a\alpha_s + b\beta)$, which in turn implies relation (5.9). We may therefore assume that $\alpha, \beta \neq \alpha_s$.

We now have four cases to consider:

(1) $\beta < \alpha_s$: Then the result follows immediately from relation (5.7) and the definition of $<^s$.

(2) $a\alpha + b\beta < \alpha_s < \beta$: Then, by relation (5.7), $\alpha < a\alpha + b\beta < \alpha_s < \beta$ and relation (5.8) follows.

(3) $\alpha < \alpha_s < a\alpha + b\beta$: Then, from Proposition 5.2.2, relation (5.7), and the definition of reflection ordering, there follows that

$$s(\alpha) < as(\alpha) + bs(\beta) < s(\beta)$$

(note that $s(\alpha)$, $s(\beta)$, $s(a\alpha + b\beta) \in \Phi^+$). On the other hand, $\alpha_s < as(\alpha) + bs(\beta)$ by Proposition 5.2.2, so

$$\alpha < \alpha_s < s(a\alpha + b\beta) < s(\beta),$$

and relation (5.8) again follows.

(4) $\alpha_s < \alpha$: Then, by the definition of reflection ordering, we have that either

$$s(\alpha) < s(a\alpha + b\beta) < s(\beta)$$

or

$$s(\beta) < s(a\alpha + b\beta) < s(\alpha),$$

so again either relation (5.8) or (5.9) follows. $\square$

Let $<$ be a reflection ordering on Φ^+. Since there is a canonical bijection between Φ^+ and T (see Proposition 4.4.5), we will from now on equivalently consider $<$ as a total ordering on T.

5.3 R-polynomials

In this section, we give two combinatorial interpretations for the R-polynomials of a Coxeter system: one in terms of certain paths in the Bruhat graph, and one in terms of reduced decompositions and subexpressions.

The reader may be surprised at the term "combinatorial interpretation" since even the simplest examples computed in Section 5.1 show that the R-polynomials do not have non-negative coefficients. The explanation lies in the following consequence of Theorem 5.1.1.

Proposition 5.3.1 *Let $u, v \in W$. Then, there exists a unique polynomial* $\widetilde{R}_{u,v}(q) \in \mathbb{N}[q]$ *such that*

$$R_{u,v}(q) = q^{\frac{\ell(u,v)}{2}} \widetilde{R}_{u,v}(q^{\frac{1}{2}} - q^{-\frac{1}{2}}). \tag{5.10}$$

Proof. The result is trivially true if $u \not\leq v$, so we may assume that $u \leq v$. If $f, g \in \mathbb{R}[q]$ are such that

$$f(q^{\frac{1}{2}} - q^{-\frac{1}{2}}) = g(q^{\frac{1}{2}} - q^{-\frac{1}{2}})$$

for all $q \in \mathbb{R}^+$, then, since $\lim_{q \to +\infty}(q^{\frac{1}{2}} - q^{-\frac{1}{2}}) = +\infty$, $\lim_{q \to 0+}(q^{\frac{1}{2}} - q^{-\frac{1}{2}}) = -\infty$, and $q^{\frac{1}{2}} - q^{-\frac{1}{2}}$ is a continuous function for $q > 0$, we conclude that f and g are identical as polynomials. This proves the uniqueness statement.

To prove the existence, we proceed by induction on $\ell(v)$, the existence being clear if $v = e$ by Theorem 5.1.1. So, let $u, v \in W$, $u \leq v$, be such that $\ell(v) > 0$ and choose $s \in D_R(v)$. If $s \in D_R(u)$, then by Theorem 5.1.1 and our induction hypothesis, we conclude that

$$R_{u,v}(q) = R_{us,vs}(q) = q^{\frac{\ell(us,vs)}{2}} \widetilde{R}_{us,vs}(q^{\frac{1}{2}} - q^{-\frac{1}{2}})$$

and the result follows since $\ell(us, vs) = \ell(u, v)$ in this case. If $s \notin D_R(u)$, then we conclude in a similar way that

$$
\begin{aligned}
R_{u,v}(q) &= qR_{us,vs}(q) + (q-1)R_{u,vs}(q) \\
&= q^{1+\frac{\ell(us,vs)}{2}} \widetilde{R}_{us,vs}(q^{\frac{1}{2}} - q^{-\frac{1}{2}}) + (q-1)q^{\frac{\ell(u,vs)}{2}} \widetilde{R}_{u,vs}(q^{\frac{1}{2}} - q^{-\frac{1}{2}}) \\
&= q^{\frac{\ell(u,v)}{2}} (\widetilde{R}_{us,vs}(q^{\frac{1}{2}} - q^{-\frac{1}{2}}) + (q^{\frac{1}{2}} - q^{-\frac{1}{2}})\widetilde{R}_{u,vs}(q^{\frac{1}{2}} - q^{-\frac{1}{2}}))
\end{aligned}
$$

since $\ell(us, vs) = \ell(u, v) - 2 = \ell(u, vs) - 1$ in this case, and the result again follows. $\square$

For example, from the computations done in connection with Example 5.1.2, we conclude that

$$
\begin{aligned}
\widetilde{R}_{123,123}(q) &= 1, \\
\widetilde{R}_{123,213}(q) = \widetilde{R}_{123,132}(q) &= q, \\
\widetilde{R}_{123,312}(q) = \widetilde{R}_{123,231}(q) &= q^2, \\
\widetilde{R}_{123,321}(q) &= q^3 + q.
\end{aligned}
$$

The crucial advantage of the $\widetilde{R}$-polynomials over the R-polynomials is that they have non-negative integer coefficients. The combinatorial interpretations that we derive in this section are all for the coefficients of the $\widetilde{R}$-polynomials.

From Proposition 5.3.1, Theorem 5.1.1, and Proposition 5.1.3, we conclude the following.

Proposition 5.3.2 *Let* $u, v \in W$, $u \leq v$. *Then,* $\widetilde{R}_{u,v}(q)$ *is a monic polynomial of degree* $\ell(u, v)$. *Furthermore, if* $s \in D_R(v)$, *then*

$$\widetilde{R}_{u,v}(q) = \begin{cases} \widetilde{R}_{us,vs}(q), & \text{if } s \in D_R(u), \\ \widetilde{R}_{us,vs}(q) + q\widetilde{R}_{u,vs}(q), & \text{if } s \notin D_R(u). \end{cases} \square$$

Note that $R_{u,v}(0) = (-1)^{\ell(u,v)}$ is automatically encoded in formula (5.10), by the preceding proposition. Some other basic properties of the $\widetilde{R}$-polynomials are given in Exercises 5 and 6.

The following property of the Bruhat graph is used repeatedly in what follows.

Lemma 5.3.3 *Let* $u, v \in W$ *be such that* $u \rightarrow v$, *and* $s \in S \setminus \{u^{-1}v\}$. *Then,* $us \rightarrow vs$.

Proof. Since $vs = us(sts)$ (if $v = ut$, $t \in T$), it suffices to show that $\ell(us) < \ell(vs)$. This is clear if $\ell(u, v) \geq 3$. If $\ell(u, v) = 1$, it follows from Proposition 2.2.7. $\square$

We begin with the interpretation of $\widetilde{R}$-polynomials in terms of Bruhat paths. For this, we need to introduce some notation. Given a path $\Delta = (a_0, \ldots, a_r)$ in the Bruhat graph and a reflection ordering $<$ on Φ^+ (hence, on T), we let

$$E(\Delta) \stackrel{\text{def}}{=} \{a_{i-1}^{-1}a_i : i = 1, \ldots, r\} \subseteq T$$

and

$$D(\Delta; <) \stackrel{\text{def}}{=} \{i \in [r-1] : a_{i-1}^{-1}a_i > a_i^{-1}a_{i+1}\}.$$

We call $E(\Delta)$ the *edge set* of Δ, and $D(\Delta; <)$ its *descent set with respect to* $<$. For convenience, we define an element $R_<$ in the incidence algebra $I(W, <)$ of W, partially ordered by Bruhat order, by letting

$$R_<(u, v) \stackrel{\text{def}}{=} \sum_{\{\Delta \in B(u,v):\, D(\Delta,<)=\emptyset\}} q^{\ell(\Delta)} \tag{5.11}$$

for all $u, v \in W$, $u \leq v$. Here, $B(u, v)$ denotes the set of all the directed paths, in the Bruhat graph of W, from u to v.

Theorem 5.3.4 *Let* $<$ *be a reflection ordering, and* $u, v \in W$, $u \leq v$. *Then,*

$$\widetilde{R}_{u,v}(q) = R_<(u, v).$$

Proof. We proceed by induction on $\ell(v)$, the result being clearly true if $\ell(v) = 0$. Let $v \in W$ be such that $\ell(v) > 0$ and choose $s \in D_R(v)$. Let

$$f(x,y) \overset{\text{def}}{=} \sum_{\Delta \in B_s(x,y)} q^{\ell(\Delta)} \tag{5.12}$$

and

$$g(x,y) \overset{\text{def}}{=} \sum_{\Delta \in B'_s(x,y)} q^{\ell(\Delta)}$$

for all $x, y \in W$, where

$$B_s(x,y) \overset{\text{def}}{=} \{\Delta \in B(x,y) : D(\Delta, <) = \emptyset, \ s \leq E(\Delta)\} \tag{5.13}$$

and

$$B'_s(x,y) \overset{\text{def}}{=} \{\Delta \in B(x,y) : D(\Delta, <^s) = \emptyset, \ s \leq E(\Delta)\}.$$

(Note that f and g depend on s and on $<$, although, for simplicity, we omit this dependence from the notation.)

We claim that

$$f(x,v) = \begin{cases} g(xs, vs), & \text{if } s \in D_R(x), \\ g(xs, vs) + q\,g(x, vs), & \text{if } s \notin D_R(x) \end{cases} \tag{5.14}$$

and

$$g(x,v) = \begin{cases} f(xs, vs) + q(g(x, vs) - f(x, vs)), & \text{if } s \in D_R(x), \\ f(xs, vs) + q\,g(x, vs), & \text{if } s \notin D_R(x) \end{cases} \tag{5.15}$$

for all $x \in W$. We prove two of these equations (namely equations (5.15)), the proof of the other two (equations (5.14)) being entirely similar (in fact, slightly simpler).

Suppose that $s \in D_R(x)$. Let $\Delta = (a_0, \ldots, a_r) \in B'_s(x,v)$. If $s \in E(\Delta)$, then necessarily $s = a_{r-1}^{-1} a_r$ (since $D(\Delta, <^s) = \emptyset$). Hence, $\Delta' \overset{\text{def}}{=} (a_0, \ldots, a_{r-1})$ is a Bruhat path from x to vs such that $s < E(\Delta')$, and $D(\Delta', <^s) = \emptyset$, so $\Delta' \in B'_s(x, vs)$. Furthermore, every path in $B'_s(x, vs)$ does not contain s in its edge set and, hence, arises in this way from a path $\Delta \in B'_s(x, v)$ such that $s \in E(\Delta)$. Hence,

$$\sum_{\{\Delta \in B'_s(x,v): \ s \in E(\Delta)\}} q^{\ell(\Delta)} = q\,g(x, vs). \tag{5.16}$$

If $\Delta \in B'_s(x,v)$ is such that $s \notin E(\Delta)$, then, by Lemma 5.3.3 and the definition of $<^s$, $\Delta s \overset{\text{def}}{=} (a_0 s, \ldots, a_r s) \in B_s(xs, vs)$. Furthermore, every path in $B_s(xs, vs)$ that does not contain s in its edge set arises in this way from a path $\Delta \in B'_s(x, v)$ such that $s \notin E(\Delta)$. Therefore,

$$\sum_{\{\Delta \in B'_s(x,v): \ s \notin E(\Delta)\}} q^{\ell(\Delta)} = f(xs, vs) - \sum_{\{\Delta \in B_s(xs,vs): \ s \in E(\Delta)\}} q^{\ell(\Delta)}. \tag{5.17}$$

Moreover, if $\Delta = (b_0, \ldots, b_i) \in B_s(xs, vs)$ is such that $s \in E(\Delta)$, then necessarily $s = b_0^{-1} b_1$. Hence, $\Delta'' \overset{\text{def}}{=} (b_1, \ldots, b_i)$ is a Bruhat path in $B_s(x, vs)$. Furthermore, every path in $B_s(x, vs)$ does not contain s in its edge set (since $s \in D_R(x)$) and therefore arises in this way from a path $\Delta \in B_s(xs, vs)$ such that $s \in E(\Delta)$. This shows that

$$\sum_{\{\Delta \in B_s(xs, vs): \, s \in E(\Delta)\}} q^{\ell(\Delta)} = qf(x, vs),$$

and this, together with equations (5.16) and (5.17), gives the first equation in (5.15).

Suppose now that $s \notin D_R(x)$. Let $\Delta = (a_0, \ldots, a_r) \in B'_s(x, v)$. If $s \in E(\Delta)$, then necessarily $s = a_{r-1}^{-1} a_r$ and, hence, $\Delta' \overset{\text{def}}{=} (a_0, \ldots, a_{r-1}) \in B'_s(x, vs)$ and $s \notin E(\Delta')$. Furthermore, every path $\Gamma \in B'_s(x, vs)$ does not contain s in its edge set (because $s \notin D_R(vs)$) and so arises in this way. Therefore,

$$\sum_{\{\Delta \in B'_s(x, v): \, s \in E(\Delta)\}} q^{\ell(\Delta)} = q \, g(x, vs). \tag{5.18}$$

If $\Delta \in B'_s(x, v)$ is such that $s \notin E(\Delta)$, then, by Lemma 5.3.3 and the definition of $<^s$, $\Delta s \in B_s(xs, vs)$. Furthermore, every path in $B_s(xs, vs)$ does not contain s in its edge set (since $s \in D_R(xs)$) and, hence, arises in this way. Therefore,

$$\sum_{\{\Delta \in B'_s(x, v): \, s \notin E(\Delta)\}} q^{\ell(\Delta)} = f(xs, vs),$$

and this, together with equation (5.18), implies the second equation in (5.15).

Now, let

$$h(x, y) \overset{\text{def}}{=} \sum_{\{\Delta \in B(x, y): \, D(\Delta, <) = \emptyset, \, E(\Delta) < s\}} q^{\ell(\Delta)}$$

for all $x, y \in W$. Then, we clearly have that

$$R_< = hf \quad \text{and} \quad R_{<^s} = hg \tag{5.19}$$

in the incidence algebra $I(W, \leq)$, since $\{\Delta \in B(x, y) : D(\Delta, <) = \emptyset, \, E(\Delta) < s\} = \{\Delta \in B(x, y) : D(\Delta, <^s) = \emptyset, \, E(\Delta) < s\}$.

We now claim that $f(x, v) = g(x, v)$ for all $x \leq v$. In fact, by equations (5.19) and our induction hypothesis,

$$f(a, vs) = (h^{-1} R_<)(a, vs) = (h^{-1} R_{<^s})(a, vs) = g(a, vs) \tag{5.20}$$

for all $a \leq vs$, and the claim follows from equations (5.14) and (5.15).

Therefore, by equations (5.19),

$$R_<(u, v) = R_{<^s}(u, v). \tag{5.21}$$

However, equation (5.21) holds for any reflection ordering $<$. Hence, in particular, we obtain from equations (5.21) and (5.6) that

$$R_{<_s}(u,v) = R_{<^s}(u,v) = R_<(u,v). \tag{5.22}$$

Now notice that by equations (5.12) and (5.13) and definition (5.11), for the reflection ordering $<_s$,

$$f(x,y) = R_{<_s}(x,y)$$

for all $x,y \in W$, $x \leq y$ (recall that f depends on the choice of a reflection ordering). Hence, by equations (5.14) and (5.20),

$$R_{<_s}(x,v) = \begin{cases} R_{<_s}(xs, vs), & \text{if } s \in D_R(x), \\ R_{<_s}(xs, vs) + qR_{<_s}(x, vs), & \text{if } s \notin D_R(x) \end{cases}$$

for all $x \in W$. However this, by Proposition 5.3.2 and our induction hypothesis, implies that $R_{<_s}(u,v) = \widetilde{R}_{u,v}(q)$ which, by equation (5.22), concludes the induction step and, hence, the proof. $\square$

Example 5.3.5 Consider the two permutations $u = 1234$ and 4312 of S_4, and choose the reflection ordering

$$(1,2) < (1,3) < (1,4) < (2,3) < (2,4) < (3,4).$$

Then, $B(u,v)$ is the labeled directed graph depicted in Figure 5.1, and

$$\{\Delta \in B(u,v) : D(\Delta, <) = \emptyset\} = \{(1234, 2134, 4132, 4312),$$
$$(1234, 3214, 4213, 4312),$$
$$(1234, 2134, 3124, 4123, 4213, 4312)\}.$$

Therefore, $\widetilde{R}_{u,v}(q) = q^5 + 2q^3$. $\square$

Note that it is a consequence of Theorem 5.3.4 that the polynomial $\widetilde{R}_{u,v}(q)$ contains only odd powers of q or only even powers of q, depending on the parity of $\ell(u,v)$.

We now study the second combinatorial interpretation for the $\widetilde{R}$-polynomials, which is in terms of reduced decompositions and certain subexpressions. Although it can be deduced from Theorem 5.3.4, we present here a self-contained proof, both because it is interesting in its own right and because it does not require reflection orderings. The remainder of this section is not needed in the rest of this chapter, except in some of the exercises.

Let $\xi \stackrel{\text{def}}{=} (s_1, \dots, s_r) \in S^r$. A *subexpression* of ξ is a sequence $(a_1, \dots, a_r) \in (S \cup \{e\})^r$ such that $a_i \in \{s_i, e\}$ for $i = 1, \dots, r$ (clearly, there are 2^r such subexpressions). We let

$$\| (a_1, \dots, a_r) \| \stackrel{\text{def}}{=} |\{i \in [r] : a_i = s_i\}|.$$

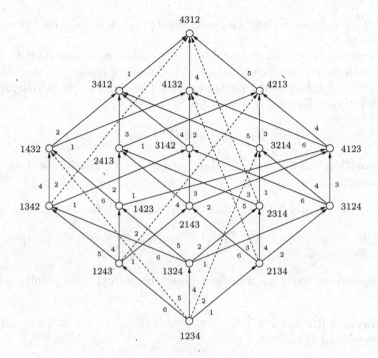

Figure 5.1. The directed graph $B(1234, 4312)$.

A subexpression $(a_1, \ldots, a_r)$ is *distinguished* if

$$s_j \notin D_R(a_1 \cdots a_{j-1})$$

for all $2 \leq j \leq r$ such that $a_j = e$. In particular, $(e, \ldots, e, s_i, \ldots, s_r)$ is always a distinguished subexpression of $(s_1, \ldots, s_r)$, for $i = 1, \ldots, r+1$. We denote by

$$\mathcal{D}(s_1, \ldots, s_r)$$

the set of all distinguished subexpressions of $(s_1, \ldots, s_r) \in S^r$.

For example, let $W = S_5$, $s_i = (i, i+1)$ for $i = 1, \ldots, 4$, and $\xi = (s_3, s_2, s_1, s_2, s_4)$. Then,

$$
\begin{aligned}
\mathcal{D}(\xi) = \{ &(3, -, -, -, -), (-, -, 1, -, -), (-, -, -, 2, -), (-, -, -, -, 4), \\
&(3, -, 1, -, -), (3, -, -, 2, -), (3, -, -, -, 4), (3, 2, 1, -, -), \\
&(3, 2, -, 2, -), (3, -, 1, 2, -), (3, -, -, 2, 4), (-, 2, 1, 2, -), \\
&(-, 2, 1, -, 4), (-, 2, -, 2, 4), (-, -, 1, 2, 4), (-, 2, 1, 2, 4), \\
&(3, -, 1, 2, 4), (3, 2, -, 2, 4), (3, 2, 1, -, 4), (3, 2, 1, 2, -), \\
&(3, 2, 1, 2, 4), (-, -, -, -, -), (-, 2, 1, -, -), (-, 2, -, 2, -), \\
&(-, -, 1, 2, -), (-, -, 1, -, 4), (-, -, -, 2, 4), (3, -, 1, -, 4) \},
\end{aligned}
$$

where, for notational simplicity, we identify "s_i" with "i" and write "$-$" instead of "e."

The combinatorics of distinguished subexpressions is important for the understanding of the R-polynomials. The following result lists two simple properties of distinguished subexpressions. They both follow trivially from the definitions. Given $u \in W$ and $\xi \in S^r$, we let

$$\mathcal{D}(\xi)_u \overset{\text{def}}{=} \{(a_1, \ldots, a_r) \in \mathcal{D}(\xi) : a_1 \cdots a_r = u\}.$$

Proposition 5.3.6 *Let* $\xi \overset{\text{def}}{=} (s_1, \ldots, s_r) \in S^r$. *Then, we have the following:*

(i) *If* $(a_1, \ldots, a_r) \in \mathcal{D}(\xi)_u$, *then* $(a_1, \ldots, a_{r-1}) \in \mathcal{D}(s_1, \ldots, s_{r-1})_{ua_r}$.

(ii) *If* $(a_1, \ldots, a_{r-1}) \in \mathcal{D}(s_1, \ldots, s_{r-1})_u$, *then* $(a_1, \ldots, a_{r-1}, s_r) \in \mathcal{D}(\xi)_{us_r}$. $\square$

We are now ready to state and prove the second main result of this section.

Theorem 5.3.7 *Let* $u, v \in W$, $u \leq v$, *and* $(s_1, \ldots, s_r) \in S^r$ *be a reduced decomposition of* v. *Then,*

$$\widetilde{R}_{u,v}(q) = \sum_{\xi \in \mathcal{D}(s_1, \ldots, s_r)_u} q^{\ell(v) - \|\xi\|}.$$

Proof. We proceed by induction on $\ell(v)$, the result being clear if $\ell(v) = 0$. So assume $r \geq 1$ and let, for convenience, $s \overset{\text{def}}{=} s_r$ and $\rho \overset{\text{def}}{=} (s_1, \ldots, s_r)$. We distinguish two cases.

Case 1: $s \in D_R(u)$. Define a map

$$\varphi : \mathcal{D}(\rho)_u \to \mathcal{D}(s_1, \ldots, s_{r-1})_{us}$$

by letting

$$\varphi((a_1, \ldots, a_r)) \overset{\text{def}}{=} (a_1, \ldots, a_{r-1})$$

for $(a_1, \ldots, a_r) \in \mathcal{D}(\rho)_u$. We claim that φ is a bijection and that

$$\| \varphi(\xi) \| = \| \xi \| - 1$$

for all $\xi \in \mathcal{D}(\rho)_u$. In fact, if $(a_1, \ldots, a_r) \in \mathcal{D}(\rho)_u$, then $a_r = s_r$ (for if $a_r = e$, then $u = a_1 \ldots a_{r-1}$ and $s_r \notin D_R(a_1 \ldots a_{r-1})$, which is a contradiction) and, hence, $(a_1, \ldots, a_{r-1}) \in \mathcal{D}(s_1, \ldots, s_{r-1})_{us}$. This shows that $\varphi(\mathcal{D}(\rho)_u) \subseteq \mathcal{D}(s_1, \ldots, s_{r-1})_{us}$, that φ is injective, and that $\| \varphi(\xi) \| = \| \xi \| - 1$ for all $\xi \in \mathcal{D}(\rho)_u$. Also, if $(a_1, \ldots, a_{r-1}) \in \mathcal{D}(s_1, \ldots, s_{r-1})_{us}$, then $(a_1, \ldots, a_{r-1}, s) \in \mathcal{D}(\rho)_u$ and, hence, φ is surjective.

Therefore, by Proposition 5.3.2 and our induction hypothesis,

$$\sum_{\xi \in \mathcal{D}(\rho)_u} q^{\ell(v)-\|\xi\|} = \sum_{\eta \in \mathcal{D}(s_1,\ldots,s_{r-1})_{us}} q^{\ell(v)-\|\eta\|-1}$$

$$= \widetilde{R}_{us,vs}(q)$$

$$= \widetilde{R}_{u,v}(q),$$

as desired.

Case 2: $s \notin D_R(u)$. Let

$$\mathcal{D}(\rho)_u^+ \stackrel{\text{def}}{=} \{(a_1,\ldots,a_r) \in \mathcal{D}(\rho)_u : a_r = s_r\}$$

and

$$\mathcal{D}(\rho)_u^- \stackrel{\text{def}}{=} \{(a_1,\ldots,a_r) \in \mathcal{D}(\rho)_u : a_r = e\}.$$

Define a map

$$\varphi : \mathcal{D}(\rho)_u \to \mathcal{D}(s_1,\ldots,s_{r-1})_u \cup \mathcal{D}(s_1,\ldots,s_{r-1})_{us}$$

by letting

$$\varphi((a_1,\ldots,a_r)) \stackrel{\text{def}}{=} (a_1,\ldots,a_{r-1})$$

for $(a_1,\ldots,a_r) \in \mathcal{D}(\rho)_u$. We claim that φ is a bijection, $\varphi(\mathcal{D}(\rho)_u^-) = \mathcal{D}(s_1,\ldots,s_{r-1})_u$, $\varphi(\mathcal{D}(\rho)_u^+) = \mathcal{D}(s_1,\ldots,s_{r-1})_{us}$, and $\| \varphi(\xi) \|=\| \xi \|$ if $\xi \in \mathcal{D}(\rho)_u^-$, whereas $\| \varphi(\xi) \|=\| \xi \| -1$ if $\xi \in \mathcal{D}(\rho)_u^+$. All of these properties are obvious (by Proposition 5.3.6), except for the surjectivity of φ. However, if $(a_1,\ldots,a_{r-1}) \in \mathcal{D}(s_1,\ldots,s_{r-1})_u$, then $(a_1,\ldots,a_{r-1},e) \in \mathcal{D}(\rho)_u^-$ (because $s_r \notin D_R(u)$), whereas if $(a_1,\ldots,a_{r-1}) \in \mathcal{D}(s_1,\ldots,s_{r-1})_{us}$, then $(a_1,\ldots,a_{r-1},s) \in \mathcal{D}(\rho)_u^+$. This proves the surjectivity.

Therefore, by Proposition 5.3.2 and our induction hypothesis,

$$\sum_{\xi \in \mathcal{D}(\rho)_u} q^{\ell(v)-\|\xi\|} = \sum_{\eta \in \mathcal{D}(s_1,\ldots,s_{r-1})_u} q^{\ell(v)-\|\eta\|} + \sum_{\eta \in \mathcal{D}(s_1,\ldots,s_{r-1})_{us}} q^{\ell(vs)-\|\eta\|}$$

$$= q\widetilde{R}_{u,vs}(q) + \widetilde{R}_{us,vs}(q)$$

$$= \widetilde{R}_{u,v}(q),$$

as desired. $\square$

Example 5.3.8 Let $W = S_4$, $u = 1234$, $v = 4321$ and take $(1,2,1,3,2,1)$ as a reduced decomposition for v. Then,

$$\mathcal{D}((1,2,1,3,2,1))_u = \{(-,-,-,-,-,-), (1,-,1,-,-,-),$$
$$(-,2,-,-,2,-), (-,-,1,-,-,1),$$
$$(1,2,-,-,2,1)\}$$

and, hence,

$$\widetilde{R}_{1234,4321}(q) = q^6 + 3q^4 + q^2.$$

Consequently,

$$R_{1234,4321}(q) = q^6 - 3q^5 + 4q^4 - 4q^3 + 4q^2 - 3q + 1.$$

$\square$

Since distinguished subexpressions play such a fundamental role in the combinatorics of the R-polynomials, it is useful to have a different characterization of this concept.

Proposition 5.3.9 *Let* $(s_1, \ldots, s_r) \in S^r$ *be a reduced decomposition and* $(a_1, \ldots, a_r)$ *be a subexpression of* $(s_1, \ldots, s_r)$. *Then, the following are equivalent:*

(i) $(a_1, \ldots, a_r) \in \mathcal{D}(s_1, \ldots, s_r)$.

(ii) $a_1 \cdots a_{j-1} s_j \cdots s_r \geq a_1 \cdots a_j s_{j+1} \cdots s_r$, *for* $j = 2, \ldots, r$.

Proof. Let $(a_1, \ldots, a_r) \in \mathcal{D}(s_1, \ldots, s_r)$ and $2 \leq j \leq r$. If $a_j = s_j$, then (ii) clearly holds, so assume that $a_j = e$. Then, since $(a_1, \ldots, a_r) \in \mathcal{D}(s_1, \ldots, s_r)$, we have that $s_j \notin D_R(a_1 \cdots a_{j-1})$. On the other hand, $s_j \notin D_L(s_{j+1} \cdots s_r)$ (since $s_j \cdots s_r$ is a reduced expression). Therefore, by Lemma 2.2.10, $(a_1 \cdots a_{j-1})(s_{j+1} \cdots s_r) \leq (a_1 \cdots a_{j-1}) s_j (s_{j+1} \cdots s_r)$, which proves (ii).

Conversely, suppose that $(a_1, \ldots, a_r)$ satisfies (ii) and $2 \leq j \leq r$ is such that $a_j = e$. We claim that then $s_j \notin D_R(a_1 \cdots a_{j-1})$. In fact, if $s_j \notin D_R(a_1 \cdots a_{j-1} s_j)$, then, since $s_j \notin D_L(s_{j+1} \cdots s_r)$, we conclude from Lemma 2.2.10 that $(a_1 \cdots a_{j-1} s_j)(s_{j+1} \cdots s_r) \leq (a_1 \cdots a_{j-1} s_j) s_j (s_{j+1} \cdots s_r) = a_1 \cdots a_j s_{j+1} \cdots s_r$, and this, by (ii), implies that $s_j = e$, which is a contradiction. $\square$

Intuitively, Proposition 5.3.9 states that we obtain all possible distinguished subexpressions of a reduced expression by deleting elements in it, from left to right, so that each deletion moves the resulting product *down* in Bruhat order. For example, the distinguished subexpression $(1, 2, -, 3, -, -, -, 3, 2, 1)$ of $(1, 2, 1, 3, 2, 1, 4, 3, 2, 1)$ in S_5 corresponds to the sequence of deletions

$$(1, 2, 1, 3, 2, 1, 4, 3, 2, 1)$$
$$(1, 2, -, 3, 2, 1, 4, 3, 2, 1)$$
$$(1, 2, -, 3, -, 1, 4, 3, 2, 1)$$
$$(1, 2, -, 3, -, -, 4, 3, 2, 1)$$
$$(1, 2, -, 3, -, -, -, 3, 2, 1)$$

In Section 2.7, we associated with every maximal chain $\mathbf{m} : v = x_0 \rhd x_1 \rhd \cdots \rhd x_k = u$ in Bruhat order a string of integers $\lambda(\mathbf{m}) = (\lambda_1, \lambda_2, \ldots, \lambda_k)$. The λ_i's are induced from a given reduced expression for v; namely they record the positions of successively deleted generators. The labeling rule given in Section 2.7 can be directly generalized from maximal chains in

Bruhat order to paths $v = x_0 \leftarrow x_1 \leftarrow \cdots \leftarrow x_j = u$ in the Bruhat graph. By the preceding paragraph, distinguished subexpressions correspond to increasingly labeled such paths.

Both Theorems 5.3.4 and 5.3.7 can be used to obtain some rather non-trivial properties of the R-polynomials. See, for example, Exercises 4, 5, 6, 10, 18, 35, and 36.

5.4   Lattice paths

In this section, we define and study a family of polynomials, indexed by sequences of positive integers, which plays an important role in the combinatorics of the Kazhdan-Lusztig polynomials. These polynomials are independent of W and are defined in terms of lattice paths.

Recall that a *composition* of n ($n \in \mathbb{P}$) is a sequence $(\alpha_1, \ldots, \alpha_s)$ (for some $s \in \mathbb{P}$) of positive integers such that $\alpha_1 + \cdots + \alpha_s = n$. When writing compositions, we will sometimes omit the parentheses (i.e., we will write $\alpha_1, \ldots, \alpha_s$ instead of $(\alpha_1, \ldots, \alpha_s)$). For $n \in \mathbb{P}$, we let C_n be the set of all compositions of n and $C \overset{\text{def}}{=} \bigcup_{n \geq 1} C_n$. Given $\beta \in C$, we denote by $\ell(\beta)$ the number of parts of β, and by β_i, for $i = 1, \ldots, \ell(\beta)$, the i-th part of β (so that $\beta = (\beta_1, \beta_2, \ldots, \beta_{\ell(\beta)})$). Furthermore, we let

$$|\beta| \overset{\text{def}}{=} \sum_{i=1}^{\ell(\beta)} \beta_i,$$

$$\overline{\beta} \overset{\text{def}}{=} (\beta_2, \beta_3, \ldots, \beta_{\ell(\beta)}) \quad (\text{if } \ell(\beta) \geq 2),$$

$$\beta^* \overset{\text{def}}{=} (\beta_{\ell(\beta)}, \ldots, \beta_2, \beta_1),$$

$$T(\beta) \overset{\text{def}}{=} \{\beta_r, \beta_r + \beta_{r-1}, \ldots, \beta_r + \cdots + \beta_2\}, \text{ where } r \overset{\text{def}}{=} \ell(\beta).$$

Given $(\alpha_1, \ldots, \alpha_s), (\beta_1, \ldots, \beta_t) \in C_n$, we say that $(\alpha_1, \ldots, \alpha_s)$ *refines* $(\beta_1, \ldots, \beta_t)$ if there exist $1 \leq i_1 < i_2 < \cdots < i_{t-1} \leq s$ such that $\sum_{j=i_{k-1}+1}^{i_k} \alpha_j = \beta_k$ for $k = 1, \ldots, t$ (where $i_0 \overset{\text{def}}{=} 0$, $i_t \overset{\text{def}}{=} s$). We then write $(\alpha_1, \ldots, \alpha_s) \preceq (\beta_1, \ldots, \beta_t)$. It is easy to see that the map $\alpha \mapsto T(\alpha)$ is an isomorphism from $(C_n, \preceq)$ to the Boolean algebra of subsets of $[n-1]$, ordered by reverse inclusion.

Let $a, b \in \mathbb{Z}$, $a \leq b$. By a *lattice path* on $[a, b]$ we mean a function $\Gamma : [a, b] \to \mathbb{Z}$ such that $\Gamma(a) = 0$ and

$$|\Gamma(i+1) - \Gamma(i)| = 1$$

for all $i \in [a, b-1]$. Given such a lattice path Γ, we let

$$
\begin{aligned}
N(\Gamma) &\overset{\text{def}}{=} \{i \in [a+1, b-1] : \Gamma(i) < 0\}, \\
d_+(\Gamma) &\overset{\text{def}}{=} |\{i \in [a, b-1] : \Gamma(i+1) - \Gamma(i) = 1\}|, \\
\ell(\Gamma) &\overset{\text{def}}{=} b - a, \\
\Gamma_{\geq 0} &\overset{\text{def}}{=} \ell(\Gamma) - 1 - |N(\Gamma)|.
\end{aligned}
$$

We call $N(\Gamma)$ the *negative set* of Γ and $\ell(\Gamma)$ the *length* of Γ. Note that $b \notin N(\Gamma)$ and that

$$
d_+(\Gamma) = \frac{\Gamma(b) + b - a}{2}. \tag{5.23}
$$

For example, if Γ is the lattice path illustrated in Figure 5.2, then $N(\Gamma) = \{3, 4, 5\}$, $d_+(\Gamma) = 2$, $\ell(\Gamma) = 6$, and $\Gamma_{\geq 0} = 2$.

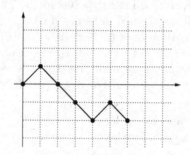

Figure 5.2. A lattice path on $[0, 6]$.

For $n \in \mathbb{P}$, we denote by $\mathcal{L}(n)$ the set of all the lattice paths on $[0, n]$. Given $S \subseteq [n-1]$, we let

$$
H(S, n) \overset{\text{def}}{=} \{\Gamma \in \mathcal{L}(n) : N(\Gamma) \supseteq S\}
$$

and

$$
E(S, n) \overset{\text{def}}{=} \{\Gamma \in \mathcal{L}(n) : N(\Gamma) = S\}.
$$

For example, the four lattice paths in $E(\{5\}, 6)$ are illustrated in Figure 5.3.

For $\alpha \in C$, we define two polynomials $\Psi_\alpha(q), \Upsilon_\alpha(q) \in \mathbb{Z}[q]$ by letting

$$
\Psi_\alpha(q) = (-1)^{|\alpha|} \sum_{\Gamma \in H(T(\alpha), |\alpha|)} (-q)^{d_+(\Gamma)} \tag{5.24}
$$

and

$$
\Upsilon_\alpha(q) = (-1)^{|\alpha| - \ell(\alpha)} \sum_{\Gamma \in E(T(\alpha), |\alpha|)} (-q)^{d_+(\Gamma)}.
$$

For example, let $\alpha = (1,5)$. Then, $E(\{5\},6)$ consists of the four lattice paths illustrated in Figure 5.3 and, hence, $\Upsilon_{1,5}(-q) = 2q^2 + 2q^3$.

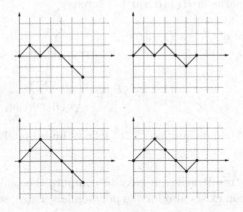

Figure 5.3. The four lattice paths in $E(\{5\},6)$.

Note that the definitions imply that

$$\Psi_\beta(q) = \sum_{\alpha \preceq \beta} (-1)^{\ell(\alpha)} \Upsilon_\alpha(q).$$

Hence, by the Principle of Inclusion-Exclusion,

$$\Upsilon_\beta(q) = \sum_{\alpha \preceq \beta} (-1)^{\ell(\alpha)} \Psi_\alpha(q). \tag{5.25}$$

For $j \in \mathbb{Q}$, we define an operator $L_j : \mathbb{R}[q] \to \mathbb{R}[q]$ by letting

$$L_j \left(\sum_{i \geq 0} a_i q^i \right) \overset{\text{def}}{=} \sum_{0 \leq i \leq j} a_i q^i.$$

Note that L_j is linear and idempotent, and that $L_j = L_{\lfloor j \rfloor}$ for all $j \in \mathbb{Q}$.

Although the polynomials $\Psi_\beta(q)$ have several interesting properties (see, for example, Exercises 28 through 34), we only need here the following recurrence relation.

Proposition 5.4.1 Let $\alpha \in C$. Then,

$$\Psi_\alpha(q) = (q-1)^{\alpha_1} L_{\frac{|\overline{\alpha}|-1}{2}} (\Psi_{\overline{\alpha}}(q)) \tag{5.26}$$

if $\ell(\alpha) \geq 2$, and

$$\Psi_\alpha(q) = (q-1)^{|\alpha|} \tag{5.27}$$

if $\ell(\alpha) = 1$.

Proof. If $\ell(\alpha) = 1$, then $T(\alpha) = \emptyset$ and equation (5.27) follows immediately from equation (5.24). So assume $\ell(\alpha) \geq 2$ and let $\Gamma \in H(T(\alpha), |\alpha|)$. Then,

$\Gamma_{|_{[0,|\overline{\alpha}|]}} \in H(T(\overline{\alpha}), |\overline{\alpha}|)$ and $\Gamma(|\overline{\alpha}|) < 0$. Conversely, if $\Gamma' \in H(T(\overline{\alpha}), |\overline{\alpha}|)$ and $\Gamma'(|\overline{\alpha}|) < 0$, then extending Γ' to $[0, |\alpha|]$ by adding any α_1 steps yields 2^{α_1} different lattice paths in $H(T(\alpha), |\alpha|)$. Therefore,

$$\sum_{\Gamma \in H(T(\alpha),|\alpha|)} q^{d_+(\Gamma)} = (1+q)^{\alpha_1} \sum_{\{\Gamma' \in H(T(\overline{\alpha}),|\overline{\alpha}|): \ \Gamma'(|\overline{\alpha}|) < 0\}} q^{d_+(\Gamma')}$$

$$= (1+q)^{\alpha_1} L_{\frac{|\overline{\alpha}|-1}{2}} \left(\sum_{\Gamma \in H(T(\overline{\alpha}),|\overline{\alpha}|)} q^{d_+(\Gamma)} \right),$$

by equation (5.23), and equation (5.26) follows immediately from equation (5.24). $\square$

The preceding result makes it easy to compute the polynomials $\Psi_\alpha(q)$ (a similar recursion exists also for the polynomials $\Upsilon_\beta(q)$; see Exercise 28). For example,

$$\begin{aligned} \Psi_{1,5}(q) &= (q-1) L_{\frac{5-1}{2}}(\Psi_5(q)) \\ &= (q-1) L_2((q-1)^5) \\ &= -10q^3 + 15q^2 - 6q + 1. \end{aligned}$$

5.5   Kazhdan-Lusztig polynomials

As we have seen in Section 5.1, once the R-polynomials have been computed, then one can use Theorem 5.1.4 to compute the Kazhdan-Lusztig polynomials. This, however, is a recursive procedure, as is the one based on Theorem 5.1.7. In this section, we derive a nonrecursive formula for the computation of the Kazhdan-Lusztig polynomials.

Our first step is that of "solving" the recurrence relation given in part (iv) of Theorem 5.1.4. More precisely, we wish to find a nonrecursive formula for $P_{u,v}(q)$ in terms of the R-polynomials. To do this, it is convenient to introduce the following concept. Given a chain $a_0 < a_1 < \cdots < a_i$ in W, we define

$$R_{a_0,\ldots,a_i}(q) \stackrel{\text{def}}{=} R_{a_0,a_1}(q) L_{\frac{d-1}{2}}(R_{a_1,\ldots,a_i}(q)) \tag{5.28}$$

(where $d \stackrel{\text{def}}{=} \ell(a_1, a_i)$) if $i \geq 2$, and

$$R_{a_0,\ldots,a_i}(q) \stackrel{\text{def}}{=} R_{a_0,a_1}(q) \tag{5.29}$$

if $i = 1$.

Example 5.5.1 In $W = S_4$, we have that

$$
\begin{aligned}
R_{2134,2431,4321}(q) &= R_{2134,2431}(q)\, L_{\frac{2-1}{2}}(R_{2431,4321}(q)) \\
&= (q^3 - 2q^2 + 2q - 1)\, L_{\frac{1}{2}}((q-1)^2) \\
&= q^3 - 2q^2 + 2q - 1
\end{aligned}
$$

and, hence,

$$
\begin{aligned}
R_{1234,2134,2431,4321}(q) &= R_{1234,2134}(q)\, L_{\frac{5-1}{2}}(R_{2134,2431,4321}(q)) \\
&= (q-1)\, L_2(q^3 - 2q^2 + 2q - 1) \\
&= -2q^3 + 4q^2 - 3q + 1.
\end{aligned}
$$

$\square$

The polynomial $R_{a_0,\ldots,a_i}(q)$ just defined is called the *R-polynomial* of the chain $a_0 < a_1 < \cdots < a_i$. This terminology is consistent with the one introduced in Section 5.1 since the R-polynomial of a chain of size two coincides with the usual R-polynomial of the two elements in increasing order. Note that although definition (5.28) is recursive, it can be formulated in an entirely explicit way (see Exercises 21 and 22) so that one can compute the R-polynomial of a given chain without having to compute that of any other chain.

The R-polynomial of a chain is important in the nonrecursive computation of the Kazhdan-Lusztig polynomials. Recall that a chain from u to v is a chain whose first element is u and whose last element is v.

Theorem 5.5.2 *Let* $u, v \in W$, $u < v$. *Then,*

$$
P_{u,v}(q) - q^{\ell(u,v)} P_{u,v}\left(\frac{1}{q}\right) = \sum_{C \in C(u,v)} (-1)^{\ell(C)} R_C(q),
$$

where $C(u,v)$ *is the set of all the chains from* u *to* v.

Proof. We prove the result by induction on $\ell(u,v)$. If $\ell(u,v) = 1$, then

$$
\sum_{C \in C(u,v)} (-1)^{\ell(C)} R_C(q) = -R_{u,v}(q) = 1 - q = P_{u,v}(q) - qP_{u,v}\left(\frac{1}{q}\right).
$$

Now, let $u < v$ be such that $\ell(u,v) \geq 2$. Then, from Theorem 5.1.4, definitions (5.28) and (5.29), and our induction hypothesis, we have that

$$q^{\ell(u,v)}P_{u,v}\left(\frac{1}{q}\right) - P_{u,v}(q)$$

$$= R_{u,v}(q) + \sum_{u<a<v} R_{u,a}(q)L_{\frac{\ell(a,v)-1}{2}}\left(\sum_{\mathcal{C}\in C(a,v)}(-1)^{\ell(\mathcal{C})}R_{\mathcal{C}}(q)\right)$$

$$= R_{u,v}(q) + \sum_{u<a<v}\sum_{\mathcal{C}\in C(a,v)}(-1)^{\ell(\mathcal{C})}R_{u,a}(q)L_{\frac{\ell(a,v)-1}{2}}(R_{\mathcal{C}}(q))$$

$$= R_{u,v}(q) + \sum_{u<a<v}\sum_{\mathcal{C}\in C(a,v)}(-1)^{\ell(\mathcal{C})}R_{u,\mathcal{C}}(q)$$

$$= R_{u,v}(q) + \sum_{\{\mathcal{C}'\in C(u,v):\, \ell(\mathcal{C}')\geq 2\}}(-1)^{\ell(\mathcal{C}')-1}R_{\mathcal{C}'}(q)$$

$$= -\sum_{\mathcal{C}'\in C(u,v)}(-1)^{\ell(\mathcal{C}')}R_{\mathcal{C}'}(q),$$

as desired. $\square$

As an immediate consequence, we derive the following.

Corollary 5.5.3 *Let $u,v \in W$, $u < v$. Then,*

$$P_{u,v}(q) = L_{\frac{\ell(u,v)-1}{2}}\left(\sum_{\mathcal{C}\in C(u,v)}(-1)^{\ell(\mathcal{C})}R_{\mathcal{C}}(q)\right). \qquad \square$$

Now that we have a nonrecursive formula for the Kazhdan-Lusztig polynomials in terms of the R-polynomials, the strategy is that of using one of the combinatorial interpretations obtained in Section 5.3 and "transfering" it to the Kazhdan-Lusztig polynomials using Theorem 5.5.2. We will do this using the one in terms of Bruhat paths (Theorem 5.3.4). Another nonrecursive formula (involving multichains, instead of chains) for computing the Kazhdan-Lusztig polynomials in terms of the R-polynomials is given in Exercise 25.

To state and prove the results that follow, it is convenient to introduce some notation. Given $u,v \in W$, and $k \in \mathbb{N}$, we denote by $B_k(u,v)$ the set of all Bruhat paths from u to v of length k, so that $B(u,v) = \bigcup_{k\geq 0} B_k(u,v)$.

Let Δ be a Bruhat path and $<$ be a reflection ordering. We define the *descent composition* of Δ with respect to $<$ to be the unique composition $\mathcal{C}(\Delta,<) \in C$ such that $|\mathcal{C}(\Delta,<)| = \ell(\Delta)$ and

$$T(\mathcal{C}(\Delta,<)^*) = D(\Delta,<).$$

In other words, if Δ has length i (say) and $\mathcal{C}(\Delta,<) = (b_1,\ldots,b_j)$, then $b_1 + \cdots + b_j = i$ and the descent set of Δ with respect to $<$ is $\{b_1, b_1+b_2, \ldots, b_1 + \cdots + b_{j-1}\}$. For example, if $\Delta = (2147563, 2147653, 6147253, 6157243)$ and $<$ is the reflection ordering used in Figure 5.1, then $\mathcal{C}(\Delta;<) = (1,2)$.

For $u, v \in W$, and $\alpha \in C$, we let

$$c_\alpha(u, v) \overset{\text{def}}{=} |\{\Delta \in B_{|\alpha|}(u, v) : \mathcal{C}(\Delta, <) \succeq \alpha\}| \tag{5.30}$$

and

$$b_\alpha(u, v) \overset{\text{def}}{=} |\{\Delta \in B_{|\alpha|}(u, v) : \mathcal{C}(\Delta, <) = \alpha\}|. \tag{5.31}$$

Note that these definitions imply that

$$c_\alpha(u, v) = \sum_{\beta \succeq \alpha} b_\beta(u, v) \tag{5.32}$$

for all $u, v \in W$ and $\alpha \in C$, and that

$$c_\alpha(u, v) = b_\alpha(u, v) = |\{\Delta \in B_{|\alpha|}(u, v) : D(\Delta, <) = \emptyset\}| \tag{5.33}$$

if $l(\alpha) = 1$.

From now on and until the end of this section, we fix once and for all a reflection ordering $<$.

Proposition 5.5.4 *Let $u, v \in W$, $u \leq v$, and $\alpha_1, \ldots, \alpha_r \in \mathbb{P}$, $r \geq 2$. Then,*

$$c_{\alpha_1, \ldots, \alpha_r}(u, v) = \sum_{u \leq a \leq v} c_{\alpha_1, \ldots, \alpha_{r-1}}(u, a) c_{\alpha_r}(a, v).$$

Proof. Let $\Delta \overset{\text{def}}{=} (a_0, \ldots, a_{\beta_r}) \in B(u, v)$ be such that $D(\Delta, <) \subseteq \{\beta_1, \ldots, \beta_{r-1}\}$, where $\beta_i \overset{\text{def}}{=} \alpha_1 + \cdots + \alpha_i$, for $i = 1, \ldots, r$. Then, clearly, $u \leq a_{\beta_{r-1}} \leq v$, $\Delta' \overset{\text{def}}{=} (a_0, \ldots, a_{\beta_{r-1}}) \in B(u, a_{\beta_{r-1}})$, $\Delta'' \overset{\text{def}}{=} (a_{\beta_{r-1}}, \ldots, a_\alpha) \in B(a_{\beta_{r-1}}, v)$, $D(\Delta', <) \subseteq \{\beta_1, \ldots, \beta_{r-2}\}$, and $D(\Delta'', <) = \emptyset$. It is easy to see that this correspondence $\Delta \leftrightarrow (\Delta', \Delta'')$ is a bijection, and the result follows from definition (5.30). $\square$

The following result is crucial for the proof of the main theorem of this section.

Proposition 5.5.5 *Let $u, v \in W$, $u \leq v$, and $\alpha \in C$. Then,*

$$c_\alpha(u, v) = \sum_{(a_0, \ldots, a_r) \in C_r(u, v)} \prod_{j=1}^{r} [q^{\alpha_j}](\widetilde{R}_{a_{j-1}, a_j}),$$

where $C_r(u, v)$ denotes the set of all chains of length r from u to v and $r \overset{\text{def}}{=} \ell(\alpha)$.

Proof. We proceed by induction on $r \in \mathbb{P}$. If $r = 1$ then, by equation (5.33) and Theorem 5.3.4, we have that

$$c_{\alpha_1}(u, v) = [q^{\alpha_1}](\widetilde{R}_{u, v}), \tag{5.34}$$

as desired. If $r \geq 2$, then we have from Proposition 5.5.4, equation (5.34), and our induction hypothesis that

$$
\begin{aligned}
c_{\alpha_1,\dots,\alpha_r}(u,v) &= \sum_{u \leq a \leq v} c_{\alpha_r}(a,v) c_{\alpha_1,\dots,\alpha_{r-1}'}(u,a) \\
&= \sum_{u \leq a \leq v} [q^{\alpha_r}](\widetilde{R}_{a,v}) \sum_{(a_0,\dots,a_{r-1}) \in C_{r-1}(u,a)} \prod_{j=1}^{r-1} [q^{\alpha_j}](\widetilde{R}_{a_{j-1},a_j}) \\
&= \sum_{u \leq a < v} \sum_{(a_0,\dots,a_{r-1}) \in C_{r-1}(u,a)} [q^{\alpha_r}](\widetilde{R}_{a,v}) \prod_{j=1}^{r-1} [q^{\alpha_j}](\widetilde{R}_{a_{j-1},a_j})
\end{aligned}
$$

and the thesis follows. $\square$

The next result extends Proposition 5.3.1 to the R-polynomial of a chain.

Proposition 5.5.6 *Let $a_0 < a_1 < \cdots < a_i$ be a chain in W. Then,*

$$
R_{a_0,\dots,a_i}(q) = \sum_{\alpha \in \mathbb{P}^i} q^{\frac{\ell(a_0,a_i)-|\alpha|}{2}} \Psi_\alpha(q) \prod_{r=1}^{i} [q^{\alpha_r}](\widetilde{R}_{a_{r-1},a_r}). \tag{5.35}
$$

Proof. If $i = 1$, then equation (5.35) follows from equations (5.27), (5.29) and (5.10). We now proceed by induction on $i \in \mathbb{P}$. Since $i \geq 2$, we have from definition (5.28) and our induction hypothesis that

$$
R_{a_0,\dots,a_i}(q)
$$

$$
\begin{aligned}
&= R_{a_0,a_1}(q) L_{\frac{d-1}{2}}(R_{a_1,\dots,a_i}(q)) \\
&= \sum_{\alpha_0 > 0} q^{\frac{\ell(a_0,a_1)-\alpha_0}{2}} \Psi_{\alpha_0}(q)\, [q^{\alpha_0}](\widetilde{R}_{a_0,a_1}) L_{\frac{d-1}{2}}(R_{a_1,\dots,a_i}(q)) \\
&= \sum_{\alpha_0 > 0} \sum_{\alpha \in \mathbb{P}^{i-1}} q^{\frac{\ell(a_0,a_1)-\alpha_0}{2}} \prod_{r=0}^{i-1} [q^{\alpha_r}](\widetilde{R}_{a_r,a_{r+1}}) \Psi_{\alpha_0}(q) L_{\frac{d-1}{2}}(q^{\frac{d-|\alpha|}{2}} \Psi_\alpha(q)) \\
&= \sum_{\alpha_0 > 0} \sum_{\alpha \in \mathbb{P}^{i-1}} q^{\frac{\ell(a_0,a_i)-\alpha_0-|\alpha|}{2}} \prod_{r=1}^{i} [q^{\alpha_{r-1}}](\widetilde{R}_{a_{r-1},a_r}) \Psi_{\alpha_0}(q) L_{\frac{|\alpha|-1}{2}}(\Psi_\alpha(q)),
\end{aligned}
$$

where $d \overset{\text{def}}{=} \ell(a_1,a_i)$. Hence, equation (5.35) follows from equations (5.26) and (5.27). $\square$

We can now state and prove the main result of this section, namely a nonrecursive combinatorial formula for the Kazhdan-Lusztig polynomials.

Theorem 5.5.7 *Let $<$ be a reflection ordering, $u,v \in W$, $u < v$. Then,*

$$
P_{u,v}(q) - q^{\ell(u,v)} P_{u,v}\left(\frac{1}{q}\right) = \sum_{\Delta \in B(u,v)} q^{\frac{\ell(u,v)-\ell(\Delta)}{2}} \Upsilon_{C(\Delta,<)}(q). \tag{5.36}
$$

Proof. From Theorem 5.5.2 and Propositions 5.5.5 and 5.5.6 we have that

$$
P_{u,v}(q) - q^{\ell(u,v)} P_{u,v}\left(\frac{1}{q}\right) = \sum_{C \in \mathcal{C}(u,v)} (-1)^{\ell(C)} R_C(q)
$$

$$
= \sum_{\alpha \in C} (-1)^{\ell(\alpha)} q^{\frac{\ell(u,v)-|\alpha|}{2}} \Psi_\alpha(q)\, c_\alpha(u,v).
$$

On the other hand, from equations (5.32) and (5.25), we have that

$$
\sum_{\alpha \in C_n} (-1)^{\ell(\alpha)} \Psi_\alpha(q)\, c_\alpha(u,v) = \sum_{\alpha \in C_n} (-1)^{\ell(\alpha)} \Psi_\alpha(q) \sum_{\beta \succeq \alpha} b_\beta(u,v)
$$

$$
= \sum_{\beta \in C_n} b_\beta(u,v) \sum_{\alpha \preceq \beta} (-1)^{\ell(\alpha)} \Psi_\alpha(q)
$$

$$
= \sum_{\beta \in C_n} b_\beta(u,v)\, \Upsilon_\beta(q)
$$

for all $n \in \mathbb{P}$. Therefore we conclude that

$$
P_{u,v}(q) - q^{\ell(u,v)} P_{u,v}\left(\frac{1}{q}\right) = \sum_{\beta \in C} q^{\frac{\ell(u,v)-|\beta|}{2}} \Upsilon_\beta(q)\, b_\beta(u,v),
$$

which, by definition (5.31), is equivalent to equation (5.36). □

The preceding result shows that the Kazhdan-Lusztig polynomials can be computed in a way similar to that given in Theorem 5.3.4 for the R-polynomials, and it also makes clear why they are considerably more difficult to compute. In fact, by Theorem 5.3.4, to compute $R_{u,v}(q)$, we just have to consider the Bruhat paths from u to v that have an empty descent set with respect to a given reflection order. To compute $P_{u,v}(q)$, by Theorem 5.5.7, we have to consider *all* of the Bruhat paths from u to v, and for each such path, we have to know the descent set with respect to the given reflection order. In Theorem 5.5.7, we also have to know the polynomials $\Upsilon_\beta(q)$ studied in the previous section. This, however, is not a big problem since these polynomials are defined explicitly as counting certain lattice paths, can be computed through simple recursions, and admit explicit formulas in terms of Catalan numbers (see Exercises 28, 30, and 31).

Example 5.5.8 Let $u = 2147563$ and $v = 6157243$. Considering all of the directed paths from 2147563 to 6157243 in Figure 5.1 (there are 62 of them) and computing the descent composition of each one, we obtain that

$P_{2147563,6157243}(q) - q^5 P_{2147563,6157243}(q^{-1})$

$$= q(2\Upsilon_3 + 3\Upsilon_{2,1} + 3\Upsilon_{1,2} + 2\Upsilon_{1,1,1})$$
$$+ q^0(\Upsilon_5 + 2\Upsilon_{4,1} + 4\Upsilon_{3,2} + 4\Upsilon_{2,3} + 2\Upsilon_{1,4} + 3\Upsilon_{3,1,1}$$
$$+ 4\Upsilon_{1,3,1} + 3\Upsilon_{1,1,3} + 6\Upsilon_{2,2,1} + 4\Upsilon_{2,1,2} + 6\Upsilon_{1,2,2} + 2\Upsilon_{2,1,1,1}$$
$$+ 4\Upsilon_{1,2,1,1} + 4\Upsilon_{1,1,2,1} + 2\Upsilon_{1,1,1,2} + \Upsilon_{1,1,1,1,1})$$
$$= q[2(-q^3 + 2q^2 - q) + 3(-q^2 + q) + 3(0) + 2(1 - q)]$$
$$+ q^0[(-q^5 + 4q^4 - 5q^3 + 2q^2) + 2(-q^4 + 2q^3 - q^2) + 4(0)$$
$$+ 4(q^3 - q^2) + 2(0) + 3(0) + 4(0) + 3(q^2 - q) + 6(-q^3 + q^2) + 4(0)$$
$$+ 6(0) + 2(q^3 - q^2) + 4(0) + 4(-q^2 + q) + 2(0) + (2q^2 - 3q + 1)]$$
$$= 1 - q^5.$$

$\square$

Theorem 5.5.7 has the following immediate consequence. Given $n \in \mathbb{Z}$ and $A \subseteq \mathbb{Z}$, we let $n - A \stackrel{\text{def}}{=} \{n - a : a \in A\}$.

Corollary 5.5.9 *Let $<$ be a reflection ordering, $u, v \in W$, $u < v$. Then,*

$$P_{u,v}(q) = \sum_{(\Gamma, \Delta)} (-1)^{\Gamma_{\geq 0} + d_+(\Gamma)} q^{\frac{\ell(u,v) + \Gamma(\ell(\Gamma))}{2}},$$

where the sum is over all pairs (Γ, Δ) such that Γ is a lattice path, $\Delta \in B(u, v)$, $\ell(\Gamma) = \ell(\Delta)$, $N(\Gamma) = \ell(\Delta) - D(\Delta, <)$, and $\Gamma(\ell(\Gamma)) < 0$. $\square$

Theorems 5.5.2 and 5.5.7 can be used to deduce some rather nontrivial properties of the Kazhdan-Lusztig polynomials; see Exercises 12, 13, and 9.

5.6   Complement: Special matchings

Recall that a *matching* of a graph $G = (V, E)$ is an involution $M : V \to V$ such that $\{v, M(v)\} \in E$ for all $v \in V$. (Here we assume that G has no loops or multiple edges.)

Let P be a graded partially ordered set. A matching M of the Hasse diagram of P is a *special matching* if, for all $x, y \in P$ such that $M(x) \neq y$, we have that

$$x \lhd y \implies M(x) \leq M(y).$$

Note that this implies, in particular, that if $x \lhd y$ and $M(x) \rhd x$, then $M(y) \rhd y$ and $M(y) \rhd M(x)$ and, dually, that if $x \lhd y$ and $M(y) \lhd y$, then $M(x) \lhd x$ and $M(x) \lhd M(y)$.

The motivation for this definition is given by the next result.

Proposition 5.6.1 *Let (W, S) be a Coxeter system, $u, v \in W$, $u \leq v$, and $s \in D(v) \setminus D(u)$. Let*

$$M(x) \overset{\text{def}}{=} xs$$

for all $x \in [u, v]$. Then, M is a special matching of $[u, v]$.

Proof. This follows immediately from the definition of a special matching and the Lifting Property. $\square$

Corollary 5.6.2 *Every lower Bruhat interval $[e, v]$ has a special matching.*

There is, of course, a left version of Proposition 5.6.1. Note that the converse is not true. Namely there are special matchings of Bruhat intervals which are not given by right or left multiplication by a simple reflection. For example, let (W, S) be a Coxeter system such that $|S| \geq 3$ and there are $a, b, c \in S$ such that $m(a, b), m(b, c) \geq 3$. Then, $M = \{\{e, b\}, \{a, ab\}, \{c, bc\}, \{ac, abc\}\}$ is a special matching of $[e, abc]$.

It is a surprising fact that Kazhdan-Lusztig and R-polynomials can be computed using only special matchings. This can be done by a recursive procedure based on the following result, which was first proved in [99] for the symmetric group and then in [100] for the general case.

Theorem 5.6.3 *Let (W, S) be a Coxeter system, $v \in W$, and let M be a special matching of $[e, v]$. Then,*

$$R_{u,v}(q) = q^c R_{M(u),M(v)}(q) + (q^c - 1) R_{u,M(v)}(q)$$

for all $u \leq v$, where $c \overset{\text{def}}{=} 1$ if $M(u) \rhd u$ and $c \overset{\text{def}}{=} 0$ otherwise.

Thus, despite the fact that by definition the R-polynomials, and hence the Kazhdan-Lusztig polynomials, depend heavily on descent sets, they can in fact be computed only in terms of the abstract poset structure of a lower interval. We illustrate this process on an example.

Example 5.6.4 Let $P = [e, v]$ be the lower Bruhat interval whose Hasse diagram is depicted in Figure 5.4. The elements of P are labeled with the integers from 1 to 18. In order to use Theorem 5.6.3, we need to find a special matching M of P.

Suppose $M(1) = 2$. We have two possible choices for $M(3)$, namely 7 and 8, and two for $M(4)$, namely 5 and 6. Suppose we choose $M(3) = 7$ and $M(4) = 5$. These choices force

$$M = \{\{1, 2\}, \{3, 7\}, \{4, 5\}, \{6, 11\}, \{8, 10\},$$
$$\{9, 12\}, \{13, 15\}, \{14, 16\}, \{17, 18\}\}.$$

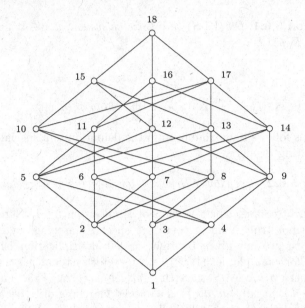

Figure 5.4. A lower Bruhat interval.

Applying Theorem 5.6.3, we obtain that[1]

$$R_{1,18} = q\,R_{M(1),M(18)} + (q-1)R_{1,M(18)} = qR_{2,17} + (q-1)R_{1,17}.$$

We therefore need to compute the polynomials $R_{u,17}$ for all $u \leq$ 17. Since M does not restrict to a special matching of $[1,17]$ ($= \{1,2,3,4,5,6,7,8,9,10,11,13,14,17\}$), we need to repeat the above procedure to find a special matching, N, of $[1,17]$.

Suppose that $N(1) = 2$. This forces $N(3) \in \{7,8\}$ and $N(4) \in \{5,6\}$. Suppose we choose $N(3) = 7$ and $N(4) = 6$. Then, our choices force

$$N = \{\{1,2\},\{3,7\},\{4,6\},\{5,11\},\{8,10\},\{9,14\},\{13,17\}\}.$$

Applying Theorem 5.6.3, we get[2]

$$R_{1,17} = q\,R_{2,13} + (q-1)R_{1,13},$$
$$R_{2,17} = R_{1,13}.$$

---

[1]As well as $R_{2,18} = R_{1,17}$, $R_{3,18} = qR_{7,17}+(q-1)R_{3,17}$, $R_{4,18} = qR_{5,17}+(q-1)R_{4,17}$, $R_{5,18} = R_{4,17}$, $R_{6,18} = qR_{11,17} + (q-1)R_{6,17}$, $R_{7,18} = R_{3,17}$, $R_{8,18} = qR_{10,17} + (q-1)R_{8,17}$, $R_{9,18} = qR_{12,17} + (q-1)R_{9,17} = (q-1)R_{9,17}$, and similarly $R_{10,18} = R_{8,17}$, $R_{11,18} = R_{6,17}$, $R_{12,18} = R_{9,17}$, $R_{13,18} = (q-1)R_{13,17}$, $R_{14,18} = (q-1)R_{14,17}$, $R_{15,18} = R_{13,17}$, $R_{16,18} = R_{14,17}$, $R_{17,18} = (q-1)R_{17,17}$.

[2]As well as $R_{3,17} = (q-1)R_{3,13}$, $R_{4,17} = (q-1)R_{4,13}$, $R_{5,17} = (q-1)R_{5,13}$, $R_{6,17} = R_{4,13}$, $R_{7,17} = R_{3,13}$, $R_{8,17} = (q-1)R_{8,13}$, $R_{9,17} = (q-1)R_{9,13}$, $R_{10,17} = R_{8,13}$, $R_{11,17} = R_{5,13}$, $R_{13,17} = (q-1)R_{13,13}$, $R_{14,17} = R_{9,13}$.

We now need to compute the polynomials $R_{u,13}$ for all $u \in [1,13]$ ($= \{1,2,3,4,5,8,9,13\}$). The poset $[1,13]$ is a 3-crown, so $R_{u,13} = (q-1)^{l(u,13)}$ for all $u \leq 13$ (see Exercise 5). However, since no outside result is needed by the procedure suggested by Theorem 5.6.3, and for completeness, we conclude the example using only special matchings.

We now need a special matching, L, of $[1,13]$. If $L(1) = 4$, then this forces $L = \{\{1,4\},\{2,5\},\{3,9\},\{8,13\}\}$. So by Theorem 5.6.3, we get[3]

$$R_{1,13} = (q-1)R_{1,8},$$
$$R_{2,13} = (q-1)R_{2,8}.$$

A special matching of $[1,8]$ ($= \{1,2,3,8\}$) is given by $\{\{1,2\},\{3,8\}\}$, and from Theorem 5.6.3 we get[4]

$$R_{1,8} = (q-1)R_{1,3},$$
$$R_{2,8} = R_{1,3}.$$

Finally, $\{\{1,3\}\}$ is a special matching of $[1,3]$ ($= \{1,3\}$) and so, again by Theorem 5.6.3, we obtain $R_{1,3} = (q-1)R_{1,1}$.

Putting all these relations together, we get

$$
\begin{aligned}
R_{1,18} &= q\,R_{2,17} + (q-1)R_{1,17} \\
&= q\,R_{1,13} + (q-1)(q\,R_{2,13} + (q-1)R_{1,13}) \\
&= q(q-1)R_{1,8} + q(q-1)^2 R_{2,8} + (q-1)^3 R_{1,8} \\
&= q(q-1)^2 R_{1,3} + q(q-1)^2 R_{1,3} + (q-1)^4 R_{1,3} \\
&= 2q(q-1)^3 + (q-1)^5,
\end{aligned}
$$

and similarly for all the other polynomials $R_{u,18}$. Clearly, in the same way (and in fact without much additional effort since we already have a special matching of $[e,w]$ for all $w \in P$) we may compute all the polynomials $R_{x,y}$ for $x,y \in P$, $x \leq y$.

The computation of the Kazhdan-Lusztig polynomials $P_{x,y}$ for $x,y \in P$, $x \leq y$, now proceeds using Theorem 5.1.4 and induction on $\ell(x,y)$, as explained in Section 5.1. $\square$

As a matter of fact, essentially all of the results presented in this chapter can be generalized using special matchings, as shown in [100].

Surprising as this result is, even more may be true. In fact, there is a tantalizing conjecture regarding the Kazhdan-Lusztig polynomials, the so-called *combinatorial invariance conjecture*. This conjecture, made by Lusztig [364] and independently by Dyer [200], states that if W_1 and W_2 are two Coxeter groups, $u,v \in W_1$, $x,y \in W_2$, and the two Bruhat intervals

---

[3]As well as $R_{3,13} = (q-1)R_{3,8}$, $R_{4,13} = R_{1,8}$, $R_{5,13} = R_{2,8}$, $R_{8,13} = (q-1)R_{8,8}$, $R_{9,13} = R_{3,8}$.

[4]As well as $R_{3,8} = (q-1)R_{3,3}$.

$[u, v]$ and $[x, y]$ are isomorphic as posets, then $P_{u,v}(q) = P_{x,y}(q)$. In other words, the Kazhdan-Lusztig polynomial of u, v supposedly depends only, on the unlabeled abstract poset $[u, v]$. This conjecture is known to be true if $[u, v]$ is a lattice (see Exercise 36) and holds for intervals of rank ≤ 4 (see Exercises 5 and 6).

Note that, by Theorem 5.1.4, the combinatorial invariance conjecture is equivalent to the analogous statement for the R-polynomials. Thus, Theorem 5.6.3 implies that the conjecture holds also for the case that both u and x are the identity.

In connection with the combinatorial invariance conjecture, it is worth noting that the relation $u \to v$ is known to depend only on the poset $[u, v]$, as shown by Dyer [203]. Regarding single coefficients, the combinatorial invariance conjecture is open even for the coefficient of q. Some further special cases are treated in Exercises 9, 23, and 35.

Exercises

In Exercises 1-14, let $u, v \in W$ and $u \leq v$.

1. Show that

$$(-q)^{\ell(u,v)} R_{u,v} \left(\frac{1}{q} \right) = R_{u,v}(q).$$

2. Suppose $u < v$. Show that $R_{u,v}(1) = 0$.

3. Suppose $\ell(u, v) \leq 2$. Show that $R_{u,v}(q) = (q-1)^{\ell(u,v)}$.

4. Suppose that $\ell(u, v) = 3$. Show that

$$R_{u,v}(q) = \begin{cases} q^3 - 2q^2 + 2q - 1, & \text{if } [u, v] \text{ is a 2-crown,} \\ (q-1)^3, & \text{otherwise.} \end{cases}$$

5. (a) Show that if $u < v$, then $\widetilde{R}_{u,v}(0) = 0$.
   (b) Show that if $\ell(u, v) \leq 3$, then

$$\widetilde{R}_{u,v}(q) = \begin{cases} q^3 + q, & \text{if } [u, v] \text{ is a 2-crown,} \\ q^{\ell(u,v)}, & \text{otherwise.} \end{cases}$$

6. Show that if $\ell(u, v) = 4$, then we have the following:
   (a)

$$\widetilde{R}_{u,v}(q) = q^4 + \frac{B_2(u, v)}{2} q^2,$$

   where $B_2(u, v)$ is the number of paths of length 2 from u to v in the Bruhat graph of W.
   (b) $B_2(u, v) \leq 4$.

(c) $B_2(u, v)$ equals the number of subintervals of $[u, v]$ isomorphic to 2-crowns.

7. (a) Show that if $\ell(u, v) \leq 2$, then $P_{u,v}(q) = 1$.
    (b) Show that if $\ell(u, v) = 3$, then

$$P_{u,v}(q) = \begin{cases} 1, & \text{if } [u, v] \text{ is a 2-crown,} \\ 1 + (a(u, v) - 3)q, & \text{otherwise,} \end{cases}$$

where $a(u, v)$ denotes the number of atoms of $[u, v]$.

8. Suppose that $\ell(u, v) = 4$. Show that

$$P_{u,v}(q) = 1 + \left(c(u, v) + \frac{B_2(u, v)}{2} - 4\right)q,$$

where $c(u, v)$ is the number of coatoms in $[u, v]$ and $B_2(u, v)$ is as in Exercise 6.

9. Let $a(u, v)$, $c(u, v)$ have the same meaning as in Exercises 7 and 8.

    (a) Show that $(-1)^{\ell(v)-1}[q](R_{e,v}) = a(e, v)$.
    (b) Show that $[q](P_{e,v}) = c(e, v) - a(e, v)$.
    (c) Show that

$$[q](P_{u,v}) = (-1)^{\ell(u,v)}[q](R_{u,v}) + c(u, v).$$

    (d) Show that

$$[q](P_{u,v}) = - \sum_{\{x,y \in [u,v]: \, \ell(x,v) \geq 3\}} [q](R_{x,y}).$$

    (e) Show that

$$[q^2](P_{u,v}) = - \sum_{\{x,y \in [u,v]: \, \ell(x,v) \geq 5\}} ([q^2](R_{x,y}) + [q](R_{x,y})[q](P_{y,v})).$$

    (f) Show that

$$[q]\left(\sum_{x,y \in [u,v]} R_{x,y} \right) = (-1)^{\ell(u,v)-1}[q](R_{u,v}).$$

Conclude from this that $[q](\sum_{x,y \in [u,v]} R_{x,y}) \geq 0$.

10. (a) Prove that $R_{u,v}(q) = R_{u^{-1},v^{-1}}(q)$.
     (b) Show, using Theorem 5.1.1, that if W is finite and w_0 is its longest element, then $R_{u,v}(q) = R_{w_0v,w_0u}(q) = R_{vw_0,uw_0}(q)$ for all $u, v \in W$.
     (c) Give a direct proof of (b) using the combinatorial interpretations for the R-polynomials described in Theorems 5.3.4 and 5.3.7.

11. Show that
$$\sum_{w \in [u,v]} (-1)^{\ell(u,w)} R_{u,w}(q) R_{w,v}(q) = \delta_{u,v}.$$

12. Show that $P_{u,v}(q) = P_{u^{-1},v^{-1}}(q)$.

13. Let W be a finite Coxeter group with longest element w_0.
    (a) Show that
    $$\sum_{a \in [u,v]} (-1)^{\ell(u,a)} P_{u,a}(q) P_{w_0v,w_0a}(q) = \delta_{u,v}.$$
    (b) Deduce from part (a) that $\bar{\mu}(u,v) = \bar{\mu}(w_0v, w_0u)$.
    (c) Show that $P_{u,v}(q) = P_{w_0uw_0,w_0vw_0}(q)$.

14. (a) Show that $\sum_{x \in [y,v]} R_{y,x}(q) = q^{\ell(y,v)}$ for all $y \in [u,v]$ if and only if $P_{y,v}(q) = 1$ for all $y \in [u,v]$.
    (b) Show that if $P_{x,y}(q) = 1$ for all $x,y \in [u,v]$, $x \leq y$, then $R_{x,y}(q) = (-1)^{\ell(x,y)} \sum_{a \in [x,y]} (-q)^{\ell(x,a)}$ for all $x,y \in [u,v]$.
    (c) Use part (a) to show that the converse of the statement in part (b) also holds.

15. Compute $R_{124356,564312}(q)$. Conclude that it is not true that the coefficients of the R-polynomials alternate in sign.

16. For $v \in W$, let
    $$F_v(q) \overset{\text{def}}{=} \sum_{u \leq v} q^{\ell(u)} P_{u,v}(q).$$

    Show that $q^{\ell(v)} F_v\left(\frac{1}{q}\right) = F_v(q)$.

17. Show that $\sum_{u \leq v} (-1)^{\ell(u)} P_{u,v}(q) = 0$ for all $v \in W \setminus \{e\}$.

18. Let $u \in W$. Show that if $R_{e,u}(q) = (q-1)^{\ell(u)}$, then $[e,u]$ is isomorphic to a Boolean algebra.

19. Let (W,S) be a Coxeter system of rank n, $S = \{s_1, \ldots, s_n\}$, and $\sigma \in S_n$. Show that there exists a reflection order $<$ on T such that
    $$s_{\sigma(1)} < s_{\sigma(2)} < \cdots < s_{\sigma(n)}.$$

20. Let W be a finite Coxeter group and let w_0 be its longest element. Let $<$ be a total order on T. Suppose that $T = \{t_1, \ldots, t_N\}$ with $t_1 < t_2 < \cdots < t_N$. Show that the following conditions are equivalent:
    (i) $<$ is a reflection order.
    (ii) There exists a reduced decomposition $s_1 \ldots s_N = w_0$ such that
    $$t_i = s_N \ldots s_{i+1} s_i s_{i+1} \ldots s_N$$
    for $i = 1, \ldots, N$.

21. Let $k \in \mathbb{Z}$ and $a_0 < a_1 < \cdots < a_i$ be a chain in W. Show that

$$[q^k](R_{a_0,\ldots,a_i}) = \sum_\alpha \prod_{r=1}^i [q^{\alpha_r}](R_{a_{r-1},a_r})$$

where the sum is over all $\alpha = (\alpha_1, \ldots, \alpha_i) \in \mathbb{N}^i$ such that $\alpha_1 + \cdots + \alpha_i = k$, $\alpha_r \leq \ell(a_{r-1}, a_r)$ for $r = 1, \ldots, i$, and $\alpha_{r+1} + \cdots + \alpha_i \leq \frac{1}{2}(\ell(a_r, a_i) - 1)$ for $r = 1, \ldots, i-1$.

22. Let $a_0 < a_1 < \cdots < a_i$ be a chain in W and $k \in \mathbb{Z}$. Show that

$$[q^k](R_{a_0,\ldots,a_i}) = (-1)^{\ell(a_0,a_i)} \sum_\alpha \prod_{r=1}^i [q^{\alpha_r}](R_{a_{r-1},a_r}) \qquad (5.37)$$

where the sum is over all $\alpha = (\alpha_1, \ldots, \alpha_i) \in \mathbb{N}^i$ such that $\alpha_1 + \cdots + \alpha_i = \ell(a_0, a_i) - k$, $\alpha_r \leq \ell(a_{r-1}, a_r)$ for $r = 1, \ldots, i$, and $\alpha_{r+1} + \cdots + \alpha_i \geq \frac{1}{2}(\ell(a_r, a_i) + 1)$ for $r = 1, \ldots, i-1$.
Note that this formula is simpler than the one in Exercise 21 if $k > \frac{1}{2}\ell(a_0, a_i)$.

23. Let $u, v \in S_n$, $u \leq v$.
    (a) Show that if $u \to v$, then $R_{u,v}(q) = (q-1)(q^2-q+1)^{\frac{1}{2}(\ell(u,v)-1)}$.
    (b) Is the statement in part (a) true if $W \neq S_n$?
    (c) Show that if $D_R(v), D_R(u) \supseteq [2, n-2]$, then there exists $a \in \mathbb{N}$ such that

    $$R_{u,v}(q) = (q-1)^a(q^2-q+1)^{\frac{1}{2}(\ell(u,v)-a)}.$$

    (d) Is the converse of the statement in part (a) true?

24. Let $u, v \in S_n$, $u \leq v$, be such that $D_R(v) \supseteq [2, n-2]$. Show that

$$P_{u,v}(q) = \begin{cases} 1 + q^{v(1)-v(n)}, & \text{if } u(n) > v(1) \geq v(n) > u(1), \\ 1, & \text{otherwise.} \end{cases}$$

25. For $j \in \mathbb{Q}$, define an operator $U_j : \mathbb{R}[q] \to \mathbb{R}[q]$ by letting

$$U_j \left(\sum_{i \geq 0} a_i q^i \right) \overset{\text{def}}{=} \sum_{i \geq j} a_i q^i.$$

Given a multichain $a_0 \leq a_1 \leq \cdots \leq a_{r+1}$ ($r \in \mathbb{N}$) in W, define a polynomial $\mathcal{R}_{a_0,a_1,\ldots,a_{r+1}}(q)$ inductively as follows:

$$\mathcal{R}_{a_0,a_1,\ldots,a_{r+1}}(q) \overset{\text{def}}{=} \mathcal{R}_{a_0,a_1}(q)U_{\frac{d+1}{2}}\left(q^d \mathcal{R}_{a_1,\ldots,a_{r+1}}\left(\frac{1}{q}\right)\right)$$

(where $d \overset{\text{def}}{=} \ell(a_1, a_{r+1})$) if $r \in \mathbb{P}$, and

$$\mathcal{R}_{a_0,a_1,\ldots,a_{r+1}}(q) \overset{\text{def}}{=} (-1)^{\ell(a_0,a_1)} \mathcal{R}_{a_0,a_1}(q)$$

if $r = 0$. $\mathcal{R}_{a_0, a_1, \ldots, a_{r+1}}(q)$ is called the R-*polynomial* of the multichain $a_0 \leq a_1 \leq \cdots \leq a_{r+1}$.

(a) Show that if $r \in \mathbb{P}$ and $a_0 \leq a_1 \leq \cdots \leq a_{r+1}$ is a multichain in W such that $\mathcal{R}_{a_0, \ldots, a_{r+1}} \neq 0$, then $\ell(a_r, a_{r+1}) \geq 1$, and $\ell(a_i, a_{i+1}) \geq 2$ for $i = 1, \ldots, r-1$. In particular, $\ell(a_1, a_{r+1}) \geq 2r - 1$.

(b) Prove that if $u, v \in W$, then

$$P_{u,v}(q) = \sum_{\mathcal{C} \in M(u,v)} \mathcal{R}_{\mathcal{C}}(q),$$

where $M(u, v)$ denotes the set of all multichains from u to v.

26. For $n \in \mathbb{P}$, consider the q-analog $F_n(q)$ of the Fibonacci number F_n defined by

$$F_n(q) \overset{\text{def}}{=} F_{n-1}(q) + qF_{n-2}(q),$$

where $F_n(q) \overset{\text{def}}{=} 0$ if $n < 0$ and $F_0(q) \overset{\text{def}}{=} 1$. It is easy to verify that for $n \geq 0$,

$$F_n(q) = \sum_{i=0}^{\lfloor n/2 \rfloor} \binom{n-i}{i} q^i.$$

(a) Let $n \geq 3$. Show that $P_{e,\ 3\ 4 \ldots n\ 1\ 2}(q) = F_{n-2}(q)$.

(b) Let $n \geq 5$. Show that $P_{e,\ 3\ 4 \ldots n-2\ n\ n-1\ 1\ 2}(q) = F_{n-3}(q)$.

27. (a) Let $3 \leq e_1 < e_2 < \cdots < e_a \leq n - 1$ be integers (so, $n \geq 4$ and $a \in [n-3]$). Furthermore, define $u, v \in S_n$ in complete notation by

$$u \overset{\text{def}}{=} 1\, e_1 \ldots e_a\, f_1 \ldots f_{n-3-a}\, 2\, n,$$
$$v \overset{\text{def}}{=} e_1 \ldots e_a\, n\, f_1 \ldots f_{n-3-a}\, 1\, 2,$$

where $\{f_{n-3-a}, \ldots, f_2, f_1\}_< \overset{\text{def}}{=} [n-3] \setminus \{e_1, \ldots, e_a\}$. Show that

$$P_{u,v}(q) = 1 + \sum_{i=1}^{a} q^{e_i - i - 1}.$$

(b) Conclude from part (a) that given $P \in \mathbb{N}[q]$ such that $P(0) = 1$, there exist $n \in \mathbb{P}$ and $u, v \in S_n$ such that $P(q) = P_{u,v}(q)$.

28. Let $\beta \in C$ (in this and the following exercises, we use the notation introduced in Section 5.4). Show that

$$\Upsilon_\beta(q) = (q-1)U_{\frac{|\beta|-1}{2}}(\Upsilon_{\beta^{(1)}}(q))$$

if $\beta_1 \geq 2$,

$$\Upsilon_\beta(q) = (1-q)L_{\frac{|\beta|-2}{2}}(\Upsilon_{\overline{\beta}}(q))$$

if $\beta_1 = 1$ and $\ell(\beta) \geq 2$, and $\Upsilon_1(q) = 1 - q$.

29. Let $\beta \in C$ and $\tilde{\beta} \in C$ be the unique composition such that $|\tilde{\beta}| = |\beta|$ and $T(\tilde{\beta}) = [|\beta| - 1] \setminus T(\beta)$. Show that

$$-q^{|\beta|} \Upsilon_\beta \left(\frac{1}{q} \right) = \Upsilon_{(\tilde{\beta},1)}(q).$$

In particular, $-q^n \Upsilon_n(1/q) = \Upsilon_{(1^{n+1})}(q)$ for $n \in \mathbb{P}$.

30. (a) Let $\beta \in C$, $\ell(\beta) \geq 2$. Show that there exist $A, B \in \mathbb{Z}$ such that

$$\Upsilon_\beta(q) = \begin{cases} Aq^{\frac{|\beta| - \beta_1 + 1}{2}} \Upsilon_{\beta_1 - 1}(q), & \text{if } \beta_1 \geq 2, \\ Bq^{\frac{|\beta| - \tilde{\beta}_1}{2}} \Upsilon_{(1^{\tilde{\beta}_1})}(q), & \text{if } \beta_1 = 1. \end{cases}$$

(b) Compute A and B explicitly as a product of Catalan numbers.

31. Exercises 29 and 30 reduce the problem of computing any Υ-polynomial to that of computing $\Upsilon_n(q)$ for $n \in \mathbb{P}$. These polynomials, in turn, can be computed in a completely explicit way. For $n \in \mathbb{P}$, let

$$B_n(q) \overset{\text{def}}{=} \sum_{i=0}^{n} \binom{2n}{n-i} \frac{2i+1}{1+n+i} q^{n+i}.$$

Show that

$$\Upsilon_{2n}(q) = -B_n(-q)$$

and

$$\Upsilon_{2n+1}(q) = (1 - q)B_n(-q).$$

32. Characterize the $\beta \in C$ such that $\Upsilon_\beta(q) \neq 0$.

33. A polynomial $\sum_{i=0}^{d} a_i x^i$ is called *log-concave* if $a_i^2 \geq a_{i-1} a_{i+1}$ for $i = 1, \ldots, d - 1$. Let $\beta \in C$ be such that $\Upsilon_\beta(q) \neq 0$. Show the following:

(a) $(-1)^{|\beta| - \ell(\beta)} \Upsilon_\beta(-q)$ is a log-concave polynomial.

(b) The maximum power of q that divides $\Upsilon_\beta(q)$ is $\left\lfloor \frac{|\beta| - \tilde{\beta}_1 + 1}{2} \right\rfloor$.

(c) $\deg(\Upsilon_\beta) = \beta_1 + \max\left(\left\lfloor \frac{|\tilde{\beta}| - 1}{2} \right\rfloor, 0 \right)$.

34. Let $\beta \in C$ be such that $\ell(\beta), \ell(\tilde{\beta}) \geq 2$ (where $\tilde{\beta}$ has the same meaning as in Exercise 29) and $\Upsilon_\beta \neq 0$. Show that $\Upsilon_{\tilde{\beta}} = 0$. Show that the converse statement does not hold.

35. (a) Show that if $u, v \in W$, $u \leq v$, then

$$\frac{d}{dq} (R_{u,v}(q))|_{q=1} = \begin{cases} 1, & \text{if } u \to v, \\ 0, & \text{otherwise.} \end{cases}$$

(b) Use part (a) to show that if $u, v \in W$, $u \leq v$, then

$$\left. \frac{d}{dq} \left(q^{\ell(u,v)} P_{u,v} \left(\frac{1}{q^2} \right) \right) \right|_{q=1} = \sum_{\{x \in [u,v]:\ u \to x\}} P_{x,v}(1).$$

(c) For $u, v \in W$, $u \leq v$, define the *defect* of the interval $[u, v]$ by

$$\mathrm{df}(u, v) \overset{\mathrm{def}}{=} |\{x \in [u, v] : u \to x\}| - \ell(u, v).$$

Use part (b) to show that if $P_{x,v}(q) = 1$ for all $x \in [u, v]$, then $\mathrm{df}(x, v) = 0$ for all $x \in [u, v]$.

(d) Does the converse of the statement in part (c) also hold?

36. Let $u, v \in W$, $u \leq v$.

(a) Show that if $[q^i](\tilde{R}_{u,v}) \neq 0$ for some $i < \ell(u, v)$, then $[q^{i+2}](\tilde{R}_{u,v}) \neq 0$.

(b) Use part (a) to show that $R_{u,v}(q) = (q-1)^{\ell(u,v)}$ if and only if $[q](R_{u,v}) = (-1)^{\ell(u,v)}(-\ell(u, v))$.

(c) Show that if $\ell(u, v) = 3$, then $u \to v$ if and only if $[u, v]$ is a 2-crown.

(d) Use parts (a) and (c) to show that $R_{u,v}(q) = (q-1)^{\ell(u,v)}$ if and only if $[u, v]$ does not contain any 2-crown as a subinterval.

(e) Deduce from part (d) that if $[u, v]$ is isomorphic to a Boolean algebra, then $P_{u,v}(q) = 1$.

37. Let (W, S) be a universal Coxeter system (see Example 1.2.2). Show that $P_{u,v}(q) \in \mathbb{N}[q]$ for all $u, v \in W$.

38. Show that $[q](P_{u,v}) \geq 0$ for all $u, v \in W$.

39. A permutation $u \in S_n$ is said to be 3412-*avoiding* if there are no $1 \leq i_1 < i_2 < i_3 < i_4 \leq n$ such that $u(i_3) < u(i_4) < u(i_1) < u(i_2)$. A permutation $v \in S_n$ is called *bigrassmannian* if $|D_R(v)| = |D_L(v)| = 1$. It is easy to see that v is bigrassmannian if and only if there exist $0 \leq a < b < c \leq n$ such that

$$v = 1 \ldots a\, b+1 \ldots c\, a+1 \ldots b\, c+1 \ldots n.$$

For $u \in S_n$ let

$$\mathrm{bg}(u) \overset{\mathrm{def}}{=} \{v \in S_n : v \text{ is bigrassmannian}, v \leq u\}.$$

For $v \in \mathrm{bg}(u)$, $v = 1 \ldots a\, b+1 \ldots c\, a+1 \ldots b\, c+1 \ldots n$ for some $0 \leq a < b < c \leq n$, define the *distance*, denoted $d(v, u)$, from v to u to be

$$\max\{i \in \mathbb{N} : 1 \ldots a-i\, b+1 \ldots c+i\, a-i+1 \ldots b\, c+i+1 \ldots n \leq u\}.$$

Suppose u avoids 3412, $u \neq w_0$. Associate a tree to uw_0 as follows. Let $I = (i_1, \ldots, i_n)$ be the inversion table of uw_0 and let $\lambda(u)$ be its nondecreasing rearrangement. Associate to $\lambda(u)$ a word in the two-letter alphabet $\{(,)\}$ by associating a "(" (respectively, ")") to each vertical step (respectively, horizontal step) as you follow the boundary of the diagram of $\lambda(u)$ from southwest to northeast. Then, associate to this word a rooted tree $T(u)$ by "matching the parentheses" in this word (i.e., each vertex of the tree, except the root, corresponds to a matching pair $(\dots)$ and a vertex is a descendant of another if and only if its pair is enclosed by the other pair). Note that the leaves of the tree correspond to the corners of the partition and, therefore, to the nonzero values of $I(uw_0)$.

For example, if $uw_0 = 15764238$ then $I = (0, 3, 4, 3, 2, 0, 0, 0)$ and hence $\lambda(u) = (2, 3, 3, 4)$. The word associated to $\lambda(u)$ is therefore $(()())()$ and the rooted tree associated to this word is

(a) Show that the map $z \mapsto z(i) - i$, where i is the unique (right) descent of z, is a bijection between the maximal elements z of $\mathrm{bg}(uw_0)$ and the nonzero values of $I(uw_0)$ and, hence, the leafs of $T(u)$.

(b) Given $v \in S_n$, $v \le u$, label each leaf of $T(u)$ by $d(z, vw_0)$, where z is the maximal element of $\mathrm{bg}(uw_0)$ corresponding to the leaf under the bijection in part (a). Let $E(u)$ be the set of edges of $T(u)$ and call a map $f : E(u) \to \mathbb{N}$ v-admissible if the following hold:

  (i) f weakly increases along any path from the root.

  (ii) The value of f at any leaf edge does not exceed the label of the leaf.

  Given such an f let $|f| \overset{\text{def}}{=} \sum_{x \in E(u)} f(x)$. Show that

$$P_{v,u}(q) = \sum_f q^{|f|},$$

where $f : E(u) \to \mathbb{N}$ runs over all v-admissible functions.

(c) Deduce from part (b) that if $u \in S_n$ is 3412-avoiding and 4231-avoiding, then $P_{v,u}(q) = 1$ for all $v \le u$.

(d) Show that the converse of the statement in part (c) also holds.

(e) Let $u \in S_n$ be such that $D_R(u) \subseteq \{1, n-1\}$. Use part (b) to show that, for any $v \le u$,

$$P_{v,u}(q) = (1 + q)^r,$$

where $r \overset{\text{def}}{=} \left| \left\{ j \in [u(n) + 1, u(1) - 2] : \sum_{i=1}^{j} v(i) = \binom{j+1}{2} \right\} \right|$.

    (f) Deduce from part (e) that $P_{e,(1,n)}(q) = (1+q)^{n-3}$ for all $n \geq 3$.

    (g) Let V_n be the number of 3412-avoiding permutations of S_n. Show that $\lim_{n \to \infty} \frac{V_n}{n!} = 0$.

Notes

The R-polynomials and Kazhdan-Lusztig polynomials were introduced by Kazhdan and Lusztig in [322] and all the results in Section 5.1 appear there, although some of them not explicitly. Reflection orderings appear in Bourbaki [79] for finite Coxeter systems (under the name of normal orderings) and were first introduced by Dyer [205] for general Coxeter systems. All of the results in Section 5.2 are due to Dyer and appear in [205], although usually with different proofs.

    Theorem 5.3.4 appeared in Dyer [200] and then in [205], whereas Theorem 5.3.7 is due to Deodhar [180]. The polynomials in Section 5.4 were introduced by Brenti [92] and Proposition 5.4.1 appears there. The R-polynomial of a chain was also introduced in [92], and all of the results in Section 5.5 appear there, except for Proposition 5.5.5, which is from Brenti [89].

Exercises 9, 21, 22, 30, and 31. See Brenti [92]. The polynomials $B_n(q)$ in Exercise 31 are closely related to ballot problems (see, e.g., [9], [237, §III.1, p. 21], [268, §5.3)]).

Exercises 11 and 13. See Kazhdan and Lusztig [322].

Exercise 14. See Kazhdan and Lusztig [322] and Brenti [87].

Exercise 15. See Boe [69].

Exercise 20. See Dyer [205].

Exercise 23. See Brenti [88] and Shapiro, Shapiro, and Vainshtein [454].

Exercise 24. See Shapiro, Shapiro, and Vainshtein [454].

Exercises 25 and 36. See Brenti [87].

Exercise 26. See Brenti and Simion [102].

Exercise 27. See Polo [421] and Caselli [113].

Exercise 37. See Dyer [201].

Exercise 38. See Dyer [210] and Tagawa [523].

Exercise 16. See Kazhdan and Lusztig [322]. It can be shown [323] that if W is a Weyl group, the polynomial $F_v(q)$ is the Poincaré polynomial for the intersection homology of the Schubert variety X_v associated to v. So the result stated in Exercise 16 is, for Weyl groups, a special case of the general topological fact that (middle perversity) intersection homology satisfies Poincaré duality.

Exercises 35 and 39. See Carrell [109] for Exercise 35, and for Exercise 39 see Lakshmibai and Sandhya [337], Lascoux [341], and Shapiro, Shapiro, and Vainshtein [454]. These two exercises have strong connections to the geometry of Schubert varieties. In fact, the equivalent conditions in parts (c) and (d) of Exercise 39 are themselves equivalent to the smoothness of X_u [337]. Similarly, if W is a Weyl group, then the conditions in part (c) of Exercise 35 are equivalent and are themselves equivalent, in the case that $u = e$, to the statement that the Schubert variety X_v is rationally smooth [109]. For Schubert varieties in type A, smoothness and rational smoothness are equivalent (see Deodhar [181]), so all of the preceding conditions are equivalent. In general, however, rational smoothness is a weaker statement than smoothness. For a comprehensive discussion of these issues, see Billey and Lakshmibai [47].

In addition to their connection to the geometry of Schubert varieties, the Kazhdan-Lusztig polynomials also encode a good deal of topological information. In fact, it is known (see [323]) that if W is a Weyl group, then the coefficients of the Kazhdan-Lusztig polynomial $P_{u,v}(q)$ equal the dimensions of the local intersection homology spaces of the Schubert variety X_v at a point lying on the Schubert cell indexed by u. Thus, Theorem 5.6.3 has the surprising geometrical consequence that these dimensions depend only on the inclusion relation between the Schubert subvarieties of X_v. The Kazhdan-Lusztig polynomials also play an important role in the representation theory of semisimple algebraic groups (see, e.g., Andersen [8] and the references cited there) and Hecke algebras, which is also the original reason for their introduction in [322].

Aside from their importance in the fields just mentioned, there are purely combinatorial reasons that make the Kazhdan-Lusztig polynomials interesting objects to study. Perhaps the main one is the *non-negativity conjecture* [322], which simply says that all Kazhdan-Lusztig polynomials have non-negative coefficients. This conjecture has been proved in the case that W is a finite or affine Weyl group using the topological interpretation of the polynomials and also for the other finite Coxeter groups by computer verification. It has also been proved for the universal Coxeter systems (see Exercise 37). Since the constant term of the Kazhdan-Lusztig polynomials is 1, the first nontrivial coefficient is the one of the q-term, and this has been proved to be indeed always non-negative (see Exercise 38).

In fact, much more than non-negativity could conceivably be true since for the finite and affine Weyl groups, a monotonicity result holds that trivially implies the non-negativity. Namely Braden and MacPherson [83] have shown that for these groups it is true that if $u, v, w \in W$ and $u \leq v \leq w$, then $P_{u,w}(q) - P_{v,w}(q) \in \mathbb{N}[q]$. In other words, if the second index of a Kazhdan-Lusztig polynomial is fixed, and the first one moves down in Bruhat order, then all the coefficients of the polynomial weakly increase.

It is worth noting that this monotonicity property always holds for the $\tilde{R}$-polynomials [89].

If the coefficients of Kazhdan-Lusztig polynomials are indeed always non-negative, then a natural problem is that of finding a combinatorial interpretation for them, meaning a set of combinatorial objects whose number equals the given coefficient. Even in the simplest nontrivial case, that of the symmetric group, this seems to be an extremely hard problem, and only some partial results are known (see Exercises 24, 26, 27, and 39).

It is known that *any* polynomial with non-negative integer coefficients and constant term equal to 1 is the Kazhdan-Lusztig polynomial of some pair of permutations. This has been proved geometrically by Polo [421] and combinatorially by Caselli [113] (see Exercise 27).

6

Kazhdan-Lusztig representations

This chapter concerns the so-called "left cell" representations of Coxeter groups, introduced by Kazhdan and Lusztig in [322]. (Actually, what they constructed are representations of Hecke algebras, and we are here talking of the $q = 1$ specializations.) This class of representations admits an encoding into the language of labeled graphs, and the whole theory has a strongly combinatorial flavor. For example, the "Kazhdan-Lusztig graph"

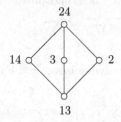

encodes the irreducible representation of the symmetric group S_5 indexed by the partition $(3, 2)$. From the information given by this graph, there is an algorithm for writing down the integral matrix representing any given element of S_5.

One of the most remarkable features of Kazhdan and Lusztig's construction, from a combinatorial point of view, is that for symmetric groups, it produces the irreducible representations. This fact is treated rather briefly in their article, and it is one of the main purposes of this chapter to work it out in detail. It will appear that this construction of the irreducible representations of S_n is in very close contact with the combinatorics of tableaux, in fact more so than the classical approaches. In particular, it fits hand-in-glove with the Robinson-Schensted correspondence and gives a more general meaning to the related tableau combinatorics. The combinatorial background is reviewed in Appendix A3.

6.1   Review of background material

To get started, we need to summarize some algebraic background material. A thorough account of this is given in Chapter 7 of Humphreys' book [306].

Let $\mathcal{H} = \mathcal{H}(W, S)$ be the Hecke algebra of a Coxeter group (W, S) over the ring $\mathbb{Z}[q^{\frac{1}{2}}, q^{-\frac{1}{2}}]$. This algebra has a linear basis $\{T_w\}_{w \in W}$ and its multiplication is determined by

$$\begin{cases} T_s T_w = q T_{sw} + (q-1)T_w, & \text{if } sw < w, \ s \in S, \\ T_w T_u = T_{wu}, & \text{if } \ell(w) + \ell(u) = \ell(wu). \end{cases}$$

In particular, if $w = s_1 s_2 \dots s_k$ is a reduced decomposition, then $T_w = T_{s_1} T_{s_2} \dots T_{s_k}$. Note that if $q^{\frac{1}{2}} = 1$, as will be assumed from Section 6.3 on, then $\mathcal{H}$ specializes to the group algebra $\mathbb{Z}[W]$.

The multiplication rule shows that the special basis elements T_s, for $s \in S$, are invertible: $T_s^{-1} = (q^{-1} - 1)T_e + q^{-1}T_s$. Consequently, all basis elements T_w are invertible in $\mathcal{H}$, namely

$$T_{w^{-1}}^{-1} = q^{-\ell(w)} \sum_{y \leq w} (-1)^{\ell(y,w)} R_{y,w}(q)\, T_y,$$

where $R_{y,w}(q) \in \mathbb{Z}[q]$ are the R-polynomials treated in Chapter 5.

There is an involution (ring homomorphism of order 2) of $\mathcal{H}$ defined by

$$\overline{\sum_{w \in W} p_w(q^{\frac{1}{2}})T_w} \overset{\text{def}}{=} \sum_{w \in W} p_w(q^{-\frac{1}{2}})T_{w^{-1}}^{-1},$$

where $p_w(q^{\frac{1}{2}})$ are Laurent polynomials in $q^{\frac{1}{2}}$. Furthermore, there is a basis $\{C_w\}_{w \in W}$ for $\mathcal{H}$ defined by

$$C_w = q^{\frac{\ell(w)}{2}} \sum_{y \leq w} (-1)^{\ell(y,w)} q^{-\ell(y)} P_{y,w}(q^{-1})T_y, \tag{6.1}$$

where $P_{y,w} \in \mathbb{Z}[q]$ are the Kazhdan-Lusztig polynomials discussed in the previous chapter. This basis satisfies

$$\overline{C_w} = C_w, \quad \text{for all } w \in W \tag{6.2}$$

and this relation is, in fact, equivalent to the recursion in part (iv) of Theorem-Definition 5.1.4.

Recall definition (5.3): $\overline{\mu}(y, w) = [q^{\frac{1}{2}(\ell(y,w)-1)}]P_{y,w}$ for $y < w$. If $y > w$, let $\overline{\mu}(y, w) = \overline{\mu}(w, y)$, and let

$$E \overset{\text{def}}{=} \left\{ \{x, y\} \in \binom{W}{2} : \ x < y \text{ or } x > y \text{ and } \overline{\mu}(x, y) \neq 0 \right\}. \tag{6.3}$$

The following multiplication rule in $\mathcal{H}$ (see [322, p. 171], or [306, p. 166]) is basic to this chapter:

$$T_s C_w = \begin{cases} -C_w, & \text{if } sw < w, \\ qC_w + q^{\frac{1}{2}} C_{sw} + q^{\frac{1}{2}} \sum \overline{\mu}(x,w)C_x, & \text{if } sw > w, \end{cases} \qquad (6.4)$$

where the sum is taken over all $x < w$ such that $\{x, w\} \in E$ and $sx < x$. It implies [306, p. 167] the following useful fact, already stated in Proposition 5.1.8 ($x, y \in W$ and $s \in S$):

$$x < y \text{ and } sy < y \quad \Rightarrow \quad P_{x,y} = P_{sx,y}. \qquad (6.5)$$

In this chapter, we use some basic facts and concepts from the representation theory of finite groups. This material will not be summarized, we refer to the texts of Lederman [358] or Sagan [450] (or any other book on the elements of representation theory). All representations will be over the complex numbers $\mathbb{C}$, and we do not distinguish notationally between representations and their characters.

6.2   Kazhdan-Lusztig graphs and cells

Let (W, S) be a Coxeter system. We define a graph with vertices labeled by subsets of S and edges labeled by nonzero integers as follows.

Definition 6.2.1 *The (left) Kazhdan-Lusztig graph $\Gamma_{(W,S)} = (W, E)$ (or K-L graph, for short) has for nodes the elements $w \in W$, labeled by $D_L(w)$, and for edges, it has the pairs $\{x, y\} \in E$ defined in definition (6.3), labeled by $\overline{\mu}(x,y)$.*

Figure 6.1 shows to the left the Bruhat order on S_3 (left descents are underlined), and to the right its K-L graph. All $\overline{\mu}$-labels are here equal to 1 and not marked in the picture, and in the vertex labels, we abbreviate $1 = (1,2)$, $2 = (2,3)$.

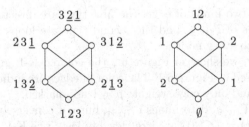

Figure 6.1. Bruhat order and K-L graph of S_3.

The K-L graph of S_4 (without labels) is the Hasse diagram of Bruhat order (see Figure 2.4) with two additional edges: $1324 - 3412$ and $2143 - 4231$, corresponding to the two length 3 intervals with $P_{y,w}(q) = 1 + q$.

Here are a few immediate observations.

Lemma 6.2.2 *Let $x, y \in W$, $x < y$. Then, the following hold:*

(i) *If $\{x, y\} \in E$, then $\ell(x, y) = \ell(y) - \ell(x)$ is odd.*

(ii) *If $\ell(x, y) = 1$, then $\{x, y\} \in E$ and $\overline{\mu}(x, y) = 1$.*

(iii) *If $\ell(x, y) > 1$ and there exists $s \in S$ such that $sy < y$ and $sx > x$, then $\{x, y\} \notin E$.*

Proof. The first two statements follow directly from the definitions. For the third, note that $P_{x,y} = P_{sx,y}$ by implication (6.5), so the degree of $P_{x,y}$ is too small. [*Remark:* This is a restatement of Proposition 5.1.9.] $\square$

Part (ii) of the lemma shows that all *Bruhat edges* (by which we mean coverings in Bruhat order) are edges of the K-L graph with label $= 1$. The remaining "non-Bruhat edges" of the K-L graph and their labels are considerably more difficult to determine. Finding them requires computation of the degree, and if maximal also of the leading coefficient, of the Kazhdan-Lusztig polynomials of all odd-length Bruhat intervals. (This task is not known to be computationally easier than the computation of the full K-L polynomials of all intervals of the group.)

It is useful for later developments to introduce a directed version of the K-L graph, with unlabeled vertices but with doubly-labeled edges. Thus, there will be two labels $s \in S$ and $\overline{\mu} \in \mathbb{Z} \setminus \{0\}$ on each edge, and we can think of s as a "color" and $\overline{\mu}$ as a "weight."

Definition 6.2.3 *The (left) colored Kazhdan-Lusztig graph is the directed graph $\widetilde{\Gamma}_{(W,S)} = (W, A)$ whose set A of labeled edges $x \xrightarrow[s]{\overline{\mu}} y$ are of the following two types ($x, y \in W$):*

(i) *$x \neq y$, $\{x, y\} \in E$, $s \in D_L(x) \setminus D_L(y)$, and $\overline{\mu} = \overline{\mu}(x, y)$.*

(ii) *$x = y$, $s \in S$, and $\overline{\mu} = \begin{cases} 1, & \text{if } s \notin D_L(x), \\ -1, & \text{if } s \in D_L(x). \end{cases}$*

Hence, there are two kinds of edge; the first type are directed versions of some of the edges of $\Gamma_{(W,S)}$ and the second type are loops (one for each $s \in S$) attached to every node.

As an example, we show in Figure 6.2 the colored K-L graph of S_3. It should be compared to Figure 6.1. The color of edges are indicated by solid or dashed lines, and only edge weights $\overline{\mu} \neq 1$ are marked.

It is clear that $\Gamma_{(W,S)}$ determines $\widetilde{\Gamma}_{(W,S)}$, but not conversely. Some edges (those with $D_L(x) = D_L(y)$) are irretrievably lost. The K-L graph $\Gamma_{(W,S)}$ is a more compact way to encode the information, but the colored version $\widetilde{\Gamma}_{(W,S)}$ is more useful for our developments. Note that $\widetilde{\Gamma}_{(W,S)}$ can have multiple (parallel or antiparallel) edges between a given pair of nodes. In such cases, these edges always have distinct labels. The down-directed edges in $\widetilde{\Gamma}_{(W,S)}$ are easy to characterize — they are the left weak order coverings:

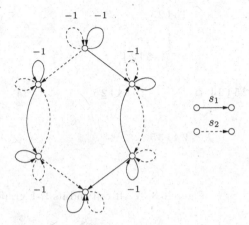

Figure 6.2. The colored K-L graph of S_3.

Lemma 6.2.4 *Suppose that $\ell(x) > \ell(y)$. Then, $x \to y$ is an edge in $\widetilde{\Gamma}_{(W,S)}$ if and only if $x \rhd_L y$.*

Proof. If $x = sy$, then clearly $x \to y$ is an s-labeled edge in $\widetilde{\Gamma}_{(W,S)}$. The converse follows from Lemma 6.2.2(iii). $\square$

The following concepts are basic to this chapter.

Definition 6.2.5 *Let $x, y \in W$. We say the following:*

(i) $x \preceq_L y$ if there exists a directed path in $\widetilde{\Gamma}_{(W,S)}$ from x to y.

(ii) The relation "$\preceq_L$" is called the left *preorder on W.*

(iii) $x \sim_L y$ if $x \preceq_L y$ and $y \preceq_L x$.

(iv) The relation "$\sim_L$" is called the left equivalence *relation on W.*

(v) An equivalence class of (W, S) under $\sim_L$ is called a left cell.

We can consider the induced subgraphs of $\Gamma_{(W,S)}$ and $\widetilde{\Gamma}_{(W,S)}$ on each left cell $\mathcal{C}$, and in fact from now on, we will usually think of left cells as the induced graphs $\Gamma_{\mathcal{C}}$ and $\widetilde{\Gamma}_{\mathcal{C}}$. [*Remark:* In graph-theoretic terminology, the left cells are the strongly connected components of $\widetilde{\Gamma}_{(W,S)}$.]

For example, $\mathcal{C} = \{24135, 25134, 34125, 35124, 45123\}$ is a left cell in S_5 (the reason for this will be easily understood from the results in Section 6.5). Figure 6.3(a) shows Bruhat order of this set and Figure 6.3(b) shows the graph $\Gamma_{\mathcal{C}}$ (there is a non-Bruhat edge because $P_{24135,45123}(q) = 1 + q$; the edge labels $\overline{\mu} = 1$ are not marked).

Figure 6.4 shows the colored K-L graph $\widetilde{\Gamma}_{\mathcal{C}}$, with all loops removed and edge weights $\overline{\mu} = 1$ not marked.

Definition 6.2.6 *Given two left cells $\mathcal{C}$ and $\mathcal{C}'$, we say that $\mathcal{C} \preceq_L \mathcal{C}'$ if $x \preceq_L y$ for some (equivalently, all) $x \in \mathcal{C}$, $y \in \mathcal{C}'$.*

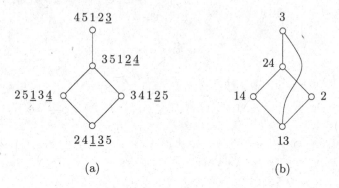

(a)   (b)

Figure 6.3. A left cell and its K-L graph Γ_C.

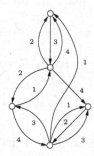

Figure 6.4. The colored K-L graph $\widetilde{\Gamma}_C$ (without loops).

Note that this is a partial order on the set of all left cells of W.

For example, the left cells of S_3 are $\{123\}, \{321\}, \{132, 231\}, \{312, 213\}$. One sees from Figure 6.1 or Figure 6.2 that their left partial order is given by

$$\{321\} \preceq_L \{132, 231\} \preceq_L \{123\},$$

$$\{321\} \preceq_L \{312, 213\} \preceq_L \{123\}.$$

There is, of course, a "right" version of everything that we have done so far, and we will use it in the sequel. From now on, we drop the word "left" everywhere, unless there is risk of confusion.

Proposition 6.2.7 *If* $x \preceq_L y$, *then* $D_R(x) \supseteq D_R(y)$. *Consequently, if* $x \sim_L y$, *then* $D_R(x) = D_R(y)$.

Proof. We may assume that $x \xrightarrow{\overline{\mu}}_s y$ is an edge in $\widetilde{\Gamma}_{(W,S)}$. There are two cases to consider:

(i) $y < x$. Then, $x = sy$, by Lemma 6.2.4, and hence $D_R(y) \subseteq D_R(x)$.

(ii) $y > x$. Assume that there exists an $s' \in D_R(y) \setminus D_R(x)$. Then, $x \xrightarrow{k}_{s'} y$ would be an edge in the *right* colored K-L graph for some

k, which implies by the "right" version of the above that $y = xs$ and, hence, $D_L(x) \subseteq D_L(y)$. However, this contradicts our hypothesis that $x \xrightarrow{s} y$ is an edge in the left colored K-L graph. $\square$

As a consequence, we conclude that the (right) *descent set* $D_R(\mathcal{C})$ of a left cell $\mathcal{C}$ is well defined by

$$D_R(\mathcal{C}) \stackrel{\text{def}}{=} D_R(x), \quad \text{for any } x \in \mathcal{C}.$$

Now, suppose that (W, S) is finite. Then, translation by the top element w_0 induces various symmetries of K-L graphs and of the left partial order on cells.

Proposition 6.2.8 *If* $x \overset{\overline{\mu}}{\text{---}} y$ *is an edge in* $\Gamma_{(W,S)}$, *then also* $w_0 x \overset{\overline{\mu}}{\text{---}} w_0 y$, $xw_0 \overset{\overline{\mu}}{\text{---}} yw_0$, *and* $w_0 x w_0 \overset{\overline{\mu}}{\text{---}} w_0 y w_0$ *are edges (all with the same label).*

Proof. This follows from Corollary 2.3.3 and Exercise 5.13. $\square$

Although the labels of edges of $\Gamma_{(W,S)}$ are preserved under these maps, the labels of vertices, of course, change.

Proposition 6.2.9 *If* $x \xrightarrow[s]{\overline{\mu}} y$, $x \neq y$, *is a directed edge in* $\widetilde{\Gamma}_{(W,S)}$, *then so also are the following:*

(i) $w_0 y \xrightarrow[w_0 s w_0]{\overline{\mu}} w_0 x.$

(ii) $y w_0 \xrightarrow[s]{\overline{\mu}} x w_0.$

(iii) $w_0 x w_0 \xrightarrow[w_0 s w_0]{\overline{\mu}} w_0 y w_0.$

Furthermore, if $x \xrightarrow[s]{\overline{\mu}} x$ *is a loop, then so is* $xw_0 \xrightarrow[s]{-\overline{\mu}} xw_0.$

Proof. Immediate from Exercise 2.10 and the preceding proposition. $\square$

It follows from these facts that for each left cell $\mathcal{C}$ also $w_0\mathcal{C}$, $\mathcal{C}w_0$ and $w_0\mathcal{C}w_0$ are left cells and that they all are isomorphic as edge-labeled graphs.

Corollary 6.2.10 $\Gamma_{\mathcal{C}} \cong \Gamma_{w_0\mathcal{C}} \cong \Gamma_{\mathcal{C}w_0} \cong \Gamma_{w_0\mathcal{C}w_0}.$ $\square$

The vertex labels, however, change according to Exercise 2.10. Similarly, the colored graphs $\widetilde{\Gamma}_{\mathcal{C}}$, $\widetilde{\Gamma}_{w_0\mathcal{C}}$, $\widetilde{\Gamma}_{\mathcal{C}w_0}$, $\widetilde{\Gamma}_{w_0\mathcal{C}w_0}$ are isomorphic as edge-labeled directed graphs, except for a permutation of the set S of "edge colors."

It also follows that the $x \mapsto w_0 x$ and $x \mapsto xw_0$ maps induce fixed-point-free involutions on the set of left cells, which reverse the $\preceq_L$ and $\preceq_R$ orders. Similarly, $x \mapsto w_0 x w_0$ induces an involution on the left cells, which preserves these orders.

6.3   Left cell representations

It is now necessary to briefly return to the Hecke algebra setup reviewed in Section 6.1, but we immediately specialize the discussion, namely

Assume from now on that $q^{\frac{1}{2}} = 1$ and that (W, S) is finite.

The first assumption implies that the Hecke algebra specializes to the group algebra of W : $\mathcal{H}(W, S) \cong \mathbb{Z}[W]$. Also, the involution on $\mathcal{H}$ becomes the identity, since $T_w^{-1} = T_{w^{-1}}$ in $\mathbb{Z}[W]$. Finally, the multiplication formula (6.4) can now be more succinctly written as

$$T_s C_w = \sum_x \overline{\mu}(x, w) C_x, \qquad (6.6)$$

with summation over all $x \in W$ having an s-colored arrow to w in $\widetilde{\Gamma}_{(W,S)}$. The simple form of this rule is the motivation behind the definition of the colored graph $\widetilde{\Gamma}_{(W,S)}$.

The (left) *regular representation* of W is the mapping $\mathrm{Reg}_W : W \to \mathrm{End}_{\mathbb{Z}}(\mathcal{H})$ defined by

$$\mathrm{Reg}_W(w) : D \mapsto T_w D$$

for all $D \in \mathcal{H}$. Our first goal is to express the regular representation in the Kazhdan-Lusztig basis $\{C_x\}_{x \in W}$.

Let $A(w) = (a_{x,y}(w))_{x,y \in W}$ be the matrix of $\mathrm{Reg}_W(w)$ with respect to the Kazhdan-Lusztig basis (so that $T_w C_y = \sum_{x \in W} a_{x,y}(w) C_x$).

Lemma 6.3.1 *If $w = s_1 \ldots s_k$ (with $s_i \in S$), then*

$$a_{x,y}(w) = \sum \overline{\mu}(x_0, x_1)\, \overline{\mu}(x_1, x_2) \ldots \overline{\mu}(x_{k-1}, x_k),$$

where the sum is over all paths $x = x_0 \xrightarrow{s_1} x_1 \xrightarrow{s_2} \cdots \xrightarrow{s_k} x_k = y$ in $\widetilde{\Gamma}_{(W,S)}$.

Proof. Equation (6.6) can be restated as

$$a_{x,y}(s) = \begin{cases} \overline{\mu}(x, y), & \text{if } (x \xrightarrow{s} y) \in A, \\ 0, & \text{otherwise.} \end{cases} \qquad (6.7)$$

Now use the fact that $A(w) = A(s_1) \cdots A(s_k)$. $\square$

Lemma 6.3.2 *Let $\mathcal{C}^1, \ldots, \mathcal{C}^k$ be the left cells of W, labeled so that if $\mathcal{C}^i \preceq_L \mathcal{C}^j$, then $i < j$. Arrange the columns and rows of $A(w)$ cellwise in this order. Then, $A(w)$ has upper-triangular block form.*

Proof. This is clear from Lemma 6.3.1, the definition of left cells, and the definition of $\preceq_L$ for left cells. $\square$

The block form of $A(w)$ is indicated in the following diagram:

	$\mathcal{C}^1$	$\mathcal{C}^2$	$\mathcal{C}^3$		$\mathcal{C}^k$
$\mathcal{C}^1$	$A_{\mathcal{C}^1}(w)$	$*$	$*$	$*$	$*$
$\mathcal{C}^2$	0	$A_{\mathcal{C}^2}(w)$	$*$	$*$	$*$
$\mathcal{C}^3$	0	0	$A_{\mathcal{C}^3}(w)$	$*$	$*$
	0	0	0	$\ddots$	$*$
$\mathcal{C}^k$	0	0	0	0	$A_{\mathcal{C}^k}(w)$

For each left cell $\mathcal{C}$ and $w \in W$, let

$$A_{\mathcal{C}}(w) = (a_{x,y}(w))_{x,y \in \mathcal{C}}$$

be the diagonal-block submatrix. It is clear from the form of the $A(w)$ matrices that $A_{\mathcal{C}}(u) A_{\mathcal{C}}(v) = A_{\mathcal{C}}(uv)$ for all $u, v \in W$; hence, they give a representation.

Definition 6.3.3 *The representation of W given by $(A_{\mathcal{C}}(w))_{w \in W}$ is the Kazhdan-Lusztig representation $KL_{\mathcal{C}}$ determined by the left cell $\mathcal{C}$.*

From now on, we consider all representations to be over the field $\mathbb{C}$ of complex numbers.

Proposition 6.3.4 *Let (W, S) be a finite Coxeter system. Then,*

$$\operatorname{Reg}_W \cong \bigoplus_{\mathcal{C}} KL_{\mathcal{C}},$$

where $\mathcal{C}$ runs over all the left cells of W.

Proof. This is an immediate consequence of Lemma 6.3.2 and Maschke's theorem (see, e.g., [358, p. 21]). $\square$

The Kazhdan-Lusztig representations are, in general, reducible. For example, let $W = B_2$ (the dihedral group of order 8) and $S = \{a, b\}$. Then, the left cells are $\{a, ba, aba\}$, $\{b, ab, bab\}$, $\{abab\}$, and $\{e\}$; but B_2 has five irreducible representations, four of degree 1 and one of degree 2. Hence, the K-L representations corresponding to the two left cells of size 3 must be reducible.

The same pattern holds for *all* dihedral groups; namely $I_2(m)$ has two irreducible representations of degree 1 if m is odd and four such if m is even, and all remaining irreducibles are of degree 2 (see, e.g.,[358, pp. 65-66]). However, $I_2(m)$ always has exactly four left cells, see Exercise 1.

It is easy to see that $\{e\}$ and $\{w_0\}$ are left (and right) cells and that they give the *trivial representation*

$$KL_{\{e\}} = 1 \tag{6.8}$$

and the *alternating representation*

$$KL_{\{w_0\}} = \varepsilon, \tag{6.9}$$

defined by $\varepsilon(w) = (-1)^{\ell(w)}$ (cf. Lemma 1.4.1). We will frequently write ε_w instead of $\varepsilon(w)$.

The w_0-induced symmetries have the following effect on Kazhdan-Lusztig representations.

Proposition 6.3.5 *Let C be a left cell. Then, the following hold:*

(i) $KL_{Cw_0} \cong \varepsilon\, KL_C$.

(ii) $KL_{w_0 C} \cong \varepsilon KL_C$.

(iii) $KL_{w_0 C w_0} \cong KL_C$.

Proof. We begin with part (i). The key fact is that $\widetilde{\Gamma}_{Cw_0}$ is obtained from $\widetilde{\Gamma}_C$ by reversing the direction of all arrows, keeping their color s and weight $\overline{\mu}$, except that the ± 1 weights on loops are switched (cf. Proposition 6.2.9(ii)). This implies on the character level that

$$KL_{Cw_0}(x) = \varepsilon_x\, KL_C(x^{-1}),$$

for all $x \in W$; namely the trace of $A_{Cw_0}(x)$ is the sum of the weights of all directed circuits in $\widetilde{\Gamma}_{Cw_0}$ beginning and ending in some $y \in Cw_0$ and whose color sequence is $(s_1, s_2, \ldots, s_k)$ for some fixed expression $x = s_1 s_2 \ldots s_k$. Similarly, $KL_C(x^{-1})$ is the sum of the weights of directed circuits in $\widetilde{\Gamma}_C$ whose color sequence is $(s_k, s_{k-1}, \ldots, s_1)$. By the previous remark, these quantities are equal, except possibly for the sign. Whether the sign will change depends on the distribution of the number of $(+1)$-labeled and (-1)-labeled loops traversed. However, since $\widetilde{\Gamma}_{(W,S)}$ without its loops is a bipartite graph (edges connect elements of even length with elements of odd length), and hence every circuit with its loops removed is of even length, a change of sign will take place for each individual path if and only if k is odd (i.e., if $\varepsilon_x = -1$).

Now use that

$$KL_C(x^{-1}) = \overline{KL_C(x)} = KL_C(x),$$

where the last equality is true because the matrices, and hence character values, are real. Consequently, the characters agree, and part (i) is proved.

For part (iii), one observes (using Proposition 6.2.9(iii)) that $\widetilde{\Gamma}_{w_0 C w_0}$ is obtained from $\widetilde{\Gamma}_C$ by applying the operator $x \mapsto w_0 x w_0$ to all nodes and all edge colors, keeping the edge weights $\bar{\mu}$. Therefore, the directed $(s_1, s_2, \ldots, s_k)$-colored circuits based at $y \in \mathcal{C}$ are in bijection with the $(w_0 s_1 w_0, w_0 s_2 w_0, \ldots, w_0 s_k w_0)$-colored circuits based at $w_0 y w_0 \in w_0 \mathcal{C} w_0$, and this bijection preserves weight. Hence, the trace of $A_{\mathcal{C}}(x)$ equals that of $A_{w_0 \mathcal{C} w_0}(w_0 x w_0)$, implying (since characters are class functions)

$$KL_{\mathcal{C}}(x) = KL_{w_0 \mathcal{C} w_0}(w_0 x w_0) = KL_{w_0 \mathcal{C} w_0}(x).$$

Part (ii) is implied by parts (i) and (iii) jointly, since $w_0 \mathcal{C} = w_0(\mathcal{C} w_0) w_0$. $\square$

In the sequel, for any representation χ of a parabolic subgroup W_J we introduce the following simplified notation for the induced representation of W:

$$\mathrm{Ind}_J^S[\chi] \overset{\mathrm{def}}{=} \mathrm{Ind}_{W_J}^W[\chi].$$

We need only the most basic facts about induced representations, as can be found, for example, in [358, Section 3.1]. In particular, the character formula

$$\mathrm{Ind}_J^S[\chi](w) = \frac{1}{|W_J|} \sum_{y \in W} \dot{\chi}(y^{-1} w y) \tag{6.10}$$

is useful. Here, $w \in W$, and $\dot{\chi}(u)$ is declared equal to $\chi(u)$ if $u \in W_J$ and equal to zero otherwise.

For any subset $A \subseteq W$, define two elements T_A and $\overline{T}_A$ of $\mathcal{H} \cong \mathbb{Z}[W]$ by

$$T_A = \sum_{w \in A} T_w \qquad \text{and} \qquad \overline{T}_A = \sum_{w \in A} \varepsilon_w T_w. \tag{6.11}$$

Let $A_1, A_2, \ldots, A_k$ be the left cosets of W_J (for some $J \subseteq S$). Then, W acts via left multiplication $x : T_A \mapsto x T_A$ on the submodule $\langle T_{A_1}, \ldots, T_{A_k} \rangle$, permuting its basis elements. This representation, the "left coset action," coincides with the induced representation $\mathrm{Ind}_{W_J}^W[1]$.

Similarly, W acts via $x : \overline{T}_A \mapsto \varepsilon_x x \overline{T}_A$ on the submodule $\overline{\mathcal{H}}_J = \langle \overline{T}_{A_1}, \ldots, \overline{T}_{A_k} \rangle$ by permuting and sign-changing its basis elements. This "signed left coset action" is in the same way identified with the representation $\mathrm{Ind}_J^S[\varepsilon] \cong \varepsilon \cdot \mathrm{Ind}_J^S[1]$. This is all easily seen from formula (6.10).

For example, let $W = S_3$, $S = \{a, b\}$ and $J = \{b\}$. Then, $\overline{T}_{A_1} = T_e - T_b$, $\overline{T}_{A_2} = T_{ab} - T_a$, and $\overline{T}_{A_3} = T_{ba} - T_{aba}$. Two examples of the action of W on $\overline{\mathcal{H}}_J$ expressed in this basis are

$$ab \longmapsto \begin{pmatrix} 0 & 0 & 1 \\ 1 & 0 & 0 \\ 0 & 1 & 0 \end{pmatrix}, \qquad aba \longmapsto \begin{pmatrix} 0 & 0 & -1 \\ 0 & -1 & 0 \\ -1 & 0 & 0 \end{pmatrix}.$$

The remainder of this section is devoted to showing how the induced representations $\mathrm{Ind}_J^S[1]$ and $\mathrm{Ind}_J^S[\varepsilon]$ are related to Kazhdan-Lusztig representations. We first need some preparatory lemmas.

Lemma 6.3.6 *Let $y \in \mathcal{D}_J^S$ and $x \le y$. Then, the following hold:*

(i) $a \le y$, *for all $a \in xW_J$. In particular, $[e, y]$ is a union of left cosets of W_J.*

(ii) $P_{a,y}(q) = P_{x,y}(q)$, *for all $a \in xW_J$.*

Proof. Use that the upper projection $\overline{P}^J : W \to \mathcal{D}_J^S$ is order-preserving (Exercise 2.16). Then, $a \le \overline{P}^J(a) = \overline{P}^J(x) \le \overline{P}^J(y) = y$, for all $a \in xW_J$.

For part (ii), let $z = \overline{P}^J(x) = \overline{P}^J(a)$. Then $x < xs_1 < xs_1s_2 < \cdots < xs_1s_2 \ldots s_k = z$ with all $s_i \in J$, and $J \subseteq D_R(z)$. Hence, $P_{x,y}(q) = P_{z,y}(q)$ follows by repeated use of property (6.5). Similarly, $P_{a,y}(q) = P_{z,y}(q)$. $\square$

Lemma 6.3.7   (i) *If $y \in \mathcal{D}_J^S$, then $C_y \in \overline{\mathcal{H}}_J$.*

(ii) *$\{C_y\}_{y \in \mathcal{D}_J^S}$ is a basis for $\overline{\mathcal{H}}_J$.*

Proof. Since $q = 1$, we have from equation (6.1) that $C_y = \sum_{x \in [e,y]} \varepsilon_x \varepsilon_y P_{x,y}(1) T_x$. Now, $[e, y]$ is a (disjoint) union of left cosets of W_J, say $A_1, \ldots, A_r$. Therefore, by Lemma 6.3.6,

$$C_y = \varepsilon_y \sum_{i=1}^{r} P_{A_i,y}(1) \sum_{x \in A_i} \varepsilon_x T_x$$

$$= \varepsilon_y \sum_{i=1}^{r} P_{A_i,y}(1) \overline{T}_{A_i} \in \overline{\mathcal{H}}_J, \qquad (6.12)$$

where $P_{A_i,y}(1)$ denotes the common value of $P_{x,y}(1)$ for $x \in A_i$.

For part (ii), the relation (6.12) is invertible over $\mathbb{Z}$, since it is triangular with 1s on the diagonal with respect to any ordering $\{y_1, \ldots, y_r\}$ of $\mathcal{D}_J^S$ such that if $y_i < y_j$, then $i < j$. (Note here that the index sets can be considered to be the same, since $\mathcal{D}_J^S$ consists precisely of the maximal elements of $A_1, \ldots, A_r$.) $\square$

We are now ready to state the promised result on induced characters. The two sums run over left cells $\mathcal{C}$.

Theorem 6.3.8 *Let $J \subseteq S$. Then,*

$$\mathrm{Ind}_J^S[1] \cong \bigoplus_{D_R(\mathcal{C}) \supseteq J} \varepsilon \, KL_{\mathcal{C}} \cong \bigoplus_{D_R(\mathcal{C}) \subseteq S \setminus J} KL_{\mathcal{C}}.$$

Proof. The map $\mathcal{C} \mapsto w_0\mathcal{C}$ is a bijection between those left cells whose right descent set contains J and those whose right descent set is contained in $S \setminus J$. Hence, the second equivalence is directly implied by Proposition 6.3.5(ii).

We will prove the first equivalence, namely

$$\text{Ind}_J^S[\varepsilon] \cong \bigoplus_{D_R(\mathcal{C}) \supseteq J} KL_{\mathcal{C}},$$

by expressing the action of W on $\overline{\mathcal{H}}_J$ in the $\{C_y\}_{y \in \mathcal{D}_J^S}$ basis. The argument runs parallel to that by which Proposition 6.3.4 is obtained from Lemma 6.3.2. Therefore, we will only sketch it, leaving the details to the reader.

Order the cells $\mathcal{C}$ satisfying $D_R(\mathcal{C}) \supseteq J$ by a linear extension of the left order $\preceq_L$ of cells. Write the matrices of the $\text{Ind}_J^S[\varepsilon]$ representation in the $\{C_y\}_{y \in \mathcal{D}_J^S}$ basis with rows and columns ordered cellwise, according to the chosen linear ordering of cells. Then, as in the proof of Proposition 6.3.4, the matrices assume upper-triangular block form. Thus, $\text{Ind}_J^S[\varepsilon]$ decomposes as claimed. $\square$

Note that for $J = \emptyset$, the theorem specializes to Proposition 6.3.4, and for $J = S$, it specializes to formulas (6.8) and (6.9). [*Remark:* The smaller sums $\oplus_{D_R(\mathcal{C})=J} KL_{\mathcal{C}}$ arise as homology representations from W's action on the type-selected subcomplexes of its Coxeter complex; see Exercise 12(c).]

6.4   Knuth paths

Here and in the next two sections, the discussion is specialized to the symmetric groups. The main goal is the proof in the following section that the Kazhdan-Lusztig construction gives the irreducible representations for S_n. In preparation for that, we need to work out some of the connections between edges in K-L graphs and (dual) Knuth equivalence. We will assume familiarity with the material on tableaux reviewed in Appendix A3.

Let $W = S_n$ with the standard Coxeter generators $s_i = (i, i+1)$, $i = 1, \ldots, n-1$. Let $x, y \in S_n$ and $1 < i < n$. The following defines a refinement of elementary Knuth equivalence:

$$x \overset{i}{\underset{K}{\approx}} y \quad \overset{\text{def}}{\iff} \quad x \underset{K}{\approx} y \text{ and } x_j = y_j \text{ if } |j - i| \geq 2. \tag{6.13}$$

This means that x and y differ by a Knuth relation, $bac = bca$ or $acb = cab$ for $a < b < c$, occurring in positions $i - 1$, i and $i + 1$. We call this a *Knuth step of type i*. For example, $4735162 \overset{4}{\underset{K}{\approx}} 4731562$.

Lemma 6.4.1 *Let $x, y \in S_n$, $\ell(x) < \ell(y)$, and $1 < i < n$. Then, the following are equivalent:*

(i) $x \overset{i}{\underset{K}{\approx}} y$.

(ii) $xs < x < xs' = y < ys$, with $\{s, s'\} = \{s_{i-1}, s_i\}$; see Figure 6.5(a).

(iii) *There exist antiparallel edges between x and y, colored by s_{i-1} and s_i, in the right colored K-L graph of S_n; see Figure 6.5(b).*

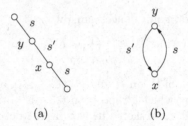

Figure 6.5. Illustration for Lemma 6.4.1.

Proof. We leave the verification to the reader. Lemma 6.2.4 is of use for showing (iii) $\Rightarrow$ (ii). $\square$

Dualizing by the left-right symmetry and dropping the reference to position i we deduce the following.

Lemma 6.4.2 *Let* $x, y \in S_n$, $\ell(x) < \ell(y)$. *Then, the following are equivalent:*

(i) $x \underset{dK}{\approx} y$.

(ii) $sx < x < s'x = y < sy$, *for some* $s, s' \in S$.

(iii) *There exist antiparallel edges between* x *and* y *in the left colored K-L graph* $\widetilde{\Gamma}_{S_n}$. $\square$

In particular, it follows that dual Knuth equivalence implies left equivalence in the sense of Kazhdan and Lusztig.

Corollary 6.4.3 $x \underset{dK}{\sim} y$ *implies* $x \underset{L}{\sim} y$. $\square$

The relation $x \underset{K}{\overset{i}{\approx}} y$ has the following influence on the right tableaux (recording tableaux) under the Robinson-Schensted correspondence.

Lemma 6.4.4 *Let* $x, y \in S_n$ *and suppose that* $x \underset{K}{\overset{i}{\approx}} y$. *Then, the following hold:*

(i) $Q(x)$ *and* $Q(y)$ *differ by a transposition* s_{i-1} *or* s_i.

(ii) $Q(x)$ *uniquely determines* $Q(y)$.

Before embarking on the proof, let us have a look at a small example. Let $x = 4735162$ and $y = 4731562$. Then, $x \underset{K}{\overset{4}{\approx}} y$ and

$$Q(x) = \begin{array}{|c|c|c|} \hline 1 & 2 & 6 \\ \hline 3 & 4 \\ \cline{1-2} 5 & 7 \\ \cline{1-2} \end{array} \qquad Q(y) = \begin{array}{|c|c|c|} \hline 1 & 2 & 6 \\ \hline 3 & 5 \\ \cline{1-2} 4 & 7 \\ \cline{1-2} \end{array}$$

These tableaux differ by the transposition $s_4 = (4,5)$. Note that $s_3 = (3,4)$ is illegal in both tableaux, so knowing that $i = 4$, one sees that either tableau uniquely determines the other.

Proof. Since $x \underset{K}{\overset{i}{\approx}} y$ and using that x and $Q(x)$ have the same descent sets, we see that $|D(Q(x)) \cap \{i-1, i\}| = 1$, and similarly for y. This shows that (i) implies (ii).

To prove (i), we have to distinguish two cases.

Case 1: $y = xs_i$. This means that we have a Knuth relation of the form $bac = bca$. Let $P(x_1 x_2 \ldots x_j)$ denote the left tableau obtained after insertion of the first j entries. Then, $P(x_1 \ldots x_j) = P(y_1 \ldots y_j)$ for $j = 1, \ldots, i-1$, and also (since $x \underset{K}{\approx} y$ implies $P(x) = P(y)$) for $j = i+1, \ldots, n$. Hence, by the formation rule for right tableaux, $Q(x)$ and $Q(y)$ differ by the transposition s_i.

Case 2: $y = xs_{i-1}$. Then, the Knuth relation is $acb = cab$. If the "bumping paths" of the insertion of a in $P(x_1 \ldots x_{i-2})$ and of the insertion of c in $P(x_1 \ldots x_{i-2})$ intersect in some row, then clearly $i-1$ is in the same position in $Q(x)$ as in $Q(y)$. So reasoning as in Case 1, we conclude that $Q(x)$ and $Q(y)$ differ by the transposition s_i. If the "bumping paths" of a and c do not intersect, then the insertions of a and c commute and, hence, $Q(x)$ and $Q(y)$ differ by the transposition s_{i-1}. $\square$

Corollary 6.4.5. *Let $x, y \in S_n$. If there exists a labeled Knuth path*

$$x \underset{K}{\overset{i_1}{\approx}} x_1 \underset{K}{\overset{i_2}{\approx}} x_2 \underset{K}{\overset{i_3}{\approx}} \cdots \underset{K}{\overset{i_k}{\approx}} x_k = y,$$

then $Q(x)$ uniquely determines $Q(y)$. $\square$

For $1 < i < n$, define

$$\mathcal{DES}_R(i) \overset{\text{def}}{=} \{x \in S_n : |D_R(x) \cap \{i-1, i\}| = 1\}. \qquad (6.14)$$

Note the following equivalent characterizations:

$$\mathcal{DES}_R(i) = \{x \in S_n : x \underset{K}{\overset{i}{\approx}} y \text{ for some } y \in S_n\}$$

$$= \{x \in S_n : x_{i-1} x_i x_{i+1} \text{ is not monotone}\}.$$

Definition 6.4.6 *If $x \in \mathcal{DES}_R(i)$, then we denote by x^* (or by x^{*_i}) the unique permutation such that $x^* \underset{K}{\overset{i}{\approx}} x$.*

This defines an involution of the set $\mathcal{DES}_R(i)$ with the following important property for the K-L graph.

Lemma 6.4.7 (Edge Transport) *Suppose $x, y \in \mathcal{DES}_R(i)$ and $\{x, y\}$ is an edge in Γ_{S_n}. Then, $\{x^*, y^*\}$ is also an edge in Γ_{S_n} and $\overline{\mu}(x^*, y^*) = \overline{\mu}(x, y)$.*

Proof. We omit the proof, which amounts to somewhat tedious case-by-case checking. See Exercise 5 or [322, pp. 175–176]. □

Corollary 6.4.8 *If* $x \in \mathcal{DES}_R(i)$ *and* $x \underset{L}{\sim} y$, *then* $y \in \mathcal{DES}_R(i)$ *and* $x^* \underset{L}{\sim} y^*$.

Proof. Since $x \underset{L}{\sim} y$, we have that $D_R(x) = D_R(y)$ by Proposition 6.2.7 and, hence, that $y \in \mathcal{DES}_R(i)$. Also, we have a circular path

$$x \to x_1 \to \cdots \to y \to y_1 \to \cdots \to x$$

in $\widetilde{\Gamma}_{S_n}$. All elements of this path are left equivalent and, hence, by what we have just shown, all of these elements are in $\mathcal{DES}_R(i)$. Hence, by Lemma 6.4.7, there is a corresponding circular path $x^*—x_1^*—\cdots— y^*— y_1^*— \cdots — x^*$ in the (undirected) K-L graph Γ_{S_n}. However, $x \overset{i}{\underset{K}{\approx}} x^*$ implies that $P(x) = P(x^*)$, and hence $D_L(x) = D(P(x)) = D(P(x^*)) = D_L(x^*)$, and similarly for all x_i and y_j. So there is also a directed path

$$x^* \to x_1^* \to \cdots \to y^* \to y_1^* \to \cdots \to x^*$$

in $\widetilde{\Gamma}_{S_n}$. Hence, $x^* \underset{L}{\sim} y^*$. □

6.5 Kazhdan-Lusztig representations for S_n

Let us begin by summarizing some basic facts about the irreducible representations of the symmetric group S_n. As earlier all representations are over $\mathbb{C}$. Proofs and further discussion of the classical theory can be found in the books by James [314], James and Kerber [315], or Sagan [450].

The following is a list of the facts that we need:

(1) *The irreducible representations* ρ_λ *are naturally indexed by partitions* $\lambda \vdash n$.

(2) *The dimension of* ρ_λ *is* $f^\lambda \overset{\text{def}}{=} \#SYT_\lambda$.

(3) *Young's rule:* $\text{Ind}_J^S[1] \cong \bigoplus_{\lambda \vdash n} \#\{T \in SYT_\lambda : D(T) \subseteq S \setminus J\} \rho_\lambda$.

The word "natural" is used in fact (1) for the following reason. Since the conjugacy classes of S_n are in bijection with the partitions of n via cycle decomposition, it is a consequence of general theory that the irreducible representations are equinumerous with the partitions of n. However, this does not lead to any particular favored way of matching these objects. There are several classical constructions of irreducible representations of S_n, due to Frobenius, Schur, Young, Specht, and others, and they all turn out to produce the same matching of representations to partitions. Thus, it seems motivated to talk about the existence of a canonical "natural" indexing of these representations by partitions.

The key fact of the three mentioned ones is Young's rule; namely, setting $J = \emptyset$, it specializes to $\operatorname{Reg}_{S_n} \cong \bigoplus_\lambda f^\lambda \rho_\lambda$, which implies fact (2). The numbers $K_{\lambda, J} = \#\{T \in SYT_\lambda : D(T) \subseteq S - J\}$ are called *Kostka numbers*. Subsets J corresponding to partitions can be chosen so that the Kostka matrix $(K_{\lambda, J})$ is upper-triangular with ones on the diagonal. Hence, the relation stated in fact (3) is invertible and determines ρ_λ as a function of the characters $\operatorname{Ind}_J^S[1]$. This can be taken as a definition of the "natural" labeling of irreducible representations by partitions.

We now return to Kazhdan-Lusztig representations and show that for S_n, each left cell can be associated with a partition in such a way that the corresponding representation is the irreducible one belonging to that partition. The construction proceeds in close contact with the Robinson-Schensted correspondence.

Theorem 6.5.1 *For each left cell C in S_n there exists a tableau T such that $C = \{x \in S_n : Q(x) = T\}$.*

Proof. The theorem says that the left cells are the dual Knuth classes. We already know from Corollary 6.4.3 that $Q(x) = Q(y)$ implies that $x \underset{L}{\sim} y$, so it remains to prove the converse. In graph-theoretic language, it is to be shown that a strongly connected component of $\widetilde{\Gamma}_{S_n}$ is actually connected already by sequences of antiparallel edges. To have some idea of what kind of situation we must deal with, we suggest taking a look at Figure 6.4.

Suppose that $x \underset{L}{\sim} y$. The first observation to be made is that the Knuth classes of x and of y produce the same collections of right descent sets:

$$\{D_R(z) : z \underset{K}{\sim} x\} = \{D_R(z) : z \underset{K}{\sim} y\}. \tag{6.15}$$

This is seen as follows. Let $x \overset{i_1}{\underset{K}{\approx}} x_1 \overset{i_2}{\underset{K}{\approx}} x_2 \overset{i_3}{\underset{K}{\approx}} \cdots \overset{i_j}{\underset{K}{\approx}} x_j$ be a Knuth path. Since $x \in \mathcal{DES}_R(i_1)$ and $x \underset{L}{\sim} y$, then (by Corollary 6.4.8) $y \in \mathcal{DES}_R(i_1)$ and $x_1 = x^{*i_1} \underset{L}{\sim} y^{*i_1}$. Continuing in this way, we see that there exists a unique Knuth path $y = y_0 \overset{i_1}{\underset{K}{\approx}} y_1 \overset{i_2}{\underset{K}{\approx}} y_2 \overset{i_3}{\underset{K}{\approx}} \cdots \overset{i_j}{\underset{K}{\approx}} y_j$, where $y_l \overset{\text{def}}{=} y_{l-1}^{*i_l}$. Furthermore, $x_l \underset{L}{\sim} y_l$, for all $l \in [j]$. Hence, by Proposition 6.2.7, $D_R(x_j) = D_R(y_j)$, which proves the claim.

Let A be the lexicographically last set of $\{D_R(z) : z \underset{K}{\sim} x\}$, and choose x_A such that $x_A \underset{K}{\sim} x$ and $D_R(x_A) = A$. We claim that

$Q(x_A)$ *is a row superstandard tableau and x_A is uniquely determined.*

To prove this, let $\lambda = \text{shape}(P(x))$. We have, using Fact A3.4.1, that $\{D_R(z) : z \underset{K}{\sim} x\} = \{D(T) : T \in SYT_\lambda\}$. Clearly, the lexicographically last set in the family $\{D(T) : T \in SYT_\lambda\}$ is that of the row superstandard tableau of shape λ, and having lexicographically last descent set within SYT_λ characterizes this tableau.

Now, let

$$x \overset{i_1}{\underset{K}{\approx}} x_1 \overset{i_2}{\underset{K}{\approx}} x_2 \overset{i_3}{\underset{K}{\approx}} \cdots \overset{i_k}{\underset{K}{\approx}} x_A \qquad (6.16)$$

be a Knuth path. Arguing as in the proof of equation (6.15), we see that there is a unique Knuth path

$$y \overset{i_1}{\underset{K}{\approx}} y_1 \overset{i_2}{\underset{K}{\approx}} y_2 \overset{i_3}{\underset{K}{\approx}} \cdots \overset{i_k}{\underset{K}{\approx}} y_A, \qquad (6.17)$$

with $D_R(y_A) = D_R(x_A) = A$. Then, equation (6.15) and the claim imply that $Q(y_A) = Q(x_A)$ is the same superstandard tableau. Hence, Corollary 6.4.5 applied to paths (6.16) and (6.17) shows that $Q(x) = Q(y)$. $\square$

The preceding result implies that we can associate a partition called *shape* with each left cell by setting

$$\text{shape}(\mathcal{C}) \overset{\text{def}}{=} \text{shape}(Q(x)), \quad \text{for any } x \in \mathcal{C}. \qquad (6.18)$$

Theorem 6.5.2 *If $\mathcal{C}_1$ and $\mathcal{C}_2$ are left cells of the same shape, then $\Gamma_{\mathcal{C}_1} \cong \Gamma_{\mathcal{C}_2}$ (as labeled graphs).*

Proof. Suppose that $\mathcal{C}_i = \{x \in S_n : Q(x) = Q_i\}$, for $i = 1, 2$, where Q_1 and Q_2 are two tableaux of equal shape. We will show that the map $(P, Q_1) \mapsto (P, Q_2)$ induces an isomorphism $\Gamma_{\mathcal{C}_1} \cong \Gamma_{\mathcal{C}_2}$. Permutations are here denoted by the corresponding pair of tableaux under the Robinson-Schensted correspondence $x \leftrightarrow (P, Q)$. It is immediately clear that the map is a bijection of the vertices, and vertex labels are preserved since $D_L(x) = D(P(x))$. We must show that also edges and their labels are preserved.

Suppose that $(P_1, Q_1) - (P_2, Q_1)$ is an edge in $\Gamma_{\mathcal{C}_1}$ with label $\bar{\mu}$. Since (P_1, Q_1) and (P_1, Q_2) have the same left tableau, we can connect them with a Knuth path:

$$(P_1, Q_1) \overset{i_1}{\underset{K}{\approx}} (P_1, Q^{(1)}) \overset{i_2}{\underset{K}{\approx}} (P_1, Q^{(2)}) \overset{i_3}{\underset{K}{\approx}} \cdots \overset{i_k}{\underset{K}{\approx}} (P_1, Q_2).$$

Since (P_1, Q_1) and (P_2, Q_1) have the same right descent set and $(P_1, Q_1) \in \mathcal{DES}_R(i_1)$, it follows from the Edge Transport Lemma 6.4.7 that $(P_1, Q^{(1)})$ $- (P_2, Q_1)^{*i_1}$ is an edge in Γ_{S_n} with label $\bar{\mu}$. However, $(P_2, Q_1)^{*i_1} = (P_2, Q^{(1)})$, as can be seen from Lemma 6.4.4. The argument can now be repeated, and after k steps, we reach the conclusion that $(P_1, Q_2) - (P_2, Q_2)$ is an edge in $\Gamma_{\mathcal{C}_2}$ with label $\bar{\mu}$. $\square$

By the preceding, it is well defined to associate K-L graphs Γ_λ and $\widetilde{\Gamma}_\lambda$ to a partition λ and, hence, also to associate a K-L representation KL_λ, by taking any left cell $\mathcal{C}$ of shape λ and defining

$$\Gamma_\lambda \stackrel{\text{def}}{=} \Gamma_{\mathcal{C}}, \quad \widetilde{\Gamma}_\lambda \stackrel{\text{def}}{=} \widetilde{\Gamma}_{\mathcal{C}}, \quad KL_\lambda \stackrel{\text{def}}{=} KL_{\mathcal{C}}. \tag{6.19}$$

We have finally reached the main conclusion: that the Kazhdan-Lusztig representations for S_n are the irreducible ones and that the supporting combinatorics induces the correct labeling by partitions.

Theorem 6.5.3 $KL_\lambda \cong \rho_\lambda$, for all $\lambda \vdash n$.

Proof. It follows from Theorems 6.3.8, 6.5.1, and 6.5.2 that

$$\operatorname{Ind}_J^S[1] = \bigoplus_{\lambda \vdash n} \#\{Q \in SYT_\lambda : D(Q) \subseteq S \setminus J\} \, KL_\lambda.$$

Hence, Young's rule implies the conclusion. $\square$

6.6   Left cells for S_n

The Kazhdan-Lusztig cells provide a very compact graphical language for writing down the irreducible representations of S_n. Integer matrices representing the adjacent transpositions are read off directly from the graph, and matrices for general group elements are obtained from them via multiplication, using reduced decompositions. The rule for producing the matrix $A(s_i)$ representing the adjacent transposition $s_i = (i, i+1)$ is given by equation (6.7), which can be restated as follows. Let $x_1, \ldots, x_g$ be the nodes of the K-L graph. Then, the (j, k)-entry of $A(s_i)$ is

$$a_{j,k}(s_i) = \begin{cases} +1, & \text{if } j = k \text{ and } i \notin D_L(x_k), \\ -1, & \text{if } j = k \text{ and } i \in D_L(x_k), \\ \overline{\mu}(x_j, x_k), & \text{if } \{x_j, x_k\} \in E, \, i \in D_L(x_j) \setminus D_L(x_k), \\ 0, & \text{otherwise.} \end{cases} \tag{6.20}$$

For instance, take the K-L graph of shape $(3, 2)$, obtained from Figure 6.3; it is also displayed in the introduction to this chapter. With rows and columns ordered (in terms of the labels of the nodes) 24, 14, 3, 2, 13, we obtain

$$s_1 \longmapsto \begin{pmatrix} 1 & 0 & 0 & 0 & 0 \\ 1 & -1 & 0 & 0 & 0 \\ 0 & 0 & 1 & 0 & 0 \\ 0 & 0 & 0 & 1 & 0 \\ 0 & 0 & 1 & 1 & -1 \end{pmatrix}.$$

To construct the K-L graph of shape λ, we know from the previous section that we can take as vertex set $\{x \in S_n : Q(x) = T\}$, for any tableau

T of shape λ. However, there is a canonical choice, namely the reading tableau T_λ (defined in Appendix A3.5). With this choice, the vertices of Γ_λ become the reading words w_T, that is, in essence the tableaux T of shape λ themselves.

For instance, we can now think of the K-L graph $\Gamma_{(3,2)}$ constructed in Figure 6.3 entirely in terms of tableaux; see Figure 6.6.

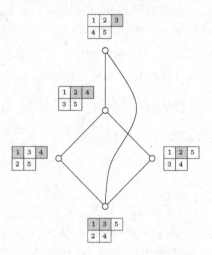

Figure 6.6. The K-L graph $\Gamma_{(3,2)}$.

The vertex labels are easily read off from the tableaux, since $D_L(w_T) = D(T)$; these elements are in shaded cells in Figure 6.6. So, a partial construction of Γ_λ can, in general, readily be done by purely combinatorial means, giving the labeled vertices (tableaux of shape λ and their descent sets) and the labeled Bruhat edges (with labels equal to 1). The non-Bruhat edges of length 3 (labeled by 1) can also be given a combinatorial description; see Exercise 2.29. The problem with the whole construction, from a practical point of view, is how to decide the general non-Bruhat edges and their labels. For this — at the present state of knowledge — the Kazhdan-Lusztig polynomials (or at least their leading coefficient) have to be computed. It is an open problem to describe the non-Bruhat edges (of length ≥ 5) and their labels in terms of tableaux.

A combinatorial description *is* possible for the special case of hook shapes (i.e., partitions $\lambda = (n-k, 1^k)$). This is because such graphs Γ_λ have no non-Bruhat edges. Recall from Section 2.4 that we denote by $L(k, m-k)$ the poset of k-element subsets of $[m]$ ordered componentwise; see, e.g., Figure 2.7.

Proposition 6.6.1 *Suppose $\lambda = (n-k, 1^k)$. Then, Γ_λ is isomorphic (as a labeled graph) to the Hasse diagram of the poset $L(k, n-1-k)$. The label of a node equals the corresponding set.*

Proof. The reading words of hook shape tableaux are $\{x \in S_n : D_R(x) = [k]\}$. These words are determined by the initial substring $x_1 > x_2 > \cdots > x_k > x_{k+1} = 1$. Hence, we get an identification $x \leftrightarrow \{x_k - 1, x_{k-1} - 1, \ldots, x_1 - 1\}$ between nodes of $\Gamma_{(n-k,1^k)}$ and k-element subsets of $[n-1]$, and one easily sees that $D_L(x) = \{x_k - 1, x_{k-1} - 1, \ldots, x_1 - 1\}$. The Bruhat order relations are, because of Proposition 2.4.8, the same as the ordering of k-sets in $L(k, n-1-k)$.

It remains only to prove that there are no non-Bruhat edges. Let $x = w_{T_1}$ and $y = w_{T_2}$ be the reading words of two tableaux T_1 and T_2 of shape λ, and assume that $x < y$. Let j be the largest entry which is differently placed in T_1 and in T_2. Then, because of the interpretation of $x < y$ given earlier, we have that $y^{-1}(j) < k+1 < x^{-1}(j)$ and also that $x^{-1}(j-1) < x^{-1}(j)$, and $y^{-1}(j) < y^{-1}(j-1)$. Hence, s_{j-1} has the property that $s_{j-1}y < y$ and $s_{j-1}x > x$, and it follows from Lemma 6.2.2(iii) that $\{x, y\}$ is an edge only if $\ell(x, y) = 1$. $\square$

For example, Figure 2.7 can now, via the isomorphism, be interpreted as showing the K-L graph $\Gamma_{(4,1,1,1)}$.

A combinatorial description of K-L graphs is also known for two-row or two-column shapes; see [325] and [347]. These, together with the hook shapes, are, as far as we know, the only classes of partitions for which an entirely combinatorial construction of K-L graphs is known.

The K-L graphs Γ_λ are easy to construct by hand calculation for $\lambda \vdash n \le 6$. For example, the graphs Γ_λ for all non-hook partitions $\lambda \vdash 6$ are shown in Figure 6.7 (for each pair of conjugate shapes λ and λ', only one is displayed — for reasons that will be explained by the next theorem). An alternative rendition of the graph in Figure 6.7(c), viewing it embedded in Bruhat order, is shown in Figure 6.8.

A glance at these small examples suggests that each K-L graph has a nontrivial symmetry of order 2. (Also, Proposition 6.6.1 supports this observation.) The vertex labels change according to a rule that is easy to guess from the examples. It is a remarkable fact that this symmetry of K-L graphs is induced by the tableau operation of evacuation (explained in Appendix A3.8). The graphs in Figure 6.7 are drawn so that the evacuation-induced involution coincides with the obvious reflection symmetry in a vertical axis. Furthermore, the tableau operation of transposition (Appendix A3.9) also has meaning for K-L graphs, as we now show.

Theorem 6.6.2 *Fix $\lambda \vdash n$ and identify the nodes of Γ_λ with SYT_λ.*

(i) *Evacuation $P \mapsto e(P)$ induces an order 2 automorphism of Γ_λ as an edge-labeled graph. The change of vertex labels is $D(e(P)) = \{n-i : i \in D(P)\}$.*

(ii) *Transposition $P \to P'$ induces an isomorphism $\Gamma_\lambda \cong \Gamma_{\lambda'}$ as edge-labeled graphs. The vertex labels are related by $D(P') = [n-1] \setminus D(P)$.*

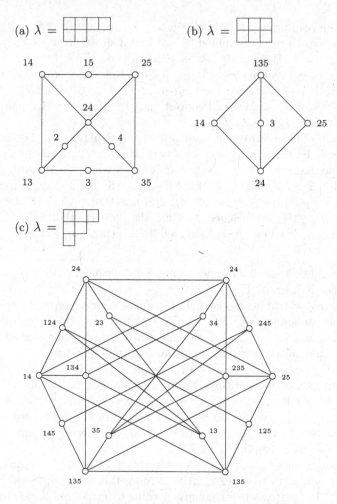

Figure 6.7. The K-L graphs Γ_λ for non-hook shapes $\lambda \vdash 6$.

Proof. *Part (i):* Fix a right tableau Q so that our left cell $\mathcal{C}$ of shape λ is realized as $\mathcal{C} = \{x \in S_n : x \leftrightarrow (P,Q), \ P \in SYT_\lambda\}$. Let $\mathcal{C}^E = \{x \in S_n : x \leftrightarrow (P, e(Q)), \ P \in SYT_\lambda\}$. The mapping $\alpha : x \longmapsto w_0 x w_0$, or in terms of Robinson-Schensted (Fact A3.9.1) $\alpha : (P,Q) \longmapsto (e(P), e(Q))$, induces an isomorphism $\alpha : \Gamma_\mathcal{C} \longrightarrow \Gamma_{\mathcal{C}^E}$ as edge-labeled graphs (Corollary 6.2.10). On the other hand, we know from the proof of Theorem 6.5.2 that $\beta : (R, e(Q)) \longmapsto (R, Q)$ induces isomorphism $\beta : \Gamma_{\mathcal{C}^E} \longrightarrow \Gamma_\mathcal{C}$. Hence, $\beta \circ \alpha : (P,Q) \longmapsto (e(P), Q)$ gives an isomorphism $\Gamma_\mathcal{C} \longrightarrow \Gamma_\mathcal{C}$ as claimed. Finally, $D(e(P)) = D_L(w_0 x w_0) = w_0 D_L(x) w_0$.

Part (ii): The mapping $\gamma : x \longmapsto x w_0$, or equivalently (by Fact A3.9.1) $\gamma : (P,Q) \longmapsto (P', e(Q'))$, gives an isomorphism of left cells by Corollary

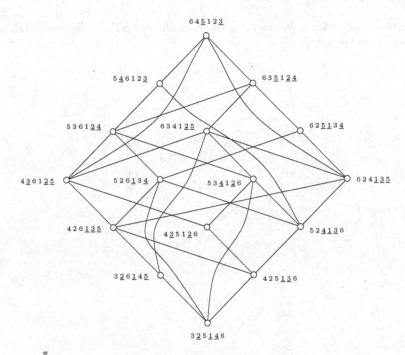

Figure 6.8. Bruhat order version of graph in Figure 6.7(c).

6.2.10. Since $\mathrm{shape}(e(Q')) = \lambda'$, this is an isomorphism $\gamma : \Gamma_\lambda \longrightarrow \Gamma_{\lambda'}$ given by $P \mapsto P'$, as claimed. That $D(P)$ and $D(P')$ are complementary sets is immediately clear. $\square$

As one more illustration, we show in Figure 6.9 the evacuation-induced involution on the K-L graph of Figure 6.6.

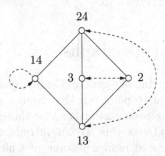

Figure 6.9. Involution on $\Gamma_{(3,2)}$.

We have seen that, on the one hand, there is a partial order "$\preceq_L$" of left cells (Definition 6.2.6) and, on the other hand, there is a bijective correspondence between left cells and standard Young tableaux. Hence, there is an induced partial order on SYT_n, the set of all standard Young tableaux of order n. We call this the *K-L order* of SYT_n. Figure 6.10 shows SYT_4 under K-L order.

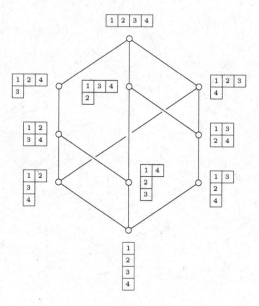

Figure 6.10. The K-L order of SYT$_4$.

It follows from the remarks made after Corollary 6.2.10 and the discussion in this section that mapping each tableau to its transpose $(x \mapsto w_0 x)$ will give an antiautomorphism of the K-L order on SYT_n. It also follows that evacuation of tableaux $(x \mapsto w_0 x w_0)$ will give an automorphism of order 2. Another combinatorial property is that tableaux of the same shape form an antichain; see Exercise 10.

6.7   Complement: W-graphs

We now return to the setting of a general Coxeter group (W, S) for one final remark. Namely we want to point out that from a combinatorial point of view, Kazhdan-Lusztig representations can be thought of as sophisticated "number-firing games." To see this, we will introduce the general notion of a W-graph with its induced representation, of which both K-L representations and the numbers game of Section 4.3 are special cases. What we present is the $q = 1$ version of a construction for Hecke algebras of Kazhdan and Lusztig [322].

Let (W, S) be a Coxeter system and R a commutative ring.

Definition 6.7.1 *A W-graph $\Gamma = (V, E)$ is an undirected graph such that the following hold:*

(i) *The nodes are labeled by subsets of S (denote the label of $x \in V$ by $I_x \subseteq S$), and the edges $\{x, y\}$ have two labels $k_{x,y}, k_{y,x} \in R - \{0\}$, one for each direction.*

(ii) *If $y \in V$ and $s \notin I_y$, then there are at most finitely many neighbors x of y such that $s \in I_x$.*

(iii) *Let M be the free R-module generated by V: For each $s \in S$, define $\tau_s : M \to M$ by*

$$\tau_s(y) = \begin{cases} -y, & \text{if } s \in I_y, \\ y + \sum_{\{x,y\} \in E,\, s \in I_x} k_{x,y} x, & \text{if } s \notin I_y, \end{cases}$$

for all $y \in V$, and extend linearly to all of M. We demand that

$$\underbrace{\tau_s \tau_{s'} \tau_s \cdots}_{m(s,s') \text{ factors}} = \underbrace{\tau_{s'} \tau_s \tau_{s'} \cdots}_{m(s,s') \text{ factors}}$$

for all $(s, s') \in S^2_{\text{fin}}$.

Note that τ_s^2 is the identity mapping for all $s \in S$ (this is immediate from the definition). Hence, $s \mapsto \tau_s \in \text{End}_R(M)$ extends to a homomorphism $W \to \text{End}_R(M)$. In particular, we see that every W-graph with $\mathbb{C}$-labeled edges determines a complex representation of W.

The action of a W-graph representation can be thought of combinatorially as a number-firing game as follows. Call a vertex y s-*labeled* if $s \in I_y$. The elements of M are distributions of ring elements to the nodes of Γ, and a "firing of type s" changes the sign at all s-labeled nodes, whereas at all other nodes y, we add the $k_{x,y}$-weighted sum of all s-labeled neighbors x.

We have encountered two examples of W-graphs:

1. **K-L graphs of left cells.** Here, a vertex $x \in \mathcal{C}$ is labeled by $D_L(x)$ and an edge $\{x, y\}$, $x < y$, by $k_{x,y} = k_{y,x} = [q^{\frac{1}{2}(\ell(x,y)-1)}](P_{x,y})$. It was shown in Section 6.3 via a suitable interpretation in the Hecke algebra that this is a W-graph.

2. **Edge-weighted Coxeter diagrams.** Let $\Gamma = (S, E)$ be the Coxeter diagram of (W, S). Let $I_s = \{s\}$ for all nodes $s \in S$, and for each edge $\{s, s'\} \in E$, let $k_{s,s'}$ and $k_{s',s}$ be positive real numbers satisfying conditions (4.3). Then, as shown by Proposition 4.1.2, Γ is a W-graph. The induced representation is that of the numbers game (i.e., the contragredient of the geometric representation), discussed in Sections 4.1 and 4.2.

Exercises

1. Show that the dihedral group $I_2(m)$ has four left cells if $m \neq \infty$ and three left cells if $m = \infty$.

2. Prove statements (6.8) and (6.9).

3. Prove Lemma 6.4.1.

4. (a) Inverting the relationship (6.13), describe combinatorially what is a "dual Knuth step of type j."
   (b) For $1 < j < n$, define $\mathcal{DES}_L(j) = \{x \in S_n : |D_L(x) \cap \{i - 1, i\}| = 1\}$ and show that for $x \in \mathcal{DES}_L(j)$, there is a unique permutation ${}^{*j}x$ such that ${}^{*j}x \overset{j}{\underset{dK}{\approx}} x$.
   (c) Show that if $x \in \mathcal{DES}_R(i) \cap \mathcal{DES}_L(j)$, then ${}^{*j}(x^{*i}) = ({}^{*j}x)^{*i}$ (Knuth steps and dual Knuth steps commute).

5. Prove the Edge Transport Lemma 6.4.7.
   [*Hint:* The case that $x^{-1}y \in \langle s_{i-1}, s_i \rangle$ is easy to handle. The case $x^{-1}y \notin \langle s_{i-1}, s_i \rangle$ requires more work. Divide it into two subcases depending on whether $x^{-1}x^* = y^{-1}y^*$ or not. Both subcases require some special properties of Kazhdan-Lusztig polynomials, such as equation (6.5) and Theorem 5.1.7.]

6. Show that every left cell in S_n contains a unique involution.

7.* Given $x, y \in S_n$, characterize $x \preceq_L y$ combinatorially in terms of $Q(x)$ and $Q(y)$ (so that $x \underset{L}{\sim} y \Rightarrow Q(x) = Q(y)$ is a direct consequence).

8.* Characterize combinatorially the K-L order of standard Young tableaux of order n. [*Remark:* Another partial order on SYT_n is defined in Exercise 2.36.]

9. Show that the Coxeter diagram of S_n, viewed as a W-graph (Example 2 of Section 6.7 with all $k_{s,s'} = 1$), is isomorphic to the K-L graph Γ_λ for a certain shape $\lambda \vdash n$. Which shape?

10. Lusztig shows in [367] and [370] that for finite and affine Weyl groups there exists a function $\mathfrak{a} : W \to \mathbb{N}$, defined in terms of multiplication in the Hecke algebra, with the following properties (among others):
    (i) $x \preceq_L y \Rightarrow \mathfrak{a}(x) \geq \mathfrak{a}(y)$.
    (ii) $x \preceq_L y$ and $\mathfrak{a}(x) = \mathfrak{a}(y) \Rightarrow x \underset{L}{\sim} y$.
    (iii) $\mathfrak{a}(x) = \mathfrak{a}(x^{-1})$.
    (iv) $\mathfrak{a}(x) = \min\{\ell(y) - 2\deg P_{e,y}(q) : y \underset{L}{\sim} x\}$, and this minimum is achieved at a unique involution y_0 in the left cell.

(a) Deduce the following combinatorial description of $\mathfrak{a}(x)$ for the case $W = S_n$ from these properties: If $\text{shape}(Q(x)') = (\lambda_1, \ldots, \lambda_k)$, then $\mathfrak{a}(x) = \sum_{i=1}^{k} \binom{\lambda_i}{2}$.

(b) Show that all tableaux of the same shape form an antichain in the K-L order of SYT_n.

(c)* Deduce properties (i)-(iv) for S_n, taking the formula for $\mathfrak{a}$ in part (a) as a definition.

11. The notions of right (colored) K-L graph and right preorder "$\preceq_R$" can be defined as in Section 6.2 by switching left to right wherever it matters. Superimposing the left and right colored K-L graphs of (W, S), we get a directed graph $\widetilde{\Gamma}^{LR}_{(W,S)}$ whose strongly connected components are called the *two-sided cells* of W.

(a) Show that the $\mathfrak{a}$-function is constant on every two-sided cell of a Weyl group (using the facts quoted in Exercise 10).

(b) Show that the two-sided cells in S_n are of the form $\mathcal{G}_\lambda \stackrel{\text{def}}{=} \{x \in S_n : \text{shape}(P(x)) = \lambda\}$. Thus, two-sided cells are matched to partitions $\lambda \vdash n$ by the Robinson-Schensted correspondence.

(c) A partial order is induced on the set of two-sided cells by the directed graph $\widetilde{\Gamma}^{LR}_{(W,S)}$, just as $\widetilde{\Gamma}_{(W,S)}$ induces an ordering of left cells (cf. Definitions 6.2.5 and 6.2.6). Show that for the case of S_n,

$$\mathcal{G}_\lambda \leq \mathcal{G}_\mu \quad \text{if and only if} \quad \lambda_1 + \cdots + \lambda_i \geq \mu_1 + \cdots + \mu_i,$$

for all $i = 1, \ldots, n$.

12. Let (W, S) be a finite Coxeter system and denote by Δ its Coxeter complex $\Delta(W, S)$ (cf. Exercise 3.16). The group W acts on the type-selected subcomplex Δ_J, for each $J \subseteq S$, and this induces an action on the simplicial homology group $\widetilde{H}_{|J|-1}(\Delta_J; \mathbb{C})$. (Recall from Exercise 3.16(j) that all other homology groups of Δ_J vanish.) Call this complex representation β_J.

Prove the following about such *homology representations:*

(a) β_J is a character of degree $|\mathcal{D}_J|$.

(b) $\beta_J = \sum_{I \subseteq J} (-1)^{|J \setminus I|} \text{Ind}^S_{S \setminus I}[1]$.

   [*Hint:* Identify $\text{Ind}^S_{S \setminus I}[1]$ as the permutation character of W's action on the set $\{F \in \Delta : \tau(F) = I\}$. Then use the Hopf trace formula.]

(c) $\beta_J = \sum_{D_R(C)=J} KL_C$.

(d) β_S equals the alternating character ε.

(e) There is a duality $\beta_J = \varepsilon \beta_{S \setminus J}$.

(f) For $I, J \subseteq S$,

$$(\beta_I, \beta_J) = \text{card}\{w \in W : D_L(w) = I, D_R(w) = J\}.$$

(g) For the symmetric group S_n the multiplicity of the irreducible character ρ_λ in β_J equals the number of standard Young tableaux of shape λ and with descent set J;

(h)* Can every homology representation β_J be realized by matrices with $\{0, +1, -1\}$-entries?

Notes

This chapter is rather narrowly focused on the symmetric group. The general theory is developed only as far as needed for understanding the case of S_n. All of the main results are due to Kazhdan and Lusztig [322], although few of the combinatorial details concerning S_n are explicitly stated in their article.

The chapter is a slightly expanded version of lectures given by one of us in 1985 [57], a summary of which appeared in Garsia and McLarnan's article [254]. Other articles contributing to a combinatorial understanding of Kazhdan-Lusztig representations for S_n are those of Kerov [325] and Lascoux and Schützenberger [347].

The reader unfamiliar with the basics of Hecke algebras is advised to study Chapter 7 of Humphreys' book [306]. See also the articles of Curtis [168] and Lehrer [362] for more information about Hecke algebras.

There is a huge literature on left cells and W-graphs containing many results of combinatorial interest. Section 7.15 of Humphreys' book [306] gives some references; additional ones can be found in our Bibliography. Let us just mention the work of Shi [455] on left cells in the affine group $\tilde{A}_n$, which has similarities with the case of A_n treated here.

Exercise 5. See Kazhdan and Lusztig [322, pp. 175–176].

Exercises 10 and 11. See Lusztig [367, 370].

Exercise 12. See Björner [56], Bromwich [105], Solomon [481], and Stanley [492]. The decomposition of the homology representation into K-L representations in part (c) has been generalized and further studied by Mathas [389, 390, 391].

7
Enumeration

Enumeration is an important part of combinatorics, so it should not come as a surprise that a book on combinatorics of Coxeter groups gives special attention to such questions. In this chapter, we look at some of the basic enumerative aspects of Coxeter groups.

We begin by considering the natural problems of enumerating a Coxeter group by length, and jointly by length and descent number. We then treat the problem of counting the number of reduced decompositions of an element of a Coxeter group. This turns out to be a difficult and deep problem. We treat only type A and let this case illustrate the basic ideas and methods. This requires a detailed analysis of the combinatorics of tableaux, beyond that reviewed in Appendix A3, and leads to the main results that reduced decompositions of a permutation can be encoded by pairs of tableaux and can be enumerated by certain symmetric functions.

7.1 Poincaré series

In this section, we look at the oldest and most basic enumerative problem concerning a Coxeter group, namely that of enumerating it by length.

For $A \subseteq W$, we let

$$A(q) \overset{\text{def}}{=} \sum_{w \in A} q^{\ell(w)},$$

and call $A(q)$ the *Poincaré series* (or *Poincaré polynomial*, if $|A| < \infty$) of A.

Our aim in this section is to compute $W(q)$ for any Coxeter group W. First, notice that it is enough to do this for irreducible Coxeter systems. In fact, we have the following result, which is a simple consequence of the definitions.

Lemma 7.1.1 *Suppose that* $W = W_1 \times W_2 \times \cdots \times W_k$, *where* $W_1, \ldots, W_k$ *are irreducible Coxeter systems. Then,*

$$W(q) = \prod_{i=1}^{k} W_i(q).$$

$\square$

Therefore, for the rest of this section we assume that (W, S) is an irreducible Coxeter system.

Lemma 7.1.1 shows that $W(q)$ factors if (W, S) is reducible. However, there are many nontrivial factorizations of $W(q)$ also if W is irreducible.

Lemma 7.1.2 *Let* $J \subseteq S$. *Then,*

$$W(q) = W^J(q)W_J(q).$$

Proof. This is an immediate consequence of Proposition 2.4.4. $\square$

Lemma 7.1.2 reduces the problem of computing $W(q)$ to that of computing $W_J(q)$ and $W^J(q)$ for some proper subset J of S. However, whereas W_J is again a Coxeter system (and we can therefore assume that we have already computed $W_J(q)$ by induction), W^J is not. We are therefore faced with the problem of computing $W^J(q)$ for some $J \subset S$ ($J \neq \emptyset$). This problem (in fact, a slightly more general one) can be solved inductively by using one of the cornerstones of enumerative combinatorics, namely the Principle of Inclusion-Exclusion.

Recall that we use the notation

$$\mathcal{D}_I^J \stackrel{\text{def}}{=} \{w \in W : I \subseteq D_R(w) \subseteq J\}$$

for $I, J \subseteq S$. Furthermore, we write $\mathcal{D}_I = \mathcal{D}_I^I$ and have that

$$W^J = \mathcal{D}_\emptyset^{S \setminus J}. \tag{7.1}$$

Proposition 7.1.3 *Let* $I \subseteq J \subseteq S$. *Then,*

$$\mathcal{D}_I^J(q) = \sum_{J \setminus I \subseteq K \subseteq J} (-1)^{|J \setminus K|} W^{S \setminus K}(q).$$

Proof. It is clear from the definitions and equation (7.1) that

$$W^{S \setminus K}(q) = \sum_{L \subseteq K} \mathcal{D}_L(q)$$

for all $K \subseteq S$. Hence,

$$\sum_{J \setminus I \subseteq K \subseteq J} (-1)^{|J \setminus K|} W^{S \setminus K}(q) = \sum_{L \subseteq J} \mathcal{D}_L(q) \sum_{(J \setminus I) \cup L \subseteq K \subseteq J} (-1)^{|J \setminus K|}. \tag{7.2}$$

However, by the Principle of Inclusion-Exclusion,

$$\sum_{(J\setminus I)\cup L\subseteq K\subseteq J} (-1)^{|J\setminus K|} = \begin{cases} 1, & \text{if } (J\setminus I)\cup L = J, \\ 0, & \text{otherwise.} \end{cases}$$

Therefore, we conclude from equation (7.2) that

$$\sum_{J\setminus I\subseteq K\subseteq J} (-1)^{|J\setminus K|} W^{S\setminus K}(q) = \sum_{I\subseteq L\subseteq J} \mathcal{D}_L(q) = \mathcal{D}_I^J(q).$$

$\square$

It is now easy to obtain the first main result of this section.

Corollary 7.1.4 *We have the following:*

(i) If W is finite, then

$$\sum_{K\subseteq S} \frac{(-1)^{|K|}}{W_K(q)} = \frac{q^{\ell(w_0)}}{W(q)}. \tag{7.3}$$

(ii) If W is infinite, then

$$\sum_{K\subseteq S} \frac{(-1)^{|K|}}{W_K(q)} = 0. \tag{7.4}$$

Proof. Part (i) follows immediately by taking $I = J = S$ in Proposition 7.1.3 and using Lemma 7.1.2. Part (ii) follows similarly using Proposition 2.3.1. $\square$

Corollary 7.1.4 allows the recursive computation of $W(q)$ for any Coxeter system (W, S). In fact, both equations (7.3) and (7.4) express $W(q)$ in terms of $W_K(q)$ for $K \subset S$.

For example, suppose that (W, S) is the Coxeter system having Coxeter matrix

$$\begin{pmatrix} 1 & 3 & 4 \\ 3 & 1 & 3 \\ 4 & 3 & 1 \end{pmatrix}.$$

Then, we have that

$$1 - \frac{1}{W_1(q)} - \frac{1}{W_2(q)} - \frac{1}{W_3(q)} + \frac{1}{W_{13}(q)} + \frac{1}{W_{23}(q)} + \frac{1}{W_{12}(q)} = \frac{1}{W(q)}$$

(where we use the shorthand notation $W_{ab...c}(q)$ instead of $W_{\{a,b,...,c\}}(q)$). It is obvious that $W_i(q) = 1+q$ for $i = 1, 2, 3$ and that $W_{12}(q) = W_{23}(q) = 1 + 2q + 2q^2 + q^3$ (this can also be computed by applying Corollary 7.1.4, of course). To compute $W_{13}(q)$, we may use Corollary 7.1.4 again to obtain

$$\frac{q^4 - 1}{W_{13}(q)} = 1 - \frac{1}{W_1(q)} - \frac{1}{W_3(q)} = 1 - \frac{1}{1+q} - \frac{1}{1+q} = \frac{q-1}{(1+q)}$$

Hence, we conclude that

$$\frac{1}{W(q)} = 1 - \frac{3}{1+q} + \frac{2}{(1+q)(1+q+q^2)} + \frac{1}{(1+q)(1+q+q^2+q^3)}$$

$$= \frac{1-q^2-q^3-q^4+q^6}{(1+q)(1+q+q^2)(1+q+q^2+q^3)}.$$

Corollary 7.1.4 solves completely the problem that was posed at the beginning of this section, and one might expect that nothing more could possibly be said. This, however, is a superficial view. In fact, computing $W(q)$ for several finite Coxeter systems (W,S), one sees that $W(q)$ always factors into a product of polynomials whose coefficients are all equal to 1. In looking for a general explanation to this phenomenon, one first observes that this is always true if W is of type A. In fact, it follows easily from Propositions 1.5.2 and 1.5.4 and classical results of enumerative combinatorics (see, e.g., [497, Corollary 1.3.10]) that

$$W(q) = \prod_{i=1}^{n}(1+q+q^2+\cdots+q^i) \tag{7.5}$$

if W is of type A_n

Is something similar to equation (7.5) true in general? Surprisingly, the answer is yes. Define

$$[i]_q \overset{\text{def}}{=} 1+q+q^2+\cdots+q^{i-1}$$

for $i \geq 1$. This is the q-*analog* of the number i.

Theorem 7.1.5 *Let* (W,S) *be a finite irreducible Coxeter system, and* $n \overset{\text{def}}{=} |S|$. *Then, there exist positive integers* $e_1,\ldots,e_n$ *such that*

$$W(q) = \prod_{i=1}^{n}[e_i+1]_q. \tag{7.6}$$

In particular, $|W| = \prod_{i=1}^{n}(e_i+1)$ *and* $|T| = \ell(w_0) = \sum_{i=1}^{n}e_i$.

The integers $e_1,\ldots,e_n$ appearing in equation (7.6) are called the *exponents* of (W,S). A table of the exponents of all the finite irreducible Coxeter systems is given in Appendix A1.

Theorem 7.1.5 can be proved in several ways. One way is to first prove it for the infinite families of finite irreducible Coxeter systems (i.e., types A, B, and D). This can be done in a way that is absolutely analogous to the type A case discussed above. Namely one uses the combinatorial descriptions of the Coxeter groups of types B and D as signed permutations and even signed permutations and of their length functions as counting certain inversions of these permutations (see Sections 8.1 and 8.2 for details). Then, formula (7.6) can be proved in essentially the same way as the classical one (equation (7.5)). One then verifies Theorem 7.1.5 for the other finite irreducible Coxeter systems directly using Corollary 7.1.4, which can be done even by hand. This proof is not entirely satisfactory because it is a case-by-case proof. A more elegant and uniform proof can be given algebraically using the invariant theory of finite reflection groups. For this proof, we refer the reader to the excellent exposition in [306, §3.15].

There is, however, a third way of proving Theorem 7.1.5 that is both combinatorial and (to a large extent) uniform. This is through the theory of normal forms developed in Section 3.4. This theory allows, through the use of a uniform algorithm (i.e., the same algorithm works for all finite Coxeter groups), the construction of a rooted labeled tree that encodes the normal forms of the elements of W^J, where J is a maximal proper subset of S. Namely each path from the root of the tree corresponds to a normal form of an element of W^J and, hence, to a unique element of W^J; see, for instance, Example 3.4.4. Thus, $W^J(q)$ is nothing but the rank generating function of the corresponding tree (seen as a graded poset). This, as has been remarked at the beginning of this section, is enough to compute $W(q)$ inductively.

Example 7.1.6 We illustrate this procedure with an example. Let (W, S) be a Coxeter system of type F_4, and suppose that $S = \{s_1, \ldots, s_4\}$ and the s_i's are numbered as in Figure 3.6 (i.e., so that $(s_1, s_2)^4 = e$ and s_3 commutes with s_1). Let $J = S \setminus \{s_4\}$. Then, the normal form algorithm produces the labeled tree τ_4 in Figure 3.7. Thus,

$$W^J(q) = 1 + q + q^2 + q^3 + 2(q^4 + \cdots + q^{11}) + q^{12} + q^{13} + q^{14} + q^{15}.$$

We then have to compute $W_J(q)$. However,

$$W_J(q) = W_{\{s_1,s_2\}}(q)(W_J)^{\{s_1,s_2\}}(q),$$

and the normal forms of the elements of $(W_J)^{\{s_1,s_2\}}$ are encoded by the tree τ_3 in Figure 3.7. Hence,

$$(W_J)^{\{s_1,s_2\}}(q) = 1 + q + \cdots + q^5,$$

and, similarly,

$$(W_{\{s_1,s_2\}})^{\{s_1\}}(q) = 1 + q + q^2 + q^3.$$

Since, obviously, $W_{\{s_1\}}(q) = 1 + q$, we conclude that

$$W(q) = [2]_q[4]_q[6]_q(1 + q^4)[12]_q$$
$$= [2]_q[6]_q[8]_q[12]_q.$$

$\square$

This example should be sufficient for the reader to carry out the computation of all the other cases by herself. Note that, for the infinite families, one seemingly has to compute infinitely many trees of normal forms. This, however, is only an optical illusion since the trees that encode the normal forms of $(A_n)^{S\setminus\{s_n\}}$, $(B_n)^{S\setminus\{s_{n-1}\}}$, and $(D_n)^{S\setminus\{s_{n-1}\}}$ (indexing as in

Appendix A1) are all extremely easy to describe for any n (in fact, even without the normal form algorithm; see Sections 8.1 and 8.2).

We now have a completely explicit description of the Poincaré polynomials of the finite Coxeter systems. Can something as explicit be said for the infinite ones also? In general, this is too much to expect. However, it is a remarkable fact that the Poincaré series of any Coxeter system can be computed in a nonrecursive and explicit way in terms of those of the finite ones.

For the remainder of this section, we assume that W is infinite. The *nerve* of (W, S) is

$$\mathcal{N}(W, S) \stackrel{\text{def}}{=} \{J \subseteq S : |W_J| < \infty\}.$$

In other words, it is the collection of all the subsets of S whose corresponding parabolic subgroup is finite. Note that $\mathcal{N}(W, S)$ is a simplicial complex. Now, consider $\mathcal{N}(W, S) \cup \{S\}$ as a partially ordered set under set inclusion, and let $\mu_{\mathcal{N}}$ denote the Möbius function of $\mathcal{N}(W, S) \cup \{S\}$.

Proposition 7.1.7 *Let (W, S) be an infinite Coxeter system. Then,*

$$\frac{1}{W(q)} = - \sum_{K \in \mathcal{N}} \frac{\mu_{\mathcal{N}}(K, S)}{W_K(q)}.$$

Proof. Let, for brevity, $\mathcal{N} \stackrel{\text{def}}{=} \mathcal{N}(W, S)$. Let $I \subseteq S, I \notin \mathcal{N}$. Then, $|W_I| = \infty$ and, hence, we conclude from Corollary 7.1.4 that

$$\sum_{K \subseteq I} \frac{(-1)^{|K|}}{W_K(q)} = 0.$$

Therefore,

$$\sum_{I \notin \mathcal{N}} \sum_{K \subseteq I} \frac{(-1)^{|I \setminus K|}}{W_K(q)} = 0$$

and, hence,

$$\sum_{K \subseteq S} \frac{1}{W_K(q)} \sum_{I \supseteq K;\, I \notin \mathcal{N}} (-1)^{|I \setminus K|} = 0. \tag{7.7}$$

However, it follows from the Principle of Inclusion-Exclusion and the definition of the Möbius function that, if $K \subset S$,

$$\sum_{I \supseteq K;\, I \notin \mathcal{N}} (-1)^{|I \setminus K|} = - \sum_{I \supseteq K;\, I \in \mathcal{N}} (-1)^{|I \setminus K|} = \begin{cases} \mu_{\mathcal{N}}(K, S), & \text{if } K \in \mathcal{N}, \\ 0, & \text{if } K \notin \mathcal{N}, \end{cases}$$

so the result follows from equation (7.7). $\square$

Proposition 7.1.7 provides a usually much faster way of computing $W(q)$ than part (ii) of Corollary 7.1.4. For example, let U_n be the universal Coxeter group on n generators (see Example 1.2.2). Then, the nerve of

(U_n, S) consists of just the subsets of S of size ≤ 1 and we conclude from Proposition 7.1.7 that

$$\frac{-1}{U_n(q)} = \frac{n-1}{1} + n\frac{(-1)}{1+q} = \frac{-1+(n-1)q}{1+q}.$$

Although this result can be obtained also by a direct enumerative argument, this is a particularly quick derivation of it.

Because the Poincaré polynomials of the finite Coxeter groups factor nicely in terms of exponents as in Theorem 7.1.5, it is possible to rephrase Proposition 7.1.7 in a somewhat more concise form. Let $R(q)$ be the least common multiple of the polynomials $\{W_J(q)\}_{J \in \mathcal{N}}$. By Theorem 7.1.5, it is clear that $R(q)$ factors as a product of polynomials of the form $[e+1]_q$. In fact, a polynomial $[e+1]_q$ is a factor of $R(q)$ if and only if e is an exponent of some W_J for $J \in \mathcal{N}$. We then have the following result.

Corollary 7.1.8 *Let (W, S) be an infinite Coxeter system. Then,*

$$W(q) = \frac{R(q)}{-\mu_{\mathcal{N}}(\emptyset, S)R(q) + P(q)} \tag{7.8}$$

for some $P(q) \in \mathbb{Z}[q]$ such that $\deg(P) < \deg(R)$. $\square$

It should be noted that $R(q)$ and $P(q)$ may very well have common factors, so that the expression in equation (7.8) is not, in general, in lowest terms. However, Corollary 7.1.8 (and Proposition 7.1.7) do show that one may write down *explicitly* $W(q)$ in terms of the (known) exponents of the finite parabolic subgroups of W, and the combinatorially computable Möbius function.

Example 7.1.9 We illustrate the difference, from a computational point of view, between Proposition 7.1.7 and Corollary 7.1.8 with an example. Let $W = \widetilde{A}_2$. Then, the nerve of (W, S) consists of all the proper subsets of S. So, $\mathcal{N} \cup \{S\}$ is a Boolean algebra of rank 3 and we obtain from Proposition 7.1.7 that

$$\frac{-1}{\widetilde{A}_2(q)} = \frac{-1}{1} + 3\frac{1}{A_1(q)} - 3\frac{1}{A_2(q)}$$

$$= \frac{-1}{1} + \frac{3}{[2]_q} - \frac{3}{[2]_q[3]_q}$$

$$= \frac{(q-1)(1-q^2)}{[2]_q[3]_q}. \tag{7.9}$$

On the other hand, $R(q) = [2]_q[3]_q$ and so from Corollary 7.1.8, we obtain that

$$\widetilde{A}_2(q) = \frac{[2]_q[3]_q}{a + bq + cq^2 + dq^3}$$

for some $a, b, c, d \in \mathbb{Z}$. It is therefore enough to compute four terms of $\widetilde{A}_2(q)$. However, it is easy to see (using, e.g., the numbers game) that

$\widetilde{A}_2(q) = 1 + 3q + 6q^2 + 9q^3 + \cdots$ (see also Figure 8.8 in Section 8.3). So, we obtain that $a = 1$, $3a + b = 2$, $6a + 3b + c = 2$, $9a + 6b + 3c + d = 1$ and, hence, that

$$a + bq + cq^2 + dq^3 = 1 - q - q^2 + q^3 = (1 - q)(1 - q^2).$$

$\square$

The elegant formula (7.9) is not a coincidence, but a general fact that holds for all the affine Coxeter groups.

Theorem 7.1.10 *Let* (W, S) *be an affine Coxeter system, and let* $e_1, \ldots, e_n$ *be the exponents of the corresponding finite group. Then,*

$$W(q) = \prod_{i=1}^n \frac{[e_i + 1]_q}{1 - q^{e_i}}.$$

Theorem 7.1.10 is due to Bott and a proof of it can be found in [78] or [295]. A combinatorial proof of Theorem 7.1.10 is possible for the exceptional groups by proceeding as we did for $\widetilde{A}_2$ in Example 7.1.9. This is not very illuminating, but works. For the infinite families $\widetilde{A}_n$, $\widetilde{B}_n$, $\widetilde{C}_n$, and $\widetilde{D}_n$, one is reduced, by Lemma 7.1.2 and Theorem 7.1.5, to showing that

$$W^J(q) = \prod_{i=1}^n \frac{1}{1 - q^{e_i}}$$

for $J \subseteq S$ such that W_J is the corresponding finite group. This has been proved, using the combinatorial descriptions discussed in Chapter 8 and bijections with appropriate sets of partitions in [224].

7.2   Descents and length generating functions

Without any doubt, the most fundamental statistic on an element v of a Coxeter group, after its length $\ell(v)$, is its *descent number*

$$d(v) = |\{t \in S : \ell(vt) < \ell(v)\}|.$$

(Note that the length is also a kind of "descent number" since $\ell(v) = |\{t \in T : \ell(vt) < \ell(v)\}|$.) It is therefore natural to define, for any Coxeter system (W, S), the bivariate generating function

$$W(t; q) \stackrel{\text{def}}{=} \sum_{v \in W} t^{d(v)} q^{\ell(v)}.$$

Note that $W(t; q)$ is a well-defined formal power series in $\mathbb{Z}[[t, q]]$ since $|\{v \in W : d(v) = i, \ \ell(v) = j\}| \leq |\{v \in W : \ \ell(v) = j\}| < \infty$ for any $i, j \in \mathbb{N}$. However, $W(t; 1)$ is *not* an element of $\mathbb{Z}[[t]]$ unless W is finite, because $d(v) \leq |S|$ for all $v \in W$. We call $W(t; 1)$ (when W is finite) the *Eulerian polynomial* of W.

For example, it is easy to compute that

$$A_2(t; q) = 1 + 2tq + 2tq^2 + t^2 q^3,$$
$$B_2(t; q) = 1 + 2tq + 2tq^2 + 2tq^3 + t^2 q^4,$$

and similarly for the other dihedral groups.

Our main aim in this section is to obtain a simple recursive rule for computing the generating function $W(t; q)$ for any Coxeter group W. This is surprisingly easy to do if one has the right idea.

Theorem 7.2.1

$$W(t; q) = \sum_{J \subseteq S} t^{|J|} (1 - t)^{|S \setminus J|} \frac{W(q)}{W_{S \setminus J}(q)}. \tag{7.10}$$

Proof. The key idea is to write the monomial $t^{d(v)}$ in a clever way, namely

$$t^{d(v)} = \sum_{D(v) \subseteq J \subseteq S} t^{|J|} (1 - t)^{|S \setminus J|}.$$

The result now follows very easily. Indeed, we have that

$$
\begin{aligned}
W(t; q) &= \sum_{v \in W} t^{d(v)} q^{\ell(v)} \\
&= \sum_{v \in W} q^{\ell(v)} \sum_{D(v) \subseteq J \subseteq S} t^{|J|} (1 - t)^{|S \setminus J|} \\
&= \sum_{J \subseteq S} t^{|J|} (1 - t)^{|S \setminus J|} \sum_{\{v \in W : D(v) \subseteq J\}} q^{\ell(v)}
\end{aligned}
$$

However, $\{v \in W : D(v) \subseteq J\} = W^{S \setminus J}$, so equation (7.10) follows from Lemma 7.1.2. $\square$

The preceding theorem reduces the computation of $W(t; q)$ to that of $W(q)$. This generating function, in turn, can be computed using any of the techniques explained in the previous section.

We illustrate Theorem 7.2.1 with an example. Let (W, S) be the Coxeter system considered after Corollary 7.1.4. Then, using Theorem 7.2.1 and the computations already carried out, we get

$$
\begin{aligned}
W(t; q) ={}& W(q)(1 - t)^3 \left[\frac{1}{W(q)} + \frac{t}{(1 - t)} \left(\frac{1}{W_{12}(q)} + \frac{1}{W_{13}(q)} + \frac{1}{W_{23}(q)} \right) \right. \\
&\left. + \frac{t^2}{(1 - t)^2} \left(\frac{1}{W_1(q)} + \frac{1}{W_2(q)} + \frac{1}{W_3(q)} \right) + \frac{t^3}{(1 - t)^3} \right] \\
={}& 1 + \frac{-q(3q^5 - 3q^3 - 5q^2 - 6q - 3)}{1 - q^2 - q^3 - q^4 + q^6} t \\
&+ \frac{q^3(3q^3 + 3q^2 + 3q + 2)}{1 - q^2 - q^3 - q^4 + q^6} t^2.
\end{aligned}
$$

The preceding example makes it clear that, in general, the following holds.

Corollary 7.2.2 $W(t; q) \in \mathbb{Z}(q)[t]$. □

Although Theorem 7.2.1 gives a very explicit way of computing $W(t; q)$ for any Coxeter group W, there are sometimes better ways of computing $W(t; q)$ if W is a member of certain infinite families of Coxeter groups. For example, using Theorem 7.2.1 to compute $S_n(t; q)$ yields explicit polynomials but they do not factor in any nice way. However, an appropriate generating function for the $S_n(t; q)$ has a very elegant and simple expression. In fact, it is a well-known result of Stanley [490] that

$$\sum_{n \geq 0} S_n(t; q) \frac{x^n}{[n]_q!} = \frac{(1-t)\exp(x(1-t); q)}{1 - t\exp(x(1-t); q)} \qquad (7.11)$$

in $(\mathbb{Z}(q)[t])[[x]]$, where

$$\exp(x; q) \overset{\text{def}}{=} \sum_{n \geq 0} \frac{x^n}{[n]_q!} \qquad (7.12)$$

(and where we use the convention that $S_0(t; q) \overset{\text{def}}{=} S_1(t; q) \overset{\text{def}}{=} 1$).

We now present a vast generalization of this result, which, as a special case, allows us to derive the analogs of Stanley's result for $B_n(t; q)$ and $D_n(t; q)$, among others.

Let (W, S) be a Coxeter system and $r \in S$. Let $B \overset{\text{def}}{=} \{s \in S : m(s, r) \geq 3\}$ and choose (once and for all) a partition of B into two blocks, B_1 and B_2. For $n \in \mathbb{N}$, let $(W^{(n)}, (S \setminus \{r\}) \cup \{s_0, \ldots, s_n\})$ be the Coxeter system whose Coxeter graph is obtained from that of (W, S) in the following way. We replace the vertex r by the path shown in Figure 7.1 and then connect s_0 to all the vertices in B_1 and s_n to all the vertices in B_2 so that $m(s_0, b) = m(r, b)$ for all $b \in B_1$ and $m(s_n, b) = m(r, b)$ for all $b \in B_2$ (the vertices $s_1, \ldots, s_{n-1}$ are, therefore, not connected to any vertices in $S \setminus \{r\}$.

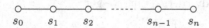

$$s_0 \quad\quad s_1 \quad\quad s_2 \quad\quad \cdots\cdots \quad\quad s_{n-1} \quad\quad s_n$$

Figure 7.1. Replace r by this path.

For example, if the Coxeter graph of (W, S) is the one shown in Figure 7.2 and $B_1 = \{a_1\}$, $B_2 = \{a_2, a_3\}$, then the Coxeter graph of $(W^{(4)}, (S \setminus \{r\}) \cup \{s_0, \ldots, s_4\})$ is the one shown in Figure 7.3. Note that $W^{(0)} = W$.

The motivation for this construction is that many important infinite families of Coxeter groups are obtained in this way, starting from a very simple

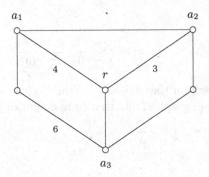

Figure 7.2. The Coxeter graph of (W, S).

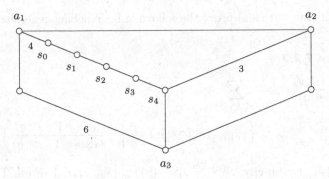

Figure 7.3. The Coxeter graph of $(W^{(4)}, (S \setminus \{r\}) \cup \{s_0, \dots, s_4\})$.

(W, S). For example, $A_n = (A_1)^{(n-1)}$, $B_n = (B_2)^{(n-2)}$, $D_n = (D_4)^{(n-4)}$, $\tilde{A}_n = (\tilde{A}_2)^{(n-2)}$, etc. (for appropriate choices of $r \in S$ and blocks B_1, B_2).

Our aim is to compute a generating function for the rational functions $W^{(n)}(t; q)$, $n \in \mathbb{N}$, assuming that we know $W(t; q)$ $(= W^{(0)}(t; q))$. Since $[n]_q! = S_n(q)$ for $n \geq 1$, Stanley's result suggests that one should consider the generating function

$$\sum_{n \geq 0} W^{(n)}(t; q) \frac{x^n}{W^{(n)}(q)}.$$

For $J \subseteq S \setminus \{r\}$ and $a, b, n \in \mathbb{N}$, we let, for brevity,

$$[W_J]^{(n)} \stackrel{\text{def}}{=} (W^{(n)})_{J \cup \{s_0, \dots, s_n\}} \tag{7.13}$$

and

$$[W_J]^{(a,b)} \stackrel{\text{def}}{=} (W^{(a+b)})_{J \cup \{s_0, \dots, \widehat{s_a}, \dots, s_{a+b}\}} \tag{7.14}$$

(s_a omitted), and we write

$$\exp_{W_J}(x; q) \stackrel{\text{def}}{=} \sum_{n \geq 0} \frac{x^n}{[W_J]^{(n)}(q)} \tag{7.15}$$

and

$$\mathrm{dex}_{W_J}(x;q) = \sum_{a,b \in \mathbb{N}} \frac{x^{a+b}}{[W_J]^{(a,b)}(q)}. \tag{7.16}$$

We adopt the convention that $W_\emptyset = \{e\}$ (the Coxeter group with $S = \emptyset$, having empty Coxeter graph). Note that from definitions (7.13) and (7.14), we deduce that

$$[W_\emptyset]^{(n)} = A_{n+1} \tag{7.17}$$

and

$$[W_\emptyset]^{(a,b)} = A_a \times A_b. \tag{7.18}$$

We can now state and prove the following far-reaching generalization of (7.11).

Theorem 7.2.3

$$\sum_{n \geq 0} \frac{W^{(n)}(t;q)}{W^{(n)}(q)} x^n = \sum_{J \subseteq S \setminus \{r\}} t^{|S \setminus J|-1}(1-t)^{|J|+1}$$

$$\left(\exp_{W_J}(x(1-t);q) + \frac{t \, \mathrm{dex}_{W_J}(x(1-t);q)}{1 - t \exp(x(1-t);q)} \right). \tag{7.19}$$

Proof. Let, for brevity, $S^{(n)} \overset{\text{def}}{=} (S \setminus \{r\}) \cup \{s_0, \ldots, s_n\}$. From Theorem 7.2.1, we obtain that

$$\frac{W^{(n)}(t;q)}{W^{(n)}(q)} = (1-t)^{|S|+n} \sum_{M \subseteq S^{(n)}} \left(\frac{t}{1-t} \right)^{|M|} \frac{1}{(W^{(n)})_{S^{(n)} \setminus M}(q)}.$$

Now, note that any $M \subseteq S^{(n)}$ can be written uniquely as $M = J \cup K$, where $J \subseteq S \setminus \{r\}$, $K \subseteq \{s_0, \ldots, s_n\}$ and $J \cap K = \emptyset$. Therefore,

$$\sum_{n \geq 0} \frac{W^{(n)}(t;q)}{W^{(n)}(q)} x^n = \sum_{J \subseteq S \setminus \{r\}} t^{|J|}(1-t)^{|S \setminus J|}$$

$$\sum_{n \geq 0} \sum_{K \subseteq \{s_0, \ldots, s_n\}} \frac{t^{|K|} x^n (1-t)^{n-|K|}}{(W^{(n)})_{S^{(n)} \setminus (J \cup K)}(q)}. \tag{7.20}$$

Observe that

$$(W^{(n)})_{S^{(n)} \setminus (J \cup K)} \cong [W_{S \setminus \{J \cup \{r\}\}}]^{(a_0, a_k)} \times S_{a_1} \times \cdots \times S_{a_{k-1}},$$

where $a_0, a_k \in \mathbb{N}$, $a_1, \ldots, a_{k-1} \in \mathbb{P}$ are uniquely determined by the condition that $K = \{s_{a_0}, s_{a_0+a_1}, s_{a_0+a_1+a_2}, \ldots, s_{a_0+\cdots+a_{k-1}}\}$ and $\sum_{i=0}^k a_i = n$. It is clear that, for each $k \geq 1$, this correspondence $K \mapsto (a_0, \ldots, a_k)$ is a bijection between subsets of $\{s_0, \ldots, s_n\}$ of size k and sequences $(a_0, \ldots, a_k) \in \mathbb{N}^{k+1}$ such that $a_1, \ldots, a_{k-1} > 0$ and $\sum_{i=0}^k a_i = n$. Hence, we may write

$$\sum_{n\geq 0}\sum_{K\subseteq\{s_0,\ldots,s_n\}}\left(\frac{t}{1-t}\right)^{|K|}\frac{(x(1-t))^n}{(W^{(n)})_{S^{(n)}\setminus(J\cup K)}(q)} = A+B,$$

where

$$A = \sum_{n\geq 0}\frac{(x(1-t))^n}{(W^{(n)})_{S^{(n)}\setminus J}(q)}$$

$$= \sum_{n\geq 0}\frac{(x(1-t))^n}{[W_{S\setminus\{J\cup\{r\}\}}]^{(n)}(q)}$$

$$= \exp_{W_{S\setminus\{J\cup\{r\}\}}}(x(1-t);q)$$

and

$$B = \sum_{k\geq 1}\left(\frac{t}{1-t}\right)^k$$

$$\sum_{a_0,a_k\geq 0}\sum_{(a_1,\ldots,a_{k-1})\in\mathbb{P}^{k-1}}\frac{(x(1-t))^{\sum_{i=0}^k a_i}}{[W_{S\setminus(J\cup\{r\})}]^{(a_0,a_k)}(q)\prod_{i=1}^{k-1}S_{a_i}(q)}$$

$$= \sum_{k\geq 1}\left(\frac{t}{1-t}\right)^k \mathrm{dex}_{W_{S\setminus(J\cup\{r\})}}(x(1-t);q)$$

$$\sum_{(a_1,\ldots,a_{k-1})\in\mathbb{P}^{k-1}}\prod_{i=1}^{k-1}\frac{(x(1-t))^{a_i}}{[a_i]_q!}$$

$$= \mathrm{dex}_{W_{S\setminus(J\cup\{r\})}}(x(1-t);q)\sum_{k\geq 1}\left(\frac{t}{1-t}\right)^k(\exp(x(1-t);q)-1)^{k-1}$$

$$= \frac{t\,\mathrm{dex}_{W_{S\setminus(J\cup\{r\})}}(x(1-t);q)}{1-t\exp(x(1-t);q)}.$$

The result then follows from equation (7.20). □

To appreciate the power and beauty of Theorem 7.2.3, it is useful to consider some examples. Let $W = A_1$ (the simplest possible Coxeter group). Then, $S = \{r\}$ and, hence, there is only one summand appearing in the right-hand side of equation (7.19), namely the one corresponding to $J = \emptyset$. We therefore obtain that

$$\sum_{n\geq 0}\frac{(A_1)^{(n)}(t;q)}{(A_1)^{(n)}(q)}x^n = (1-t)\left(\exp_{W_\emptyset}(x(1-t);q)+\frac{t\,\mathrm{dex}_{W_\emptyset}(x(1-t);q)}{1-t\exp(x(1-t);q)}\right).$$

$$(7.21)$$

Since $W_\emptyset = \{e\}$, we have from definition (7.15) and equation (7.17) that

$$\exp_{W_\emptyset}(x;q) = \sum_{n\geq 0}\frac{x^n}{A_{n+1}(q)} = \sum_{n\geq 0}\frac{x^n}{[n+2]_q!} = \frac{1}{x^2}(\exp(x;q)-1-x),$$

and from equations (7.16) and (7.18) that

$$\mathrm{dex}_{W_\emptyset}(x;q) = \sum_{a,b\geq 0} \frac{x^{a+b}}{(A_a \times A_b)(q)} = \sum_{a,b\geq 0} \frac{x^{a+b}}{[a+1]_q![b+1]_q!}$$

$$= \left(\sum_{a\geq 0} \frac{x^a}{[a+1]_q!}\right)^2 = \left(\frac{1}{x}(\exp(x;q)-1)\right)^2.$$

Substituting these into equation (7.21) we obtain, after routine algebra,

$$\sum_{n\geq 0} S_{n+2}(t;q)\frac{x^n}{S_{n+2}(q)} = \frac{1}{x^2}\left(\frac{(1+tx)\exp(x(1-t);q)-(1+x)}{1-t\exp(x(1-t);q)}\right).$$

$$(7.22)$$

However,

$$\sum_{n\geq 0} S_n(t;q)\frac{x^n}{S_n(q)} = 1 + x + \sum_{n\geq 2} S_n(t;q)\frac{x^n}{S_n(q)}$$

and comparing this with equation (7.22) yields equation (7.11).

In a similar way, one can compute analogs of equation (7.11) for the series B_n, D_n, $\tilde{A}_n$, $\tilde{B}_n$, $\tilde{C}_n$, and $\tilde{D}_n$ (see Exercises 5, 6, 7, and 8).

7.3   Dual equivalence and promotion

In this and the next section, we tackle the problem of enumerating the reduced decompositions of an element of a Coxeter group of type A. As mentioned in the introduction to the chapter, this depends on a detailed analysis of the combinatorics of tableaux. In this section, we carry out this analysis. We assume that the reader is throughly familiar with the contents of Appendix A3.

Let P and Q be two skew tableaux such that the shape of Q extends the shape of P. Then, the cells of Q, taken in the order given by the entries of Q, define a sequence of (forward) slides for P. We then denote by $j_Q(P)$ the tableau obtained by applying this sequence of slides to P, and by $v_Q(P)$, we denote the tableau formed by the cells vacated during the construction of $j_Q(P)$, which records the order in which these cells were vacated. Similarly, P determines a sequence of (backward) slides for Q and we denote by $j^P(Q)$ and $v^P(Q)$ the corresponding tableaux. It is immediate from the definitions that

$$j^{v_Q(P)}(j_Q(P)) = P$$

and

$$j_{v^P(Q)}(j^P(Q)) = Q.$$

Let T be a tableau, $t \overset{\text{def}}{=} |T|$, and $i \in [t-1]$. We define a new tableau $r_i(T)$ as follows. If i and $i+1$ are in different rows and columns in T, then $r_i(T)$ is obtained from T by exchanging i and $i+1$. Otherwise $r_i(T) \overset{\text{def}}{=} T$. We also define $T_{|+i}$ to be the tableau obtained by adding i to all the entries of T, and $T(i)$ to be the tableau consisting of the i smallest entries of T. It is clear from the definitions that if $|i-j| > 1$ $(i,j \in [t-1])$, then

$$r_i r_j(T) = r_j r_i(T). \tag{7.23}$$

The following identity is routine to check, but extremely useful.

Lemma 7.3.1 *Let T be a tableau, $t \overset{\text{def}}{=} |T|$. Then,*

$$r_1 r_2 \ldots r_{t-1}(T) = v_Q(T(t-1)) \cup j_Q(T(t-1))_{|+1},$$

where Q is the tableau consisting of the (unique) cell of T having entry t. □

As a consequence of the previous lemma, we obtain the following important fact.

Lemma 7.3.2 *Let P and Q be two tableaux such that $\mathrm{sh}(Q)$ extends $\mathrm{sh}(P)$. Then,*

$$j^P(Q) = v_Q(P) \quad and \quad v^P(Q) = j_Q(P).$$

Proof. Let $p = |P|$ and $q = |Q|$. We may clearly assume (by adding p to all the entries of Q) that $P \cup Q$ is standard. By repeated application of Lemma 7.3.1, we have that

$$
\begin{aligned}
v_{Q(i)}&(P) \cup j_{Q(i)}(P) \mid_{+i} \cup (Q \setminus Q(i)) \\
&= (r_i r_{i+1} \ldots r_{i+p-1})(r_{i-1} r_i \ldots r_{i+p-2}) \cdots (r_1 r_2 \ldots r_p)(P \cup Q)
\end{aligned} \tag{7.24}
$$

for $i = 1, \ldots, q$. Hence,

$$
\begin{aligned}
v_Q(P) \cup j_Q(P) \mid_{+q} &= r_q r_{q+1} \ldots r_{p+q-1} \\
& r_{q-1} r_q \ldots r_{p+q-2} \\
& \vdots \\
& r_1 r_2 \ldots r_p (P \cup Q).
\end{aligned}
$$

Similarly,

$$
\begin{aligned}
j^P(Q) \mid_{-p} \cup v^P(Q) \mid_{+q} &= r_q r_{q-1} \ldots r_1 \\
& r_{q+1} r_q \ldots r_2 \\
& \vdots \\
& r_{p+q-1} r_{p+q-2} \ldots r_p (P \cup Q),
\end{aligned}
$$

so the result follows from equation (7.23). □

Recall (see Appendix A3.10) that given two skew tableaux S and T, we write $S \approx T$ to mean that S and T are dual equivalent.

Proposition 7.3.3 *Let S, T, and X be tableaux such that $S \approx T$ and* sh(S) *extends* sh(X). *Then,*

$$j_S(X) = j_T(X)$$

and

$$v_S(X) \approx v_T(X).$$

Proof. It follows imediately from the definition of dual equivalence that, with our hypotheses,

$$v^X(S) = v^X(T) \text{ and } j^X(S) \approx j^X(T).$$

So, the result follows from Lemma 7.3.2. $\square$

We now define the crucial concept of this section. Let T be a skew tableau with n cells. We define a new tableau $p(T)$ as follows. Delete entry n from T and perform a forward slide into the cell that contained entry n. Now put 0 in the cell vacated by this forward slide, and finally add 1 to all the entries. We call p the *promotion step*. Note that, by Lemma 7.3.1, we have that

$$p(T) = r_1 r_2 \ldots r_{n-1}(T). \tag{7.25}$$

For example, if

$$T = \begin{array}{|c|c|c|} \hline & 1 & 3 & 6 \\ \hline 2 & 4 & 7 \\ \hline 5 \\ \hline \end{array} \tag{7.26}$$

then

$$p(T) = \begin{array}{|c|c|c|} \hline & 2 & 4 & 7 \\ \hline 1 & 3 & 5 \\ \hline 6 \\ \hline \end{array}$$

Note that T and $p(T)$ have the same shape and that p is an invertible operation. Given T as above, the tableau $p^n(T)$ is called the *total promotion* of T. For example, if T is the tableau (7.26), then its total promotion is

$$p^7(T) = \begin{array}{|c|c|c|} \hline & 1 & 3 & 4 \\ \hline 2 & 5 & 6 \\ \hline 7 \\ \hline \end{array}$$

We encourage the reader to verify, as a further example, that if

$$T = \begin{array}{|c|c|c|} \hline 1 & 2 & 3 \\ \hline 4 & 6 \\ \hline 5 \\ \hline \end{array}$$

then $p^6(T)$ is the transpose of T.

We now come to another crucial definition. A skew shape $\lambda \setminus \mu$ is called a *brick* if for all tableaux S and T of shape $\lambda \setminus \mu$, we have that $S \approx T$ implies $e^*(S) \approx e^*(T)$. (Recall that e^* is the antievacuation operator defined in Appendix A3.8). Otherwise, we call $\lambda \setminus \mu$ a *stone*. For example, the skew shape in Figure 7.4 is a brick (since there is only one tableau of this shape).

Figure 7.4. A brick.

Figure 7.5. A brick.

Figure 7.6. A stone.

Less trivially, the skew shape in Figure 7.5 is also a brick. In fact, there are only two possible tableaux S and T of this shape, namely

$$S = \begin{array}{cc} & 2 \\ \hline 1 & 3 \end{array} \quad \text{and} \quad T = \begin{array}{cc} & 1 \\ \hline 2 & 3 \end{array}$$

and, by Fact A3.10.3, they are dual equivalent. However, a simple computation shows that

$$e^*(S) = T \quad \text{and} \quad e^*(T) = S,$$

so $e^*(S) \approx e^*(T)$. As a final example, the skew shape in Figure 7.6 is a stone. In fact, if

$$S = \begin{array}{cc} 1 & 2 \\ \hline 3 & \end{array} \quad \text{and} \quad T = \begin{array}{cc} 1 & 3 \\ \hline 2 & \end{array}$$

then $S \approx T$ (by Fact A3.10.3), but

$$e^*(S) = \begin{array}{cc} 2 & 3 \\ \hline 1 & \end{array} \quad \text{and} \quad e^*(T) = \begin{array}{cc} 1 & 3 \\ \hline 2 & \end{array}$$

which are not dual equivalent, again by Fact A3.10.3.

Dually, we define a skew shape $\lambda \setminus \mu$ to be an *antibrick* if for all tableaux S and T of shape $\lambda \setminus \mu$ we have that $S \approx T$ implies $e(S) \approx e(T)$. Clearly, a shape is an antibrick if and only if its dual shape is a brick. Note that, by equation (7.25) and the definitions of e^* and e,

$$e^*(T) = (r_{n-1})(r_{n-2}r_{n-1}) \cdots (r_2r_3 \ldots r_{n-1})(r_1r_2 \ldots r_{n-1})(T) \qquad (7.27)$$

and

$$e(T) = (r_1)(r_2r_1) \cdots (r_{n-2} \ldots r_2r_1)(r_{n-1} \ldots r_2r_1)(T) \qquad (7.28)$$

for any tableau T having n cells.

It is easy to classify which miniature shapes are bricks.

Proposition 7.3.4 *Let $\lambda \setminus \mu$ be a miniature shape. Then, $\lambda \setminus \mu$ is a brick if and only if, as a partially ordered set, $\lambda \setminus \mu$ is isomorphic to one of the posets in Figure 7.7. In particular, $\lambda \setminus \mu$ is a brick if and only if it is an antibrick.*

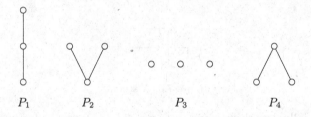

$$P_1 \qquad\qquad P_2 \qquad\qquad\qquad P_3 \qquad\qquad\qquad P_4$$

Figure 7.7. The poset isomorphism classes of miniature bricks.

Proof. We have already seen, in the examples following the definition of a brick, that the result holds if $\lambda \setminus \mu$ is isomorphic to P_1 or P_4. Similarly, one can check it for P_2 and P_3. If $\lambda \setminus \mu$ is not isomorphic to one of the posets in Figure 7.7, then it must necessarily be isomorphic to the disjoint union of two chains. Hence, $\lambda \setminus \mu$ is equivalent (with respect to antievacuation) to either $(3,1) \setminus (1)$ or $(3,2) \setminus (2)$ or their transposes. We have already shown that $(3,1) \setminus (1)$ is a stone, and the verification for the others is entirely similar. $\square$

We now come to the main result of this section. Recall that a *staircase* is a partition of the form $\delta_n \overset{\text{def}}{=} (n, n-1, \ldots, 3, 2, 1)$ for some $n \in \mathbb{P}$. Let λ be a staircase and S and T be two elementary dual equivalent tableaux of shape λ. For example, suppose

$$S = \begin{array}{|c|c|c|} \hline 1 & 3 & 6 \\ \hline 2 & 4 \\ \cline{1-2} 5 \\ \cline{1-1} \end{array} \quad \text{and} \quad T = \begin{array}{|c|c|c|} \hline 1 & 2 & 6 \\ \hline 3 & 4 \\ \cline{1-2} 5 \\ \cline{1-1} \end{array} \qquad (7.29)$$

Applying the promotion operator p to S and T, we then obtain

$$p(S) = \begin{array}{|c|c|c|}\hline 1 & 2 & 4 \\\hline 3 & 5 \\\cline{1-2} 6 \\\cline{1-1}\end{array} \quad \text{and} \quad p(T) = \begin{array}{|c|c|c|}\hline 1 & 2 & 3 \\\hline 4 & 5 \\\cline{1-2} 6 \\\cline{1-1}\end{array} \qquad (7.30)$$

and these two tableaux are still elementary dual equivalent. Note, however, that this does not go on forever. In fact, continuing, we obtain

$$p^2(S) = \begin{array}{|c|c|c|}\hline 1 & 3 & 5 \\\hline 2 & 6 \\\cline{1-2} 4 \\\cline{1-1}\end{array} \quad , \quad p^2(T) = \begin{array}{|c|c|c|}\hline 1 & 3 & 4 \\\hline 2 & 6 \\\cline{1-2} 5 \\\cline{1-1}\end{array}$$

$$p^3(S) = \begin{array}{|c|c|c|}\hline 1 & 2 & 6 \\\hline 3 & 4 \\\cline{1-2} 5 \\\cline{1-1}\end{array} \quad , \quad p^3(T) = \begin{array}{|c|c|c|}\hline 1 & 2 & 5 \\\hline 3 & 4 \\\cline{1-2} 6 \\\cline{1-1}\end{array} \qquad (7.31)$$

$$p^4(S) = \begin{array}{|c|c|c|}\hline 1 & 2 & 3 \\\hline 4 & 5 \\\cline{1-2} 6 \\\cline{1-1}\end{array} \quad , \quad p^4(T) = \begin{array}{|c|c|c|}\hline 1 & 3 & 6 \\\hline 2 & 5 \\\cline{1-2} 4 \\\cline{1-1}\end{array}$$

and these last two are not elementary dual equivalent. However, if we continue again, we obtain that

$$p^5(S) = \begin{array}{|c|c|c|}\hline 1 & 3 & 4 \\\hline 2 & 6 \\\cline{1-2} 5 \\\cline{1-1}\end{array} \quad , \quad p^5(T) = \begin{array}{|c|c|c|}\hline 1 & 2 & 4 \\\hline 3 & 6 \\\cline{1-2} 5 \\\cline{1-1}\end{array} \qquad (7.32)$$

which are elementary dual equivalent.

Staircases are important exactly because they interact well with the promotion operator and dual equivalence. The following shows that the above example is typical and is the main result of this section.

Theorem 7.3.5 *Let S and T be two elementary dual equivalent tableaux of staircase shape, with n cells. Let $\{j, j+1, j+2\}$ be the entries of S and T involved in an elementary dual equivalence. Then, if $-t \not\equiv j, j+1 \pmod{n}$, $p^t(S)$ is elementary dual equivalent to $p^t(T)$ and the entries involved in the elementary dual equivalence are $\{j+t, j+t+1, j+t+2\} \pmod{n}$.*

Proof. If $t = 1$ and $j+2 < n$, or if $t = -1$ and $j > 1$, then the result is clear from the definitions of dual equivalence and elementary dual equivalence. Thus, the result holds for any interval of allowed values of t if it holds for one element of the interval. However, the allowed values of t fall into intervals of size $n-2$ separated by "gaps" of size 2. So, we only have to show that if the result holds for the values of t in one of these intervals, then it also holds for those in the next interval, and for those in the previous one.

We will show this for the next interval, the reasoning for the previous one being entirely analogous. Thus, what we have to show is that if $j = n - 2$, then $p^3(S)$ is elementary dual equivalent to $p^3(T)$ and the entries involved in the dual equivalence are $\{1, 2, 3\}$.

Let $S = X \cup Y_S$ and $T = X \cup Y_T$, where Y_S and Y_T are the final segments of S and T, respectively, that contain the entries $\{n - 2, n - 1, n\}$. Let us compute $p^3(S)$. From equations (7.25) and (7.23), we have that

$$p^3(S) = (r_1 r_2 \ldots r_{n-1})(r_1 r_2 \ldots r_{n-1})(r_1 r_2 \ldots r_{n-1})(S)$$
$$= (r_1 r_2 r_1)(r_3 \ldots r_{n-1})(r_2 \ldots r_{n-2})(r_1 \ldots r_{n-3})(r_{n-1} r_{n-2} r_{n-1})(S).$$

However, it follows from equation (7.27) that

$$(r_{n-1} r_{n-2} r_{n-1})(S) = (r_{n-1} r_{n-2} r_{n-1})(X \cup Y_S)$$
$$= X \cup (r_{n-1} r_{n-2} r_{n-1})(Y_S)$$
$$= X \cup e^*(Y_S).$$

Hence, from equation (7.24), we conclude that

$$p^3(S) = (r_1 r_2 r_1)(v_{e^*(Y_S)}(X) \cup j_{e^*(Y_S)}(X)_{|+3})$$
$$= (r_1 r_2 r_1)(v_{e^*(Y_S)}(X)) \cup j_{e^*(Y_S)}(X))_{|+3}$$
$$= e(v_{e^*(Y_S)}(X)) \cup j_{e^*(Y_S)}(X)_{|+3}, \tag{7.33}$$

by equation (7.28), and similarly for $p^3(T)$.

Now, since Y_S and Y_T are miniature final segments of S and T, which are of staircase shape, their shape is a brick by Proposition 7.3.4. Hence, by the definition of a brick, $e^*(Y_S) \approx e^*(Y_T)$. This, by Proposition 7.3.3, implies that $j_{e^*(Y_S)}(X) = j_{e^*(Y_T)}(X)$ and $v_{e^*(Y_S)}(X) \approx v_{e^*(Y_T)}(X)$. However, $v_{e^*(Y_S)}(X)$ and $v_{e^*(Y_T)}(X)$ are miniature initial segments of tableaux of staircase shape, and so, by Proposition 7.3.4, their shape is an antibrick. By the definition of an antibrick, this implies that $e(v_{e^*(Y_S)}(X)) \approx e(v_{e^*(Y_T)}(X))$ and this, by equation (7.33), implies that $p^3(S)$ and $p^3(T)$ are elementary dual equivalent, and the entries involved in the dual equivalence are $\{1, 2, 3\}$. $\square$

We illustrate the construction used in the proof of Theorem 7.3.5 with an example. Suppose T is the tableau given in (7.29). Then,

$$Y_T = \boxed{\begin{array}{ccc} & & 6 \\ & 4 & \\ 5 & & \end{array}}$$

and, hence,

$$e^*(Y_T) = \boxed{\begin{array}{ccc} & & 1 \\ & 3 & \\ 2 & & \end{array}}$$

and the computation of $j_{e^*(Y_T)}(X)$ gives the sequence of tableaux

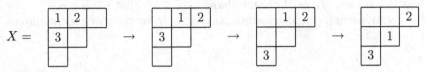

Therefore,

$$j_{e^*(Y_T)}(X) = \begin{array}{c}\boxed{2}\\\boxed{1}\\\boxed{3}\end{array} \quad \text{and} \quad v_{e^*(Y_T)}(X) = \begin{array}{cc}\boxed{1}&\boxed{3}\\\boxed{2}\end{array}.$$

Computing $e(v_{e^*(Y_T)}(X))$, we get the sequence of tableaux

$$\begin{array}{cc}\boxed{1}&\boxed{3}\\\boxed{2}\end{array} \rightarrow \begin{array}{cc}\boxed{2}&\boxed{3}\end{array} \rightarrow \boxed{3}$$

and, therefore,

$$e(v_{e^*(Y_T)}(X)) = \begin{array}{cc}\boxed{1}&\boxed{2}\\\boxed{3}\end{array}.$$

Hence,

$$e(v_{e^*(Y_T)}(X)) \cup j_{e^*(Y_T)}(X) \mid_{+3} = \begin{array}{ccc}\boxed{1}&\boxed{2}&\boxed{5}\\\boxed{3}&\boxed{4}\\\boxed{6}\end{array}$$

which agrees with our computations in (7.31).

Note that, in the statement of Theorem 7.3.5, j is not uniquely determined by S and T. For example, if S and T are as in equation (7.29), then there are two elementary dual equivalences relating S and T, namely the one involving entries $\{1,2,3\}$ and the one involving $\{2,3,4\}$. In this case we can apply Theorem 7.3.5 twice and obtain the stronger conclusion that $p^t(S) \approx p^t(T)$ for all $t \not\equiv 4 \pmod 6$.

We will use Theorem 7.3.5 over and over again. For the moment, here is an interesting and useful immediate consequence of it.

Corollary 7.3.6 *Let T be a tableau of staircase shape, with n cells. Then,*

$$p^n(T) = T'.$$

Proof. Let S be a tableau that is elementary dual equivalent to T. Let $\{j, j+1, j+2\}$ be the entries involved in an elementary dual equivalence of S and T. By Theorem 7.3.5 we then conclude that $p^n(T)$ is elementary dual equivalent to $p^n(S)$ and $\{j, j+1, j+2\}$ are the entries involved in this elementary dual equivalence. Therefore, $p^n(S) = S'$ if and only if $p^n(T) = T'$. However, all tableaux of staircase shape are dual equivalent by Fact A3.10.1 and, hence, are connected by a chain of elementary dual equivalences by Fact A3.10.4. Therefore, the result will be proved if we can find *one* tableau T_0 of the same shape as T such that $p^n(T_0) = T'_0$. This is easy to accomplish, for example, taking T_0 to be the row superstandard tableau. $\square$

7.4   Counting reduced decompositions in S_n

Although the reader can hardly be expected to be aware of this at the present point, the tableaux studied in the previous section are closely related to reduced decompositions of the Coxeter group of type A. This is explained in the present section.

Let T be a tableau of shape $\delta_n \setminus \mu$, with N cells. Assign label i to the i-th lower right corner cell of δ_n (counting from the bottom), for $i = 1, \ldots, n$. The *promotion sequence* of T is the doubly-infinite sequence $\bar{p}(T) \stackrel{\text{def}}{=} (\ldots, r_{-2}, r_{-1}, r_0, r_1, r_2, \ldots)$, where r_k is the label of the corner cell occupied by the largest entry (namely N) of $p^{N-k}(T)$, for $k \in \mathbb{Z}$. We also let $\hat{p}(T) \stackrel{\text{def}}{=} (r_1, \ldots, r_N)$ and call this the *short promotion sequence* of T. Note that if we perform promotion (and inverse promotion) on T without adding (subtracting) 1, then every integer $k \in \mathbb{Z}$ will at some point be the greatest entry in the tableau and will then occupy a corner cell of $\text{sh}(T)$, and r_k is then the label of this corner cell. Also note that if $\mu = \emptyset$, then, by Corollary 7.3.6, $\bar{p}(T) = (\ldots, r_{-1}, r_0, r_1, \ldots)$ is periodic of period $2N$ and $r_{N+k} = N + 1 - r_k$ for all $k \in \mathbb{Z}$. We let

$$\mathcal{P}(\delta_n) \stackrel{\text{def}}{=} \{\hat{p}(S) : \ S \text{ is a tableau of shape } \delta_n\}.$$

As an example, let us find the short promotion sequence of the tableau S given in (7.29). Then, from (7.30), (7.31), and (7.32), we conclude that

$$\hat{p}(S) = (2, 1, 3, 2, 1, 3) \tag{7.34}$$

and, hence,

$$\bar{p}(S) = (\ldots 2, 3, 1, 2, 3, 1, 2, 1, 3, 2, 1, 3, 2, 3, 1, 2, 3, 1, \ldots).$$

Another important property of the short promotion sequence that is convenient to note right away is that if $\hat{p}(T) = (r_1, \ldots, r_N)$, then for any $k \in [N]$, $(r_k, \ldots, r_N)$ is the short promotion sequence of the final segment of T containing the entries $\{k, \ldots, N\}$. Therefore, in particular, $(r_k, \ldots, r_N)$ depends only on this final segment. For example, if

$$X = \begin{array}{|c|c|} \hline 3 & 6 \\ \hline 4 \\ \hline \end{array}$$

then X is a final segment of S and

$$\widehat{p}(X) = (3, 2, 1, 3).$$

Before we go on, we need to verify the following simple but crucial property of the staircase shape.

Proposition 7.4.1 *Let T be a tableau of staircase shape and U be a miniature final segment of T. Then, U is uniquely determined by $\widehat{p}(U)$.*

Proof. Since U is a final segment of a tableau of staircase shape, its shape must be isomorphic (as a poset) to either P_2 or P_3 (see Figure 7.7). Suppose it is P_2. Then, U is one of the two tableaux shown in Figure 7.8.

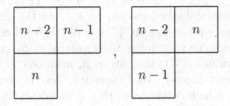

Figure 7.8. U is one of these two tableaux.

In the first case, we have that $\widehat{p}(U) = (i, i+1, i)$ for some $i \in [n-1]$, whereas in the second, $\widehat{p}(U) = (j+1, j, j+1)$ for some $j \in [n-1]$.

The case P_3 is similar. $\square$

For example, let T be a tableau of shape $(4, 3, 2, 1)$ and U be a final segment of T such that $\widehat{p}(U) = (2, 1, 4)$. Then, we can immediately conclude that U is the final segment in Figure 7.9. On the other hand, if $\widehat{p}(U) = (3, 2, 3)$, then U must necessarily be the final segment in Figure 7.10.

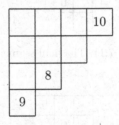

Figure 7.9. The only final segment of T such that $\widehat{p}(U) = (2, 1, 4)$.

The next result is "almost" a restatement of Theorem 7.3.5.

Theorem 7.4.2 *Let S and T be two elementary dual equivalent tableaux of staircase shape, and let $\{j, j+1, j+2\}$ be the entries involved in an elementary dual equivalence. Then, $\widehat{p}(S)$ and $\widehat{p}(T)$ differ at most in positions j, $j+1$, and $j+2$. Furthermore, given j, $\widehat{p}(S)$ is uniquely determined by $\widehat{p}(T)$.*

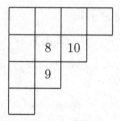

Figure 7.10. The only final segment of T such that $\widehat{p}(U) = (3, 2, 3)$.

Proof. The statements before "Furthermore" follow directly from Theorem 7.3.5. Now, let $S_1 \overset{\text{def}}{=} p^{N-j-2}(S)$ and $T_1 \overset{\text{def}}{=} p^{N-j-2}(T)$. Then, by Theorem 7.3.5, S_1 is elementary dual equivalent to T_1, and the entries involved in the elementary dual equivalence are $\{N-2, N-1, N\}$. Let U (respectively, V) be the final segment containing these entries in S_1 (respectively, T_1). Then, $\widehat{p}(T)$ determines $\widehat{p}(V)$ (the entries in positions j, $j+1$, and $j+2$ of $\widehat{p}(T)$), which determines V (by Proposition 7.4.1), which determines U (by Fact A3.10.3), which determines $\widehat{p}(U)$, which determines the entries of $\widehat{p}(S)$ in positions j, $j+1$, and $j+2$, which determines $\widehat{p}(S)$ since the rest of it coincides with $\widehat{p}(T)$. $\square$

For instance, let S and T be the tableaux given in (7.29), and $j = 2$. Then, $\widehat{p}(T) = (2, 3, 1, 2, 1, 3)$, so $\widehat{p}(V) = (3, 1, 2)$. As seen above, this implies that V is the final segment in Figure 7.11; hence, U is the final segment in Figure 7.12, and, therefore, $\widehat{p}(U) = (1, 3, 2)$. So, $\widehat{p}(S) = (2, 1, 3, 2, 1, 3)$, in accordance with equation (7.34).

Figure 7.11. The final segment V.

Figure 7.12. The final segment U.

We now come to the crucial definition of this section. Given $X \in [n]^k$ for some $k \in \mathbb{P}$, we denote by $\widehat{p}^{-1}(X)$ the set of all tableaux U such that $\text{sh}(U)$ is a final segment of δ_n, and $\widehat{p}(U) = X$. So, for example, by Proposition 7.4.1, if $X \in [n]^3$, then $|\widehat{p}^{-1}(X)| \leq 1$.

Let $k \in \mathbb{P}$ and $X, Y \in [n]^k$. The pair $\{X, Y\}$ is called a *staircase relation* (or an *s-relation*, for short) if the following two conditions are satisfied:

(i) There is a shape-preserving bijection $\phi : \widehat{p}^{-1}(X) \to \widehat{p}^{-1}(Y)$.

(ii) For any $U \in \widehat{p}^{-1}(X)$ there exists a tableau T such that $\mathrm{sh}(U)$ extends $\mathrm{sh}(T)$, $\mathrm{sh}(T) \subseteq \delta_n$, $T \cup U \approx T \cup \phi(U)$, and

$$\widehat{p}(T \cup U) = WX, \qquad \widehat{p}(T \cup \phi(U)) = WY$$

for some sequence W.

In part (i) of the definition, "shape-preserving" means that for $U \in \widehat{p}^{-1}(X)$, U and $\phi(U)$ have the *same* shape (not just isomorphic as posets). So, for example, the tableaux in Figure 7.13 *do not* have the same shape (in fact, the first one has shape $(4, 3, 2, 1) \setminus (2, 2, 2, 1)$, whereas the second one has shape $(4, 3, 2, 1) \setminus (4, 3)$).

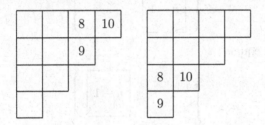

Figure 7.13. These tableaux do not have the same shape.

We illustrate the definition with some examples. Let $i \in [n-1]$; then, $\{(i+1, i, i+1), (i, i+1, i)\}$ is an *s-relation*. In fact, we know from Proposition 7.4.1 that $\widehat{p}^{-1}((i+1, i, i+1))$ equals

$$\left\{ \begin{array}{|c|c|} \hline n-2 & n \\ \hline n-1 & \\ \hline \end{array} \right\}$$

and $\widehat{p}^{-1}((i, i+1, i))$ equals

$$\left\{ \begin{array}{|c|c|} \hline n-2 & n-1 \\ \hline n & \\ \hline \end{array} \right\}$$

(where both tableaux have shape $\delta_n \setminus (n, n-1, \ldots, i+2, i-1, i-1, \ldots, 2, 1)$), so part (i) of the definition is clearly satisfied. Part (ii) can be easily satisfied by taking $T = \emptyset$, since

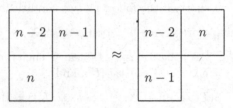

As a further example, let $i, j \in [n]$, $i + 1 < j$. Then, $\{(i,j),(j,i)\}$ is an s-relation. In fact, it is clear that $\widehat{p}^{-1}((i,j))$ equals

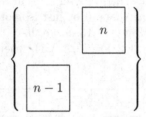

and $\widehat{p}^{-1}((j,i))$ equals

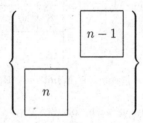

(where both tableaux have shape $\delta_n \setminus (n, \ldots, j+1, j-1, j-1, \ldots, i+1, i-1, i-1, \ldots, 2, 1)$), so part (i) of the definition is again clearly satisfied. For part (ii), take T to be the tableau consisting of the corner cell of δ_n labeled by $i + 1$ filled by the entry $N - 2$. Then, if $U \in \widehat{p}^{-1}((i,j))$, $\mathrm{sh}(U)$ extends $\mathrm{sh}(T)$, and $T \cup U \approx T \cup \phi(U)$ by Fact A3.10.3. Furthermore, it is clear that $\widehat{p}(T \cup U) = (i+1, i, j)$ and $\widehat{p}(T \cup \phi(U)) = (i+1, j, i)$, so part (ii) of the definition is also satisfied.

The examples just given can be generalized, leading to the following result.

Proposition 7.4.3 *Let μ be a miniature final segment of δ_n, and let $U \approx V$ be two tableaux of shape μ. Then, $\{\widehat{p}(U), \widehat{p}(V)\}$ is an s-relation, and any two elements of $\mathcal{P}(\delta_n)$ are related by a chain of s-relations of this form.*

Proof. The fact that $\{\widehat{p}(U), \widehat{p}(V)\}$ is an s-relation follows in exactly the same way as in the next to last example. Now, if S and T are two elementary dual equivalent tableaux of shape δ_n then by Theorem 7.4.2 (and its proof), $\widehat{p}(S)$ and $\widehat{p}(T)$ differ by an s-relation of the form $\{\widehat{p}(U), \widehat{p}(V)\}$, as in the statement of the proposition. However, by Facts A3.10.1 and A3.10.4, all tableaux of shape δ_n are connected by a sequence of elementary dual equivalences, so the proof is complete. $\square$

It is natural and useful, at this point, to understand completely the s-relations given by Proposition 7.4.3. A miniature final segment of δ_n is either of the form

where the corner cells are labeled by i and $i+1$ for some $i \in [n-1]$, or of the form

$$(7.35)$$

where all of the cells are corner cells and are labeled by i, j, and k for some $1 \le i < j < k \le n$. In the first case, we have already seen that $\{(i, i+1, i), (i+1, i, i+1)\}$ is the corresponding s-relation. In the second case, there are six possible final segments of shape (7.35) (one for each permutation of $\{N, N-1, N-2\}$) and we know from Fact A3.10.3 that they split into the four dual equivalence classes shown in Figure 7.14. The last two classes give trivial s-relations. The first two classes give the s-relations $\{(j, i, k), (j, k, i)\}$ and $\{(k, i, j), (i, k, j)\}$. We may therefore restate the last proposition in the following more explicit way.

Proposition 7.4.4 *Every two elements of $\mathcal{P}(\delta_n)$ are related by a chain of s-relations of the form $\{(a, a+1, a), (a+1, a, a+1)\}$, $\{(j, k, i), (j, i, k)\}$, and $\{(k, i, j), (i, k, j)\}$ for $1 \le i < j < k \le n$ and $a \in [n-1]$.* $\square$

The importance of s-relations comes from the following result.

Proposition 7.4.5 *Let $\{X, Y\}$ be an s-relation. Then, $|\widehat{p}^{-1}(AXB)| = |\widehat{p}^{-1}(AYB)|$ whenever $AXB, AYB \in [n]^N$. In particular, $AXB \in \mathcal{P}(\delta_n)$ if and only if $AYB \in \mathcal{P}(\delta_n)$.*

Proof. We may clearly assume that $\widehat{p}^{-1}(AXB) \ne \emptyset$ (say). We construct a map $\psi : \widehat{p}^{-1}(AXB) \to \widehat{p}^{-1}(AYB)$ as follows. Let $S \in \widehat{p}^{-1}(AXB)$, and let U be the final segment of $p^{|B|}(S)$ occupied by the entries $\{N - |X| + 1, \ldots, N-1, N\}$. Then, $U \in \widehat{p}^{-1}(X)$. Now, replace U in $p^{|B|}(S)$ by $\phi(U)$ (where ϕ is the bijection that exists by part (i) of the definition of s-relation) and apply $p^{-|B|}$. This is $\psi(S)$.

By the definition of s-relation, we know that $T \cup U \approx T \cup \phi(U)$ for some tableau T and $\widehat{p}(T \cup U) = WX$, $\widehat{p}(T \cup \phi(U)) = WY$ for some sequence W. We may assume that $\mathrm{sh}(T \cup U) = \delta_n$. Let V be the initial segment of $p^{|B|}(S)$ (and hence also of $p^{|S|}(\psi(S))$) consisting of the entries $\{1, 2, \ldots,$

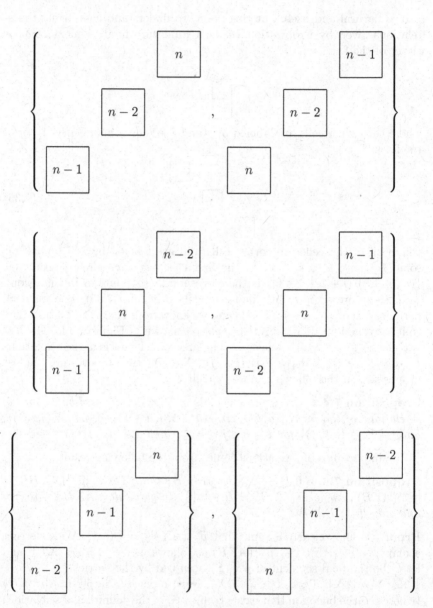

Figure 7.14. Four dual equivalence classes.

$n - |X|$}. By Facts A3.10.1 and A3.10.4, V and T are related by a chain of elementary dual equivalences. This chain transforms $p^{|B|}(S) = V \cup U$ into $T \cup U$, and $p^{|B|}(\psi(S)) = V \cup \phi(U)$ into $T \cup \phi(U)$. This, by Theorem 7.4.2, implies that $\widehat{p}(V \cup U)$ and $\widehat{p}(T \cup U)$ $(= WX)$ differ only in their first $n - |X|$ entries. Therefore, $\widehat{p}(V \cup U) = ZX$ for some sequence Z, and (again by Theorem 7.4.2) Z depends only on W (and the chain of elementary dual equivalences). Since the same chain of elementary dual equivalences transforms $V \cup \phi(U)$ into $T \cup \phi(U)$, this shows that $\widehat{p}(V \cup \phi(U)) = ZY$. Therefore,

$$\widehat{p}(p^{|B|}(\psi(S))) = ZY$$

and $\widehat{p}(p^{|B|}(S)) = ZX$ and this, since $\widehat{p}(S) = AXB$, implies that $\widehat{p}(\psi(S)) = AYB$ by Corollary 7.3.6. This shows that $\psi(\widehat{p}^{-1}(AXB)) \subseteq \widehat{p}^{-1}(AYB)$. The fact that ψ is a bijection is clear. $\square$

We are now in a position to prove the first main result of this section.

Theorem 7.4.6 *Let T be a tableau of shape δ_n. Then, T is uniquely determined by $\widehat{p}(T)$. Equivalently, $|\widehat{p}^{-1}(X)| = 1$ for all $X \in \mathcal{P}(\delta_n)$.*

Proof. It follows immediately from Propositions 7.4.4 and 7.4.5 that

$$|\widehat{p}^{-1}(X)| = |\widehat{p}^{-1}(Y)|$$

for all $X, Y \in \mathcal{P}(\delta_n)$. To conclude the proof, we therefore just have to show that there is at least one element $X_0 \in \mathcal{P}(\delta_n)$ such that $|\widehat{p}^{-1}(X_0)| = 1$, and this is easy to do (take, e.g., $X_0 = (1, 2, 1, 3, 2, 1, \ldots, n, n-1, \ldots, 2, 1)$). $\square$

The above result shows that the map $\widehat{p}$ is a bijection from the set of tableaux of shape δ_n to $\mathcal{P}(\delta_n)$. Note that, in principle, we know the set $\mathcal{P}(\delta_n)$ explicitly because we know that we can generate it by starting with a single element of it (say X_0) and applying recursively s-relations. Sounds familiar? Yes, it is not a coincidence. Recall that for $w \in W$, we denote by $\mathcal{R}(w)$ the set of all the reduced decompositions of w.

Theorem 7.4.7 *The map $\widehat{p}$ is a bijection between the set of all tableaux of shape δ_n and $\mathcal{R}(w_0)$, where w_0 denotes the longest element of S_{n+1}.*

Proof. By the preceding remarks, it is enough to show that

$$\mathcal{P}(\delta_n) = \mathcal{R}(w_0)$$

(where we identify, as done in Section 3.4, a sequence $(i_1, \ldots, i_k) \in [n]^k$ with $(s_{i_1}, \ldots, s_{i_k}) \in S^k$ and $s_i = (i, i+1)$ for $i = 1, \ldots, n$).

Let $X \in \mathcal{P}(\delta_n)$. Then, by Proposition 7.4.4, X is related to X_0 by a sequence of s-relations of the form given in the statement of Proposition 7.4.4. However, all of these relations are also Coxeter relations in S_{n+1}. Hence, since $X_0 \in \mathcal{R}(w_0)$, we conclude that $X \in \mathcal{R}(w_0)$. This shows that $\mathcal{P}(\delta_n) \subseteq \mathcal{R}(w_0)$.

Conversely, let $Y \in \mathcal{R}(w_0)$. Then, by Theorem 3.3.1, Y is related to X_0 by a sequence of braid-moves. However, all of these braid-moves are also s-relations, as we have seen in the examples preceding Proposition 7.4.3, so Y is connected to X_0 by a sequence of s-relations and this, by Proposition 7.4.5, implies that $Y \in \mathcal{P}(\delta_n)$ since $X_0 \in \mathcal{P}(\delta_n)$. This proves that $\mathcal{R}(w_0) \subseteq \mathcal{P}(\delta_n)$ and concludes the proof. $\square$

As an immediate consequence of the above theorem we obtain the following result, which was, historically, the empirical observation that started most of the enumerative research on reduced decompositions.

Corollary 7.4.8

$$|\mathcal{R}(w_0)| = \frac{\binom{n+1}{2}!}{1^n 3^{n-1} 5^{n-2} \cdots (2n-1)}.$$

Proof. This follows immediately from Theorem 7.4.7 and the well-known "hook length formula" (see, e.g., [498, Corollary 7.21.6]). $\square$

At this point, the reader may very well feel puzzled, or distressed. Have we built a whole theory just to compute the number of reduced decompositions of *one* element (and a very special one, besides) of S_{n+1}? Indeed, the answer is no. Using this theory, we can enumerate, with a little further work, the reduced decompositions of any element of S_{n+1}.

Proposition 7.4.9 Let T be a tableau of shape δ_n, $\widehat{p}(T) = (r_1, \ldots, r_N)$, and $k \in [N]$. Then, the initial segment of T consisting of the cells of T containing the entries $\{1, \ldots, k\}$ is uniquely determined by $(r_1, \ldots, r_k)$.

Proof. Let, for brevity, $A \stackrel{\text{def}}{=} (r_1, \ldots, r_k)$, and write $\widehat{p}(T) = AB$. Let S be a tableau of shape δ_n such that $\widehat{p}(S) = AC$ for some C. We wish to prove that the initial segments of S and T formed by the cells containing the entries $\{1, \ldots, k\}$ coincide. By Theorem 7.4.7, AB and AC are reduced decompositions of w_0. Hence, B and C are reduced decompositions of some element $w \in S_{n+1}$. Therefore, by Theorem 3.3.1, we may assume that B and C differ by a braid-move.

Suppose that B and C differ by a braid-move of the form $(a, a+1, a) \leftrightarrow (a+1, a, a+1)$ for some $a \in [n-1]$. Then, $\widehat{p}(S)$ and $\widehat{p}(T)$ differ by the same relation, which is an s-relation. However, for an s-relation of this form, we have (see the first example following the definition of an s-relation) that $U \approx \phi(U)$, where we use the same notation as in the proof of Proposition 7.4.5. This, by the proof of Proposition 7.4.5, implies that S and T differ only by an elementary dual equivalence on the segment involving $\{k+1, \ldots, N\}$, so the result holds in this case.

Suppose that B and C differ by a braid-move of the form $(a, b) \leftrightarrow (b, a)$ for some $a, b \in [n]$ with $|a - b| > 1$. Then, reasoning as in the previous case (namely keeping in mind the example given before Proposition 7.4.3 and the proof of Proposition 7.4.5), we conclude that S is obtained from T by applying p^t for some t (necessarily $\leq n - k - 2$), switching N and $N - 1$ (which will be in the corner cells labeled by a and b) and then applying p^{-t}. This whole process does not change the initial segment containing the entries $\{1, \ldots, k\}$, so we are done. $\square$

The preceding result is the crucial step needed to extend Theorem 7.4.7 to other elements of S_{n+1}. In fact, it allows us to define $\widehat{p}^{-1}$ in $\mathcal{R}(w)$ for any $w \in S_{n+1}$.

Let $w \in S_{n+1}$, $X \in \mathcal{R}(w)$, and $E \in \mathcal{R}(w^{-1}w_0)$. Then, by Proposition 3.1.2, $\ell(w) + \ell(w^{-1}w_0) = \ell(w_0)$ and, hence, XE is a reduced decomposition of w_0. We then define $\theta(X)$ to be the initial segment, in $\widehat{p}^{-1}(XE)$, containing the entries $\{1, \ldots, |X|\}$. This is well defined by Proposition 7.4.9. Note that $\theta(X)$ is a tableau of normal shape, contained in δ_n.

Proposition 7.4.10 *Let $w \in S_{n+1}$ and let T be a tableau with $\mathrm{sh}(T) \subseteq \delta_n$. Then,*

$$|\{X \in \mathcal{R}(w) : \theta(X) = T\}|$$

depends only on $\mathrm{sh}(T)$.

Proof. Fix $E \in \mathcal{R}(w^{-1}w_0)$. Note that $X \in \mathcal{R}(w)$ if and only if $XE \in \mathcal{R}(w_0)$.

Let $X \in \mathcal{R}(w)$ be such that $\theta(X) = T$. Let U be the final segment of $\widehat{p}^{-1}(XE)$ containing the entries $\{|X| + 1, \ldots, N\}$. Then, $\widehat{p}(U) = E$, $\widehat{p}^{-1}(XE) = T \cup U$, and, therefore, $\mathrm{sh}(U) = \delta_n \backslash \mathrm{sh}(T)$.

Conversely, if U is a tableau of shape $\delta_n \backslash \mathrm{sh}(T)$ such that $\widehat{p}(U) = E$, then

$$\widehat{p}(T \cup U|_{+|T|}) = XE$$

for some $X \in [n]^{|T|}$. Thus, by Theorem 7.4.7, $X \in \mathcal{R}(w)$ and, by definition, $\theta(X) = T$.

It is clear that this correspondence $X \leftrightarrow U$ is a bijection. Hence,

$$|\{X \in \mathcal{R}(w) : \theta(X) = T\}| = |\{U : \mathrm{sh}(U) = \delta_n \backslash \mathrm{sh}(T), \ \widehat{p}(U) = E\}|, \tag{7.36}$$

which implies the result. $\square$

Since the numbers in equation (7.36) do not depend on T and E but only on $\mathrm{sh}(T)$ and w, we define, for any partition λ, and any $w \in S_{n+1}$,

$$a_\lambda(w) \stackrel{\text{def}}{=} |\{X \in \mathcal{R}(w) : \theta(X) = T\}|, \tag{7.37}$$

where T is any tableau of shape λ. Equivalently,

$$a_\lambda(w) = |\{U : \mathrm{sh}(U) = \delta_n \backslash \lambda, \ \widehat{p}(U) = E\}|,$$

where E is any reduced decomposition of $w^{-1}w_0$. (For an alternative combinatorial interpretation of $a_\lambda(w)$, see Exercise 21.) Note that $a_\lambda(w) = 0$ unless $|\lambda| = \ell(w)$.

The following is the promised generalization of Corollary 7.4.8, and the main result of this section. Recall that f^λ denotes the number of standard Young tableaux of shape λ. As was mentioned in the proof of Corollary 7.4.8, there is an effective procedure for explicitly computing the numbers f^λ, known as the "hook length formula" (see [498, Corollary 7.21.6]).

Theorem 7.4.11 *Let $w \in S_{n+1}$. Then,*

$$|\mathcal{R}(w)| = \sum_\lambda a_\lambda(w) f^\lambda,$$

where λ runs over all partitions of $\ell(w)$ contained in δ_n.

Proof. By Proposition 7.4.10, we have that

$$|\mathcal{R}(w)| = \sum_T |\{X \in \mathcal{R}(w) : \theta(X) = T\}|$$

$$= \sum_\lambda \sum_{\{T:\, sh(T) = \lambda\}} |\{X \in \mathcal{R}(w) : \theta(X) = T\}|$$

$$= \sum_\lambda a_\lambda(w) f^\lambda,$$

where T runs over all tableaux of normal shape contained in δ_n, and the result follows. $\square$

An interesting and natural problem now is that of describing explicitly the map $\widehat{p}^{-1}$ from reduced decompositions to tableaux. This turns out to be related to some surprising variations of the Robinson-Schensted algorithm (see Exercise 21).

Another (even more natural) problem is that of extending the main results of this section to other Coxeter groups. This is a huge problem that hides some deep combinatorics. Note that the mere statement that $|\mathcal{R}(w)| = \sum_\lambda a_\lambda f^\lambda$ for *some* $a_\lambda \in \mathbb{N}$ is certainly true for any $w \in W$ in any group W, since *any* non-negative integer can be expressed in this format. What one really wants is a combinatorial interpretation for the a_λ's that gives some insight. It turns out that what we have done in this section can be carried over, with some technical complications but with essentially the same ideas, concepts, and techniques, to the Coxeter groups of types B and D (see [284] and [44]).

7.5   Stanley symmetric functions

As mentioned in Section 7.2, generating functions often provide an elegant and compact way of encoding enumerative results. In this section, we consider a multivariable generating function for the reduced decompositions of the elements of the symmetric groups. We prove the remarkable fact that this generating function is always symmetric, and we compute its expansion in terms of Schur functions. We assume that the reader is familiar with the contents of Appendix A4.

Recall that we write $S = \{s_1, \ldots, s_n\}$, where $s_i = (i, i+1)$, $i = 1, \ldots, n$, and that

$$D(s_{i_1}, \ldots, s_{i_p}) \overset{\text{def}}{=} \{j \in [p-1] : i_j > i_{j+1}\}$$

for $(s_{i_1}, \ldots, s_{i_p}) \in S^p$.

Lemma 7.5.1 *Let $w \in S_{n+1}$ and $X \in \mathcal{R}(w)$. Then,*

$$D(\theta(X)) = D(X).$$

Proof. It is a routine case-by-case exercise to verify, using equation (7.25), that if T is a tableau and k, $k+1$ are entries of T, $k+1 < |T|$, then $k \in D(T)$ if and only if $k \in D(p(T))$, where in the promotion tableau $p(T)$ we do *not* renormalize the entries.

Now, fix $E \in \mathcal{R}(w^{-1}w_0)$, so that $XE \in \mathcal{R}(w_0)$, and let $T \overset{\text{def}}{=} \widehat{p}^{-1}(XE)$. Then, by definition, $\theta(X)$ is the initial segment of T containing the entries $\{1, 2, \ldots, |X|\}$. Therefore, if $k \in [|X|-1]$, then $k \in D(\theta(X))$ if and only if $k \in D(p^{n-k-1}(T))$. However, in $p^{N-k-1}(T)$ (recall that here we do not renormalize the entries when performing p), the entry $k+1$, and hence necessarily also k, occupy corner cells of δ_n. Furthermore, since $\widehat{p}(T) = XE$, $k+1$ occupies the corner cell labeled by X_{k+1} (the $(k+1)$-st entry of X), and k occupies the one labeled by X_k. Hence, $k \in D(\theta(X))$ if and only if $X_k > X_{k+1}$, as desired. $\square$

Let $w \in S_{n+1}$, and $p \overset{\text{def}}{=} \ell(w)$. Let $\mathbf{x} \overset{\text{def}}{=} (x_1, x_2, \ldots)$ be a sequence of independent indeterminates. The *Stanley symmetric function* of w is

$$F_w(\mathbf{x}) \overset{\text{def}}{=} \sum_{E \in \mathcal{R}(w)} Q_{D(E),p}(\mathbf{x}).$$

For example, if $n = 2$ and $w = 321$, then $\mathcal{R}(w) = \{(s_1, s_2, s_1), (s_2, s_1, s_2)\}$ and, hence,

$$F_{321}(x_1, x_2, x_3) = x_1^2 x_2 + x_1^2 x_3 + x_2^2 x_3 + x_1 x_2^2 + 2x_1 x_2 x_3 + x_2 x_3^2 + x_1 x_3^2$$

(where $F_w(x_1, \ldots, x_r) \overset{\text{def}}{=} F_w(x_1, \ldots, x_r, 0, 0, \ldots)$).

Theorem 7.5.2 *Let $w \in S_{n+1}$. Then, $F_w(\mathbf{x})$ is a symmetric function and*

$$F_w(\mathbf{x}) = \sum_{\lambda \vdash \ell(w)} a_\lambda(w) s_\lambda(\mathbf{x}), \tag{7.38}$$

where $a_\lambda(w)$ has the same meaning as in Theorem 7.4.11.

Proof. From our definitions and Lemma 7.5.1, we have that

$$F_w(\mathbf{x}) = \sum_{E \in \mathcal{R}(w)} Q_{D(E),p}(\mathbf{x})$$

$$= \sum_T \sum_{\{E \in \mathcal{R}(w): \, \theta(E)=T\}} Q_{D(E),p}(\mathbf{x})$$

$$= \sum_T a_{sh(T)}(w) Q_{D(T),p}(\mathbf{x})$$

$$= \sum_{\lambda \subseteq \delta_n} a_\lambda(w) \sum_{\{T: \, sh(T)=\lambda\}} Q_{D(T),|\lambda|}(\mathbf{x}),$$

where $p = \ell(w)$ and T runs over all tableaux of shape contained in δ_n. Thus, the result follows from Fact A4.2.2. $\square$

The symmetric functions $F_w(\mathbf{x})$ were first defined by Stanley in [493] exactly for the purpose of enumerating the reduced decompositions of the elements of S_{n+1}. In fact, it is possible to prove Corollary 7.4.8 using just these symmetric functions (i.e., without using the theory developed in Sections 7.3 and 7.4; see [493]).

Note that taking the coefficient of $x_1 x_2 \ldots x_{\ell(w)}$ on both sides of equation (7.38) yields Theorem 7.4.11. However, it is not at all clear, from the definition of $F_w(\mathbf{x})$, why the coefficients $a_\lambda(w)$ are ≥ 0. Once this fact is known, then standard results from the theory of symmetric functions (see, e.g., [498, §7.18]) imply that there exists a representation $\mathcal{F}_w$ of $S_{\ell(w)}$ whose character corresponds to $F_w(\mathbf{x})$ under the characteristic map and that the integers $a_\lambda(w)$ give the multiplicity of the irreducible representation of $S_{\ell(w)}$ indexed by λ in $\mathcal{F}_w$. Such representations have been explicitly constructed by Kraśkiewicz, see [331].

The symmetric function $F_w(\mathbf{x})$ is also intimately related to the Schubert polynomial of w. The topic of Schubert polynomials is vast and is not treated here. We refer the reader to the excellent expositions in [381] and [248] and the references cited there.

Exercises

1. Let $J \subseteq S$ be such that $|W_J| = \infty$. Show that

$$\sum_{K \subseteq J} \frac{(-1)^{|K|}}{W_{S \setminus K}(q)} = 0.$$

2. Let, for $w \in W$,

$$a\ell(w) \stackrel{\text{def}}{=} \min\{k \in \mathbb{N} : \, w = t_{i_1} \cdots t_{i_k}, \text{ for some } t_{i_1}, \ldots, t_{i_k} \in T\}.$$

We call $a\ell(w)$ the *absolute length* of w. Clearly, $a\ell(w) = 1$ if and only if $w \in T$. For a finite Coxeter system (W, S), show that

$$\sum_{w \in W} t^{a\ell(w)} = \prod_{i=1}^{n}(1 + e_i t),$$

where $e_1, \ldots, e_n$ are the exponents of W.

3. Let $S = K_1 \cup K_2 \cup \cdots \cup K_k$ be the partition of S induced by the conjugacy relation (see Exercise 1.16). Define $\ell_i(w)$, $1 \le i \le k$, to be the number of letters from K_i in any reduced expression of $w \in W$, and define

$$W(x_1, \ldots, x_k) = \sum_{w \in W} x_1^{\ell_1(w)} \cdots x_k^{\ell_k(w)}.$$

(a) Show that $\ell_i(w)$ is well defined.
(b) For W of type B_n, let $K_1 = \{s_0\}$ and $K_2 = \{s_1, \ldots, s_{n-1}\}$. Show that

$$W(x_1, x_2) = \prod_{i=0}^{n-1}(1 + x_1 x_2^i)(1 + x_2 + \cdots + x_2^i).$$

4. In this and the next four exercises, we use the following notation. For a sequence of Coxeter groups $\mathcal{W} \overset{\text{def}}{=} \{W_n\}_{n=0,1,2,\ldots}$, we let

$$\exp_{\mathcal{W}}(x; q) \overset{\text{def}}{=} \sum_{n \ge 0} \frac{x^n}{W_n(q)}.$$

This extends definition (7.12) of $\exp(x; q)$. Use the q-binomial theorem (see Appendix A4.1) to show that

$$\exp(x; q) = \prod_{i \ge 0}(1 - x(1 - q)q^i)^{-1}$$

$$\exp_B(x; q) = \prod_{i \ge 0}(1 - x(1 - q)q^{2i})^{-1}$$

$$x^2 \exp_D(x; q) = \exp_B(x; q) + \prod_{i \ge 0}(1 - x(1 - q)q^{2i+1})^{-1} - 2 - x$$

$$\exp_{\tilde{B}}(x; q) = \exp_{\tilde{C}}(x; q) = \frac{-q}{[2]_q} + \frac{xq^4}{[2]_q[4]_q} + \prod_{i \ge 0}\left(\frac{1 - x(1 - q)q^{2i+1}}{1 - x(1 - q)q^{2i}}\right)$$

$$x^4 \exp_{\tilde{D}}(x; q) = \prod_{i \ge 0}\left(\frac{1 - x(1 - q)q^{2i+1}}{1 - x(1 - q)q^{2i}}\right)$$

$$= + \prod_{i \ge 0}\left(\frac{1 - x(1 - q)q^{2i}}{1 - x(1 - q)q^{2i+1}}\right) - \sum_{i=0}^{3} c_i(q)x^i$$

for some $c_i(q) \in \mathbb{Z}(q)$.

Here, $B = \{B_n\}_{n=0,1,2,...}$, $D = \{D_{n+2}\}_{n=0,1,2,...}$, $\widetilde{C} = \{\widetilde{C}_n\}_{n=0,1,2,...}$, $\widetilde{B} = \{\widetilde{B}_n\}_{n=0,1,2,...}$, and $\widetilde{D} = \{\widetilde{D}_{n+4}\}_{n=0,1,2,...}$. (We use the conventions that $B_0 = \langle e \rangle$, $B_1 = \widetilde{C}_0 = \widetilde{B}_0 = A_1$, $\widetilde{C}_1 = \widetilde{B}_1 = B_2$, $\widetilde{B}_2 = \widetilde{C}_2$, $D_2 = A_1 \times A_1$, and $D_3 = A_3$.)

5. Let $(W^{r,s}, \{s_1, s_2, s_3\})$ be the Coxeter system having Coxeter matrix

$$\begin{pmatrix} 1 & r & 2 \\ r & 1 & s \\ 2 & s & 1 \end{pmatrix}$$

$(r, s \geq 3)$. Choose $v = s_2$, $B_1 = \{s_1\}$, and $B_2 = \{s_3\}$.

(a) Use Theorem 7.2.3 to show that

$$\sum_{n \geq 0} \frac{(W^{r,s})^{(n-3)}(t; q) x^n}{(W^{r,s})^{(n-3)}(q)} = \exp_{\mathcal{W}^{r,s}}(x(1-t); q)$$

$$+ \frac{tx(1-t)\exp_{\mathcal{W}^r}(x(1-t); q)\exp_{\mathcal{W}^s}(x(1-t); q)}{1 - t\exp(x(1-t); q)},$$

where $\mathcal{W}^{r,s} \overset{\text{def}}{=} \{(W^{r,s})^{(n-3)}\}_{n=0,1,2,...}$ and $\mathcal{W}^r = \mathcal{W}^{r,3}$ (we use the conventions that $(W^{r,s})^{(-3)} \overset{\text{def}}{=} \langle e \rangle$, $(W^{r,s})^{(-2)} \overset{\text{def}}{=} A_1$, and $(W^{r,s})^{(-1)} \overset{\text{def}}{=} I_2(r)$).

(b) Deduce from part (a) that

$$\sum_{n \geq 0} \frac{B_n(t; q) x^n}{B_n(q)} = \frac{(1-t)\exp_B(x(1-t); q)}{1 - t\exp(x(1-t); q)}.$$

(c) Deduce from part (a) that

$$\frac{1}{1-t}\sum_{n \geq 0} \frac{\widetilde{C}_n(t; q) x^n}{\widetilde{C}_n(q)} = \exp_{\widetilde{C}}(x(1-t); q) + \frac{t\exp_B(x(1-t); q)^2}{1 - t\exp(x(1-t); q)}.$$

6. Deduce from Theorem 7.2.3 that

$$2tx + \sum_{n \geq 2} \frac{D_n(t; q) x^n}{D_n(q)}$$

$$= \frac{x^2(1-t)^3\exp_D(x(1-t); q) + t(2 - tx)(\exp(x(1-t); q) - 1)}{1 - t\exp(x(1-t); q)}.$$

7. (a) Use Theorem 7.2.3 to find a formula for

$$\sum_{n \geq 3} \frac{\widetilde{B}_n(t; q) x^n}{\widetilde{B}_n(q)}$$

in terms of $\exp_{\widetilde{B}}(x(1-t); q)$, $\exp_B(x(1-t); q)$, $\exp_D(x(1-t); q)$, and $\exp(x(1-t); q)$.

(b) Use Theorem 7.2.3 to find a formula for the generating function

$$\sum_{n \geq 4} \frac{\widetilde{D}_n(t; q) x^n}{\widetilde{D}_n(q)}$$

in terms of $\exp_{\widetilde{D}}(x(1 - t); q)$, $\exp_D(x(1 - t); q)$, and $\exp(x(1 - t); q)$.

8. (a) Let

$$\exp_{\widetilde{A}}(x; q) \overset{\text{def}}{=} \sum_{n \geq 1} \frac{x^n}{\widetilde{A}_{n-1}(q)}$$

(where we use the convention that $\widetilde{A}_0 = \langle e \rangle$). Show that

$$\exp_{\widetilde{A}}(x; q) = x \frac{\partial}{\partial x} \log(\exp(x; q)).$$

(b) Deduce from Theorem 7.2.3 and part (a) that

$$\sum_{n \geq 1} \frac{\widetilde{A}_{n-1}(t; q)}{1 - q^n} x^n = \frac{\exp_{\widetilde{A}}\left(x \frac{1-t}{1-q}; q\right)}{1 - t \exp\left(x \frac{1-t}{1-q}; q\right)}.$$

9. Let (W, S) be an infinite Coxeter system of rank 3. (Equivalently, the quantity d defined in Exercise 4.8 satisfies $d \leq 1$.) Show that

$$W(q) = W(q^{-1}),$$

as rational functions in $\mathbb{Z}(q)$.

10. Let (W, S) be a finite irreducible Coxeter system.

(a) Identify a subset $J \subset \{s_1, \ldots, s_n\}$ with the sequence $(n_1, \ldots, n_k)$ such that $S \setminus J = \{s_{n_1}, s_{n_1+n_2}, \ldots, s_{n_1+\ldots+n_{k-1}}\}$ and $n_1 + \cdots + n_k = n+1$ (so $k = n - |J| + 1$). Show that if $(W, \{s_1, \ldots, s_n\})$ is of type A_n, with the generators indexed in the usual way, then

$$W^J(-1) = \left(\begin{array}{c} \lfloor \frac{n+1}{2} \rfloor \\ \lfloor \frac{n_1}{2} \rfloor, \ldots, \lfloor \frac{n_k}{2} \rfloor \end{array} \right)$$

if $\lfloor \frac{n+1}{2} \rfloor = \lfloor \frac{n_1}{2} \rfloor + \cdots + \lfloor \frac{n_k}{2} \rfloor$, and

$$W^J(-1) = 0$$

otherwise.

(b) Show that if (W, S) is of type B_n, then $W^J(-1) = 0$ for all $J \subset S$.

(c) Show that if (W, S) is of type D_n (S indexed as in Appendix A1), then

$$W^J(-1) = \begin{cases} 2, & \text{if } n \text{ is odd and } J = S \setminus \{s_{n-1}\}, \\ 0, & \text{otherwise} \end{cases}$$

for all $J \subset S$.

(d) Deduce from the preceding parts that $W^J(-1) = 0$ for all $J \subset S$ if and only if every exponent of W is odd.

(e) Is there a uniform (i.e., not case by case) proof of part (d)?

11. Given a finite Coxeter group W, let $W(t; 1)$ be the Eulerian polynomial of W defined in Section 7.2.

(a) Show that

$$S_n(t; 1) = (1 + (n-1)t)S_{n-1}(t; 1) + (t - t^2)\frac{d}{dt}(S_{n-1}(t; 1))$$

for $n \in \mathbb{P}$, where $S_0(t; 1) \overset{\text{def}}{=} 1$.

(b) Deduce from part (a) that

$$\sum_{i \geq 0}(i+1)^n t^i = \frac{S_n(t; 1)}{(1-t)^{n+1}}$$

as a formal power series in $\mathbb{Z}[[t]]$.

(c) Deduce from part (a) that $S_n(t; 1)$ has only real zeros.

(d) Show that

$$B_n(t; 1) = (1 + (2n-1)t)B_{n-1}(t; 1) + 2(t - t^2)\frac{d}{dt}(B_{n-1}(t; 1))$$

for $n \in \mathbb{P}$, where $B_0(t; 1) \overset{\text{def}}{=} 1$.

(e) Deduce from part (d) that

$$\sum_{i \geq 0}(2i+1)^n t^i = \frac{B_n(t; 1)}{(1-t)^{n+1}}$$

as a formal power series in $\mathbb{Z}[[t]]$.

(f) Deduce from part (d) that the polynomial $B_n(t; 1)$ has only real zeros.

(g) For $n \in \mathbb{P}$, show that

$$D_n(t; 1) = B_n(t; 1) - n\, t2^{n-1}S_{n-1}(t; 1).$$

(h) Show that the polynomial $W(t; 1)$ has only real zeros if W is of type E_6, E_7, E_8, F_4, H_3, and H_4.

(i)* Is it true that the polynomial $D_n(t; 1)$ has only real zeros?

12. Consider the symmetric group S_n.

(a) Show that

$$\sum_{(a_1, a_2, \ldots) \in \mathcal{R}(w_0)}(x+a_1)(x+a_2)\cdots = \binom{n}{2}! \prod_{1 \leq i < j \leq n} \frac{2x + i + j - 1}{i + j - 1}.$$

(b) Deduce that

$$\sum_{(a_1,a_2,\dots)\in\mathcal{R}(w_0)} a_1 a_2 \cdots a_{\binom{n}{2}} = \binom{n}{2}!$$

[*For instance* $(n=3)$, $1\cdot 2\cdot 1 + 2\cdot 1\cdot 2 = 6$.]
  (c) Deduce the formula for $|\mathcal{R}(w_0)|$ in Corollary 7.4.8.

13. Let $S = \{s_1, s_2, \dots, s_{n-1}\}$ be the set of adjacent transpositions $s_i = (i, i+1)$ in S_n. Suppose $I \subseteq J \subseteq S$, and say that $J = \{s_{i_1}, s_{i_2}, \dots, s_{i_j}\}$, where $1 \le i_1 < i_2 < \cdots < i_j \le n-1$.
Show that

$$\mathcal{D}_I^J(q) = [n]!\ \det \begin{pmatrix} \frac{1}{[i_1]!} & \frac{1}{[i_2]!} & \frac{1}{[i_3]!} & \cdots & \frac{1}{[n]!} \\ * & \frac{1}{[i_2-i_1]!} & \frac{1}{[i_3-i_1]!} & \cdots & \frac{1}{[n-i_1]!} \\ 0 & * & \frac{1}{[i_3-i_2]!} & \cdots & \frac{1}{[n-i_2]!} \\ 0 & 0 & * & \cdots & \frac{1}{[n-i_3]!} \\ \vdots & \vdots & \vdots & & \vdots \\ 0 & 0 & 0 & \cdots & \frac{1}{[n-i_j]!} \end{pmatrix},$$

where the subdiagonal element $*$ in the p-th column equals 1 if $s_{i_p} \in I$, and is 0 otherwise. Here, $[n]! \stackrel{\text{def}}{=} [n][n-1]\cdots[2][1]$ and $[k] = [k]_q \stackrel{\text{def}}{=} 1 + q + q^2 + \cdots + q^{k-1}$.

14.* Is it true for every finite irreducible Coxeter group (W, S), and every $I \subseteq J \subseteq S$, that the polynomial $\mathcal{D}_I^J(q)$ is unimodal?
[*Remark:* This was proven for $I = \emptyset$ and essentially all cases in [491].]

15. Let T be a tableau with n cells. Show that

$$p^n(T) = e(e^*(T)).$$

16. A (rookwise) connected skew partition is a *generalized staircase* if every miniature final, and initial, segment of it is a brick. For example, $(3, 1)$ is not a generalized staircase, whereas a staircase (as the name implies) is also a generalized staircase.
Let $\lambda \setminus \mu$ be a (rookwise) connected skew partition. Show that $\lambda \setminus \mu$ is a generalized staircase if and only if it is either a staircase, an antistaircase, or a rectangle.

17. Let T be a tableau of generalized staircase shape, with n cells. Show that

$$p^n(T) = \begin{cases} T, & \text{if } \mathrm{sh}(T) \text{ is a rectangle}, \\ T', & \text{otherwise.} \end{cases}$$

18. Let $\lambda \setminus \mu$ be a connected skew partition such that $p^{|\lambda\setminus\mu|}(T) = T$ for all tableaux T of shape $\lambda \setminus \mu$. Show that then $\lambda \setminus \mu$ is a rectangle.

19. Show that $F_w(\mathbf{x}) = F_{w^{-1}}(\mathbf{x})$, for all $w \in S_n$.

20. Let $\mathcal{N}_n = \mathcal{H}_0(S_n, S)$ be the generic Hecke algebra of the symmetric group S_n (as defined in [306, §7.1]), specialized by setting parameters $a_s = b_s = 0$ for all $s \in S = \{s_1, \ldots, s_{n-1}\}$. Let $\{T_w\}_{w \in S_n}$ denote its canonical basis, and write, for convenience, $T_i \stackrel{\text{def}}{=} T_{s_i}$, for $i = 1, \ldots, n-1$. The algebra $\mathcal{N}_n$ is called the *NilCoxeter algebra* of S_n. It has generators $T_1, \ldots, T_{n-1}$ and relations

$$\begin{cases} T_i^2 = 0, & \text{if } i = 1, \ldots, n-1, \\ T_i T_j = T_j T_i, & \text{if } |i - j| > 1, \\ T_i T_{i+1} T_i = T_{i+1} T_i T_{i+1}, & \text{if } i = 1, \ldots, n-2. \end{cases}$$

(a) Show that if $r, n \in \mathbb{P}$ and $r \geq \binom{n}{2}$, then

$$\sum_{w \in S_n} F_w(x_1, \ldots, x_r) T_w = A(x_1) \cdots A(x_r)$$

in $\mathcal{N}_n \otimes_{\mathbb{Z}} \mathbb{Z}[x_1, \ldots, x_r]$, where

$$A(x_i) \stackrel{\text{def}}{=} (T_e + x_i T_1) \cdots (T_e + x_i T_{n-2})(T_e + x_i T_{n-1})$$

for $i = 1, \ldots, r$.

(b) Show that

$$A(x_i) A(x_j) = A(x_j) A(x_i).$$

for $i, j \in [r]$.

(c) Deduce from parts (a) and (b) that $F_w(x_1, \ldots, x_r)$ is a symmetric polynomial in $x_1, \ldots, x_r$.

21. Let $\mathcal{R} \stackrel{\text{def}}{=} \bigcup_{n \geq 1} \mathcal{R}(S_n)$. We say that a (not necessarily standard) tableau T is an A-*tableau* if $\rho(T) \in \mathcal{R}$ (where $\rho(T)$ denotes the reading word of T (see Appendix A3.5) and we identify i with $(i, i+1)$ for $i = 1, 2, \ldots$). Let T be an A-tableau, $R_1, \ldots, R_s$ be its rows, and $i \in \mathbb{P}$. We define a new tableau, denoted "$T \leftarrow i$," as follows. If $R_1 i$ is weakly increasing, then we add i at the end of R_1 and the algorithm stops. If $R_1 i$ is not weakly increasing, then we proceed according to the following steps:

(1) If R_1 contains the subword $i, i+1$, then we set $b_1 = i+1$.
(2) If R_1 does not contain $i, i+1$, then we let b_1 be the smallest number (in R_1) that is strictly larger than i, and we substitute b_1 by i.

At this point we repeat the above algorithm with "R_2" in place of "R_1," and "b_1" in place of "i," and so on until the algorithm stops. The result is a tableau that we denote "$T \leftarrow i$." For example, if

$$T = \begin{array}{|c|c|c|} \hline 1 & 2 & 3 \\ \hline 3 & 4 \\ \cline{1-2} 4 & 6 \\ \cline{1-2} \end{array}$$

and $i = 1$, then

$$T \leftarrow 1 = \begin{array}{|c|c|c|}\hline 1 & 2 & 3 \\\hline 2 & 4 \\\cline{1-2} 3 & 6 \\\cline{1-2} 4 \\\cline{1-1}\end{array}$$

Now, let $w = (i_1, \ldots, i_p) \in \mathcal{R}$. Define a sequence of tableaux $T_1, \ldots, T_p$ inductively by letting

$$\begin{cases} T_1 \overset{\text{def}}{=} i_1, \\ T_j \overset{\text{def}}{=} T_{j-1} \leftarrow i_j, & \text{for } j = 2, \ldots, p. \end{cases}$$

We then define

$$P_A(w) \overset{\text{def}}{=} T_p$$

and $Q_A(w)$ to be the standard tableau corresponding to the sequence

$$\operatorname{sh}(T_1) \subseteq \operatorname{sh}(T_2) \subseteq \cdots \subseteq \operatorname{sh}(T_p).$$

For example, if $w = (1,\, 2,\, 3,\, 1,\, 4,\, 3) \in \mathcal{R}(S_5)$, then

$$P_A(w) = \begin{array}{|c|c|c|c|}\hline 1 & 2 & 3 & 4 \\\hline 2 & 4 \\\cline{1-2}\end{array}$$

and

$$Q_A(w) = \begin{array}{|c|c|c|c|}\hline 1 & 2 & 3 & 5 \\\hline 4 & 6 \\\cline{1-2}\end{array}$$

We now define an equivalence relation on the set $\mathcal{R}$ as follows. Let $\sim_{AK}$ be the equivalence relation generated by the relations

$$(y, x, z) \sim_{AK} (y, z, x)$$

for $x < y < z$,

$$(z, x, y) \sim_{AK} (x, z, y)$$

for $x < y < z$, and

$$(x + 1, x, x + 1) \sim_{AK} (x, x + 1, x)$$

(note that these are exactly the s-relations of Proposition 7.4.4). In other words, we say that u is *elementary AK-equivalent* to w if and only if w can be obtained from u by changing a factor in u of length 3 according to one of the above rules. We then say that u is *AK-equivalent* to w (written $u \sim_{AK} w$) if there is a sequence of elementary AK-equivalences that transforms u to w. Note that if $u \sim_{AK} w$, then u and w are reduced decompositions of the same element.

(a) Let T be a tableau. Show that $P_A(\rho(T)) = T$.

(b) Let $w \in \mathcal{R}$ be an increasing word (so $P_A(w)$ has only one row) and $i \in \mathbb{P}$ be such that $wi \in \mathcal{R}$. Show that $wi \sim_{AK} \rho\,(P_A(w) \leftarrow i)$.

(c) Deduce from part (b) that if $w \in \mathcal{R}$, then $w \sim_{AK} \rho(P_A(w))$.

(d) Use part (c) to show that if $w \in \mathcal{R}$, then $P_A(w)$ is an A-tableau.

(e) Show that if $w, u \in \mathcal{R}$, then $w \sim_{AK} u$ if and only if $P_A(w) = P_A(u)$.

(f) Show that the map $w \mapsto (P_A(w), Q_A(w))$ is a bijection between $\mathcal{R}$ and pairs (P, Q) such that P is an A-tableau, and Q is a standard tableau of the same shape as P. Deduce that for any w and any partition λ, $a_\lambda(w)$ equals the number of tableaux P of shape λ such that $\rho(P)$ is a reduced decomposition of w. (The integer $a_\lambda(w)$ is defined by (7.37).)

(g) Show that the map $w \mapsto Q_A(w)$ is a bijection between $\mathcal{R}(w_0)$ (where w_0 denotes the longest element of S_{n+1}) and the set of all standard tableaux of shape δ_n.

(h) Show that the map in part (g) is the inverse of the map $\hat{p}$ considered in Theorem 7.4.7.

(i) Show that if $w \in \mathcal{R}$, then $D(w) = D(Q_A(w))$.

Notes

Corollary 7.1.4 is due to Steinberg [501]. Theorem 7.1.5 was first proved by Chevalley [133] for Weyl groups and then by Solomon [480] for finite Coxeter groups. Proposition 7.1.7 appears in Charney and Davis' article [125].

As mentioned in the text, Theorem 7.1.10 is due to Bott [77]. Combinatorial proofs have been given by Björner and Brenti in type $\widetilde{A}$ [62], by Bousquet-Mélou and Eriksson [80] in type $\widetilde{C}$, and by H. and K. Eriksson for the other types [224].

All the results in Section 7.2 are due to Reiner and appear in [439]. The material in Sections 7.3 and 7.4 appears in Haiman's article [284]. The symmetric functions of Section 7.5 were introduced by Stanley in [493]; again, our treatment follows [284].

Exercise 2. See Lehrer [361], Barcelo and Goupil [18], and Dyer [211].
Exercises 5, 6, and 7. See Reiner [439].
Exercise 9. See Charney and Davis [125], and Serre [453].
Exercise 10. See Tan [531].
Exercise 11. See Brenti [86].
Exercise 12. See Fomin and Kirillov [246] and Macdonald [381].
Exercise 13. See Björner and Wachs [67] and Stanley [490].
Exercises 15, 16, 17, and 18. See Haiman [284].
Exercise 20. See Fomin and Stanley [247].

Exercise 21. See Edelman and Greene [218].

Except for the classical topic of Poincaré series, most other results on Coxeter groups of an enumerative nature are very recent. The symmetric group has many beautiful enumerative properties, some of which have been known since the 1800s, and a recent trend has been to see to what extent these carry over to other (especially finite) Coxeter groups. Sections 7.1 and 7.2 contain good examples of this concerning length and descent number. Many other statistics on the symmetric group have been and are currently investigated, such as the major index, excedances, and cycle number, to cite only the most common ones. Analogs of these for other Coxeter groups have been found in some cases and are being looked for in others. The reader can see, for example, the articles of Adin, Brenti, and Roichman [1], Billey and Lam [48], Brenti [86], Clarke and Foata [140, 141, 142], Foata and Han [243], Ram [429], Reiner [433, 434, 435, 436, 437, 440], Reiner and Ziegler [443], and Steingrimsson [502], for work of this nature, where also other classical combinatorial concepts, such as partially ordered sets and tableaux, are seen as the type A case of more general ones.

Exercise 11 has connections to the geometry of certain toric varieties. In fact, it is known (see Exercise 3.16) that the Eulerian polynomial $W(t; 1)$ is also the generating function of the h-vector of the Coxeter complex associated to (W, S). Since this Coxeter complex is isomorphic to the boundary complex of a simplicial convex polytope (see, e.g., [64, Proposition 2.3.9]), there follows from standard results in the theory of toric varieties (see, e.g., [404, Chapter 2]) that this h-vector equals the sequence of even-dimensional Betti numbers of certain toric projective varieties associated to the polytope and that $W(t^2; 1)$ is the Poincaré polynomial of these varieties. It has been conjectured [86, Conjecture 5.2] that $W(t; 1)$ has only real zeros for any finite Coxeter system, and this has been verified in almost all cases; see Exercise 11. For more details on the toric varieties associated to the Eulerian polynomials $W(t; 1)$, the reader can consult Stembridge [503, 504] and Dolgachev and Lunts [191], and the references cited there.

Most of the studies on the enumeration of reduced decompositions in finite Coxeter groups have their origin in Stanley's work [493]. In it, Stanley introduced the generating functions $F_w(\mathbf{x})$, proved that they are symmetric, and used them to prove Corollary 7.4.8 (which he had conjectured himself in a previous unpublished note). This corollary had meanwhile been independently proved by Lascoux and Schützenberger in [349] (in response to Stanley's conjecture) using totally different techniques. From these early works, the subject has blossomed. In [349], and independently in articles by Edelman and Greene [217, 218], bijective proofs are given for Corollary 7.4.8. These bijections (some of which are treated in Section 7.4 and in Exercise 21) give much greater insight into the original conjecture and enable one to prove further results (such as Theorem 7.4.11), which Stanley had been unable to obtain with his symmetric function techniques.

In [493], Stanley also had some conjectures on the corresponding problems in type B, including an analog of Corollary 7.4.8. These conjectures inspired more work, and the cited bijections were generalized to type B by Kraśkiewicz [330] and independently by Haiman [284], and then to type D by Haiman [284], Lam [339], and Billey and Haiman [44]. In these last two works, generalizations of the Stanley symmetric functions (sometimes called "stable Schubert polynomials") are defined and studied also for types B and D. Another approach that has recently been followed is that of finding special classes of elements whose number of reduced decompositions admits a nice representation. We refer the reader to the articles of Fan and Stembridge [235, 236, 505, 506, 508] for work in this direction.

8
Combinatorial Descriptions

This chapter has a different flavor from the earlier ones and is intended more as a source of reference than for sequential reading. We take a detailed look at certain concrete combinatorial descriptions of the Coxeter groups of type B, D, $\tilde{A}$, $\tilde{B}$, $\tilde{C}$, and $\tilde{D}$ (as defined in Appendix A1). For each one of these families, we describe in terms of permutations the group, a set of Coxeter generators, length, descent sets, parabolic subgroups, quotients, minimal coset representatives, reflections, Bruhat graph, and (in most cases) Bruhat order. The corresponding descriptions for Coxeter groups of type A are given in Sections 1.5, 2.1, 2.4, and 2.6. Some other cases appear among the exercises.

Throughout this chapter, we use the notation for permutations introduced in Appendix A3.1.

8.1 Type B

Let S_n^B be the group of all bijections w of the set $[\pm n]$ in itself such that

$$w(-a) = -w(a)$$

for all $a \in [\pm n]$, with composition as group operation. If $w \in S_n^B$, then we write $w = [a_1, \ldots, a_n]$ to mean that $w(i) = a_i$ for $i = 1, \ldots, n$, and call this the *window* notation of w. Because of this notation, the group S_n^B is often called the group of all "signed permutations" of $[n]$.

Since the elements of S_n^B are permutations of $[\pm n]$, we can also write them in disjoint cycle form and in complete notation (as elements of $S([\pm n])$).

For example, if $v = [-3, 5, 1, -7, 2, 8, -6, -4]$, then we also write

$$v = 4\,6 - 8 - 2\,7 - 1 - 5\,3 - 3\,5\,1 - 7\,2\,8 - 6 - 4$$

and $v = (-3, -1, 3, 1)(5, 2)(8, -4, 7, -6, -8, 4, -7, 6)(-5, -2)$. We multiply elements of S_n^B "from the right." For example, if $w = [-2, 1, 4, 5, -3, 6]$ and $v = [3, -6, 2, 4, -5, -1]$, then $w^{-1} = [2, -1, -5, 3, 4, 6]$, $vw = [6, 3, 4, -5, -2, -1]$, and $wv = [4, -6, 1, 5, 3, 2]$. We identify S_n as a subgroup of S_n^B in the natural way.

As a set of generators for S_n^B we take $S_B = \{s_1^B, \ldots, s_{n-1}^B, s_0^B\}$, where

$$s_i^B = [1, \ldots, i-1, i+1, i, i+2, \ldots, n]$$

for $i \in [n-1]$, and

$$s_0^B = [-1, 2, \ldots, n].$$

Note that multiplying an element $w \in S_n^B$ on the right by s_i^B (respectively, s_0^B) has the effect of exchanging the values in position i and $i+1$ (respectively, changing the sign of the value in position 1) in the window notation of w, for $i = 1, \ldots, n-1$. On the other hand, the same operation has the effect of exchanging the values in positions i and $i+1$ *as well as* those in positions $-i$ and $-i-1$ (respectively, exchanging the values in positions 1 and -1) in the complete notation of w. This makes it clear that S_B generates S_n^B.

Figure 8.1 illustrates the action of s_0^B. Here and elsewhere in this chapter, we write the minus sign *under* an integer, for ease of notation. For the rest of this section, if there is no danger of confusion, we simply write "S" instead of "S_B" and "s_i" instead of "s_i^B," for $i = 0, \ldots, n-1$.

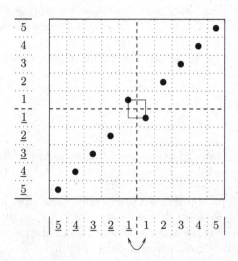

Figure 8.1. Diagram of $s_0^B \in S_5^B$ and its action on S_5^B.

As in the case of S_n, the first step is that of obtaining an explicit combinatorial description of the length function ℓ_B of S_n^B with respect to S. For $v \in S_n^B$, we let

$$\mathrm{inv}_B(v) \stackrel{\mathrm{def}}{=} \mathrm{inv}\,(v(1),\dots,v(n)) \;+\; \mathrm{neg}\,(v(1),\dots,v(n))$$
$$+\; \mathrm{nsp}\,(v(1),\dots,v(n)). \qquad (8.1)$$

For example, if $v = [-3,2,7,-1,-6,5,-4]$, then $\mathrm{inv}_B(v) = 12+4+10 = 26$.

Note that $\mathrm{inv}_B(v)$ can be interpreted combinatorially also as counting certain inversions in the complete notation of v. More precisely, we have that

$$\mathrm{inv}_B(v) \;=\; |\{(i,j) \in [n] \times [n] : \; i < j, \; v(i) > v(j)\}|$$
$$+\; |\{(i,j) \in [n] \times [n] : \; i \le j, \; v(-i) > v(j)\}|. \qquad (8.2)$$

For this reason, we call the inversions counted by the right-hand side of equation (8.2) *B-inversions*. If $v \in S_n^B$, it is not hard to see that

$$\mathrm{neg}\,(v(1),\dots,v(n)) + \mathrm{nsp}\,(v(1),\dots,v(n)) = -\!\!\!\sum_{\{j \in [n]:\; v(j)<0\}}\!\!\! v(j) \qquad (8.3)$$

and that $\mathrm{inv}_B(v) = \mathrm{inv}\,(v)$ if and only if $v \in S_n$.

Proposition 8.1.1 *Let $v \in S_n^B$. Then,*

$$\ell_B(v) = \mathrm{inv}_B(v). \qquad (8.4)$$

Proof. We prove first that

$$\mathrm{inv}_B\,(v) \le \ell_B(v) \qquad (8.5)$$

for all $v \in S_n^B$. Let $v \in S_n^B$. It is easy to see that $\mathrm{inv}\,((vs_0)(1),\dots,(vs_0)(n)) = \mathrm{inv}\,(v(1),\dots,v(n)) - v(1) + \mathrm{sgn}\,(v(1))$, and it therefore follows from definition (8.1) that

$$\mathrm{inv}_B\,(v\,s_0) = \mathrm{inv}_B\,(v) + \mathrm{sgn}\,(v(1)). \qquad (8.6)$$

On the other hand, for $i \in [n-1]$ we clearly have that

$$\mathrm{inv}_B\,(v\,s_i) = \begin{cases} \mathrm{inv}_B\,(v) + 1, & \text{if } v(i) < v(i+1), \\ \mathrm{inv}_B\,(v) - 1, & \text{if } v(i) > v(i+1). \end{cases} \qquad (8.7)$$

Since $\mathrm{inv}_B\,(e) = \ell_B(e) = 0$, equations (8.6) and (8.7) prove inequality (8.5), as claimed.

We now prove equation (8.4) by induction on $\mathrm{inv}_B\,(v)$. If $\mathrm{inv}_B\,(v) = 0$, then $v = [1,2,\dots,n] = e$ and equation (8.4) clearly holds. So, let $t \in \mathbb{N}$ and $v \in S_n^B$ be such that $\mathrm{inv}_B\,(v) = t+1$. Then, $v \ne e$ and, hence, there exists $s \in S$ such that $\mathrm{inv}_B\,(v\,s) = t$ (otherwise equations (8.6) and (8.7) would imply that $0 < v(1) < v(2) < \cdots < v(n)$ and hence that $v = e$). This, by the induction hypothesis, implies that $\ell_B(v\,s) = t$ and hence that $\ell_B(v) \le t+1$. Therefore, $\ell_B(v) \le \mathrm{inv}_B\,(v)$ and this, by inequality (8.5), concludes the induction step. $\square$

As a consequence of Proposition 8.1.1 we obtain the following simple description of the (right) descent set of an element of S_n^B.

Proposition 8.1.2 *Let* $v \in S_n^B$. *Then,*

$$D_R(v) = \{s_i \in S : v(i) > v(i+1)\}, \tag{8.8}$$

where $v(0) \overset{\mathrm{def}}{=} 0$.

Proof. By Proposition 8.1.1, we have that

$$D_R(v) = \{s \in S : \mathrm{inv}_B (v\, s) < \mathrm{inv}_B (v)\},$$

so equation (8.8) follows from equations (8.6) and (8.7). $\square$

For example, if $v = [-1, -3, 4, -2, 7, -5, -6]$, then $D_R(v) = \{s_0, s_1, s_3, s_5, s_6\}$.

Proposition 8.1.3 (S_n^B, S_B) *is a Coxeter system of type* B_n.

Proof. We proceed exactly as in the proof of Proposition 1.5.4. Let $i, i_1, \ldots, i_p \in [0, n-1]$ be such that

$$\ell_B(s_{i_1} \ldots s_{i_p} s_i) < \ell_B(s_{i_1} \ldots s_{i_p}),$$

and $w \overset{\mathrm{def}}{=} s_{i_1} \ldots s_{i_p}$. If $i \in [n-1]$, then, by Proposition 8.1.2, the reasoning goes exactly as in the proof of Proposition 1.5.4, except that $i_j \neq 0$ since $a + b \neq 0$. If $i = 0$, then, by Proposition 8.1.2, $a < 0$, where $a \overset{\mathrm{def}}{=} w(1)$. Hence, a appears in the left half of the complete notation of the identity, but in the right half of the complete notation of w. Hence, there is a $j \in [p]$ such that a is in the left half of the complete notation of $s_{i_1} \ldots s_{i_{j-1}}$ but in the right one of that of $s_{i_1} \ldots s_{i_j}$. Hence, the complete notations of $s_{i_1} \ldots s_{i_p}$ and of $s_{i_1} \ldots \widehat{s}_{i_j} \ldots s_{i_p}$ are equal except that a and $-a$ are interchanged and this, by the definitions of w and a, implies that $s_{i_1} \ldots s_{i_p} s_i = s_{i_1} \ldots \widehat{s}_{i_j} \ldots s_{i_p}$, as desired. $\square$

An alternative way of proving the above proposition is given in Exercise 1.

As was done for S_n, we now give simple combinatorial descriptions of the quotients and parabolic subgroups of S_n^B. Again, for notational simplicity, we treat only the case when $|J| = |S| - 1$. The proof of the next result is clear.

Proposition 8.1.4 *Let* $i \in [0, n-1]$ *and* $J \overset{\mathrm{def}}{=} S \setminus \{s_i\}$. *Then,*

$$(S_n^B)_J = \mathrm{Stab}\,([i+1, n])$$

and

$$(S_n^B)^J = \{v \in S_n^B : 0 < v(1) < \cdots < v(i), v(i+1) < \cdots < v(n)\}.$$

$\square$

The reader should have no trouble formulating the preceding result in the general case. For example, if $n = 6$ and $J = \{s_0, s_2, s_3, s_5\}$, then $(S_n^B)^J = \{v \in S_6^B : 0 < v(1),\ v(2) < v(3) < v(4),\ v(5) < v(6)\}$.

Also for S_n^B, there is a simple combinatorial rule for computing u^J given $u \in S_n^B$ and $J \subseteq S$, which, again, we describe only in the case that $|J| = |S| - 1$. Namely, if $i \in [0, n-1]$ and $J \overset{\text{def}}{=} S \setminus \{s_i\}$, then the window notation of u^J is obtained from that of u by rearranging the elements of $\{u(i+1), \ldots, u(n)\}$ in increasing order and then rearranging the elements of $\{|u(1)|, \ldots, |u(i)|\}$ in increasing order. For example, suppose $n = 9$, $J = \{s_1, s_3, s_4, s_5, s_6, s_8\}$, and $u = [-9, 3, 1, -5, -6, 8, 2, -4, 7]$. Then, u^J is obtained by rearranging $\{-9, 3\}$, $\{1, -5, -6, 8, 2\}$, and $\{-4, 7\}$ in increasing order; hence, $u^J = [-9, 3, -6, -5, 1, 2, 8, -4, 7]$. On the other hand, if $J = \{s_0, s_1, s_2, s_4, s_5, s_6, s_8\}$, then $u^J = [1, 3, 9, -6, -5, 2, 8, -4, 7]$.

It is easy to describe the set of reflections of S_n^B.

Proposition 8.1.5 *The set of reflections of S_n^B is*

$$\{(i,j)(-i,-j) : 1 \le i < |j| \le n\} \bigcup \{(i,-i) : i \in [n]\}.$$

In particular, S_n^B has n^2 reflections.

Proof. Let $w \in S_n^B$ and $i \in [n-1]$. Then, one computes that

$$w s_i w^{-1} = (w(i), w(i+1))(-w(i), -w(i+1)). \tag{8.9}$$

Similarly,

$$w s_0 w^{-1} = (w(1), -w(1)).$$

Since w is an arbitrary element of S_n^B, there follows that $w(i)$ and $w(i+1)$ can be any two elements of $[\pm n]$ that have different absolute value, whereas $w(1)$ can be any element of $[\pm n]$. $\square$

Proposition 8.1.5 enables us to derive a description of the Bruhat graph of S_n^B.

Proposition 8.1.6 *Let $u, v \in S_n^B$. Then, the following are equivalent:*

(i) $u \to v$.

(ii) There exist $i, j \in [\pm n]$, $i < j$, such that $u(i) < u(j)$ and either $v = u(i,j)(-i,-j)$ (if $|i| \ne |j|$) or $v = u(i,j)$ (if $|i| = |j|$).

Proof. By the definition of the Bruhat graph and Proposition 8.1.5, we only have to check that, given i, j, u, and v as in (ii), $\text{inv}_B(v) > \text{inv}_B(u)$ if and only if $u(i) < u(j)$.

If $j = -i$ (note that this implies that $i < 0$), then the complete notation of v is obtained from that of u by exchanging the values $u(i)$ and $u(-i)$. If $u(i) < 0$, then $u(i) < u(-i)$, and it is easy to see from equation (8.1) that $\text{inv}_B(u) < \text{inv}_B(v)$. If $u(i) > 0$, then, by the same reasoning (interchanging the roles of u and v), we conclude that $\text{inv}_B(u) > \text{inv}_B(v)$.

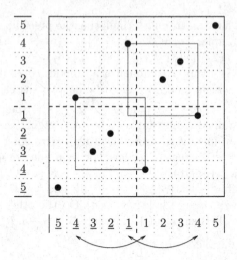

Figure 8.2. Diagram of the reflection $(-4,1)(-1,4) \in S_5^B$ and its action on S_5^B.

If $j \neq -i$ then the complete notation of v is obtained from that of u by exchanging $u(i)$ and $u(j)$ as well as $u(-i)$ and $u(-j)$. Suppose $u(i) <$ $u(j)$. The result is clear if $ij > 0$. If $ij < 0$, then, since $u(i,j)(-i,-j) = u(i,-i)(-j,i)(-i,j)(i,-i)$, the result follows from the two previous cases. Similarly, $\operatorname{inv}_B(v) < \operatorname{inv}_B(u)$ if $u(i) > u(j)$. $\square$

The preceding result implies, in particular, that if $u, v \in S_n = \operatorname{Stab}([n]) \subseteq S_n^B$, then $u \rightarrow v$ in S_n if and only if $u \rightarrow v$ in S_n^B. Thus, the Bruhat graph of S_n is the directed subgraph induced by that of S_n^B on S_n. Figure 8.3 shows the Bruhat order and the Bruhat graph of S_2^B. The reader should compare this picture to the diagram of Bruhat order of B_2 shown in Figure 2.1.

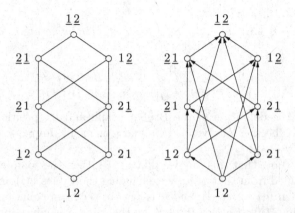

Figure 8.3. Bruhat order and Bruhat graph of S_2^B.

Proposition 8.1.6 also gives a description of the Bruhat order on S_n^B and shows, in particular, that if $u, v \in S_n^B$ and $u \leq v$ in S_n^B, then $u \leq v$ in (the Bruhat order of) $S([\pm n])$. We now show that the converse is also true.

For $v \in S_n^B$, let

$$v[i, j] \overset{\text{def}}{=} |\{a \in [-n, n] : a \leq i, v(a) \geq j\}| \qquad (8.10)$$

for $i, j \in [-n, n]$, where $v(0) \overset{\text{def}}{=} 0$. Note that identity (2.5) continues to hold if $x \in S_n^B$ and $k, i, j, l \in [\pm n]$, $k \leq i$, $j \leq l$.

The next simple identity will be used in the proof of Theorem 8.1.8.

Proposition 8.1.7 Let $v \in S_n^B$. Then,

$$v[-i - 1, -j + 1] - v[i, j] = j - i - 1,$$

for all $i \in [-n, n - 1]$, $j \in [-n + 1, n]$.

Proof. There are several cases to consider. Since they are all similar, we treat only one of them. Suppose that $i, j > 0$. Then we have that

$$\begin{aligned}
v[-i - 1, -j + 1] - v[i, j] &= |\{a < -i : -j < v(a) < j\}| \\
&\quad - |\{a \in [\pm i] : v(a) \geq j\}| \\
&= j - 1 - |\{a \in [-i, -1] : -j < v(a) < j\}| \\
&\quad - (i - |\{a \in [\pm i] : 0 < v(a) < j\}|) \\
&= j - i - 1.
\end{aligned}$$

$\square$

We can now prove the analog of Theorem 2.1.5 for the Bruhat order on S_n^B.

Theorem 8.1.8 Let $u, v \in S_n^B$. Then, the following are equivalent:

(i) $u \leq v$.

(ii) $u[i, j] \leq v[i, j]$, for all $i, j \in [-n, n]$.

Proof. Suppose that (i) holds. We may assume that $u \to v$ in S_n^B. By Proposition 8.1.6, this implies that there are $a, b \in [\pm n]$, $a < b$, such that $u(a) < u(b)$ and either $v = u(a, -a)$ (if $b = |a|$) or $v = u(a, b)(-a, -b)$ (if $b \neq |a|$). Therefore, we conclude from (8.10) that

$$v[i, j] = \begin{cases} u[i, j] + 1, & \text{if } (i, j) \in [a, -a - 1] \times [u(a) + 1, -u(a)], \\ u[i, j], & \text{otherwise,} \end{cases} \qquad (8.11)$$

if $v = u(a, -a)$, and

$$v[i, j] = \begin{cases} u[i, j] + 2, & \text{if } (i, j) \in A \cap B, \\ u[i, j] + 1, & \text{if } (i, j) \in A \triangle B, \\ u[i, j], & \text{otherwise,} \end{cases} \qquad (8.12)$$

if $v = u(a, b)(-a, -b)$, where $A = [a, b - 1] \times [u(a) + 1, u(b)]$ and $B = [-b, -a - 1] \times [-u(b) + 1, -u(a)]$. Thus, (ii) follows.

Suppose now that (ii) holds. Let, for brevity,

$$M(i,j) \stackrel{\text{def}}{=} v[i,j] - u[i,j] \tag{8.13}$$

for $i, j \in [-n, n]$. We may assume that $M(i,j) > 0$ for some i, j. Reasoning as in the proof of Theorem 2.1.5, we then conclude that there exist (a_1, b_1), $(a_0, b_0) \in [\pm n]^2$ such that $a_1 < a_0$, $v(a_1) = b_1 > b_0 = v(a_0)$, and

$$M([a_1, a_0 - 1] \times [b_0 + 1, b_1]) > 0. \tag{8.14}$$

From definition (8.13) and Proposition 8.1.7 we deduce that

$$M(i,j) = M(-i-1, -j+1) \tag{8.15}$$

for all $i \in [-n, n-1]$, $j \in [-n+1, n]$. Hence, by inequality (8.14),

$$M([-a_0, -a_1 - 1] \times [-b_1 + 1, -b_0]) > 0. \tag{8.16}$$

Also, since $v \in S_n^B$, we have that $v(-a_0) = -b_0 > -b_1 = v(-a_1)$. If the rectangles in inequalities (8.14) and (8.16) do not intersect then it follows from equation (8.12) that $v(a_1, a_0)(-a_1, -a_0)[i,j] \geq u[i,j]$ for all $i, j \in [-n, n]$ Hence, by induction we conclude that $u \leq v(a_1, a_0)(-a_1, -a_0)$, and (ii) follows since $v(a_1, a_0)(-a_1, -a_0) \rightarrow v$. If the rectangles in (8.14) and (8.16) do intersect, then $a_1 < 0 < a_0$ and $b_1 > 0 > b_0$ and, hence,

$$M([a, -a-1] \times [-v(a) + 1, v(a)]) > 0,$$

where $a \stackrel{\text{def}}{=} \max(a_1, -a_0)$. However, then, by (8.11), $v(a, -a)[i,j] \geq u[i,j]$ for all $i, j \in [-n, n]$, and we conclude again by induction. $\square$

We illustrate the preceding theorem with an example. Let $n = 4$, $v = [-4, 2, 1, -3]$, and $u = [-1, 3, -4, -2]$. Then, $u[-4, 3] = 0 < 1 = v[-4, 3]$, but $u[-3, 2] = 2 > 1 = v[-3, 2]$, so v and u are incomparable in the Bruhat order of S_4^B.

Theorem 8.1.8 also shows that Bruhat order on S_n^B is an induced subposet of Bruhat order on S_{2n} ($\cong S([\pm n])$).

Corollary 8.1.9 Let $u, v \in S_n^B$. Then, $u \leq v$ in S_n^B if and only if $u \leq v$ in S_{2n}.

Proof. This follows immediately from Theorems 2.1.5 and 8.1.8. $\square$

As for S_n, it is possible to restate Theorem 8.1.8 in a form that is sometimes known as the "tableau criterion" for S_n^B (see Exercise 6).

8.2   Type D

Let S_n^D be the subgroup of S_n^B consisting of all of the signed permutations having an even number of negative entries in their window notation. More precisely,

$$S_n^D \stackrel{\text{def}}{=} \{w \in S_n^B : \text{neg}\,(w(1),\dots,w(n)) \equiv 0 \ (\text{mod } 2)\}.$$

As a set of generators for S_n^D we take $S_D = \{s_0^D,\dots,s_{n-1}^D\}$, where $s_i^D = s_i^B$ for $i \in [n-1]$, and

$$s_0^D = [-2,-1,3,\dots,n] = (1,-2)(2,-1).$$

Figure 8.4 illustrates the action of s_0^D on S_3^D. It is clear that S_D generates S_n^D. Note that $S_n \subset S_n^D \subset S_n^B$. For the rest of this section, if there is no danger of confusion, we write simply "S" instead of "S_D" and "s_i" instead of "s_i^D," for $i = 0,\dots,n-1$.

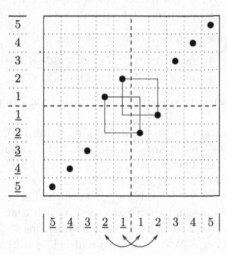

Figure 8.4. Diagram of $s_0^D \in S_5^D$ and its action on S_5^D.

Again, we begin by obtaining an explicit combinatorial description of the length function ℓ_D of S_n^D with respect to S. Let

$$\text{inv}_D(v) \stackrel{\text{def}}{=} \text{inv}_B(v) - \text{neg}\,(v(1),\dots,v(n)) \qquad (8.17)$$

for all $v \in S_n^B$. Note that, as in the case of S_n^B, $\text{inv}_D(v)$ counts certain inversions in the complete notation of v. More precisely, it follows immediately from definition (8.17) and equation (8.2) that

$$\text{inv}_D(v) = |\{(i,j) \in [n] \times [n] : i < j,\ v(i) > v(j)\}|$$
$$+ |\{(i,j) \in [n] \times [n] : i < j,\ v(-i) > v(j)\}|. \qquad (8.18)$$

For this reason, we call the inversions counted by the right-hand side of equation (8.18) D-inversions.

Proposition 8.2.1 Let $v \in S_n^D$. Then,

$$\ell_D(v) = \text{inv}_D(v). \qquad (8.19)$$

Proof. We prove first that

$$\operatorname{inv}_D(v) \le \ell_D(v) \tag{8.20}$$

for all $v \in S_n^D$. Let $v \in S_n^D$. If $i \in [n-1]$, then it follows from equation (8.7) and definition (8.17) that

$$\operatorname{inv}_D(v\, s_i) = \begin{cases} \operatorname{inv}_D(v) + 1, & \text{if } v(i) < v(i+1), \\ \operatorname{inv}_D(v) - 1, & \text{if } v(i) > v(i+1). \end{cases} \tag{8.21}$$

On the other hand, since $s_0^D = s_0^B s_1^B s_0^B$, we obtain from equations (8.6) and (8.7) (for $i = 1$) that

$$\operatorname{inv}_B(v\, s_0^D) = \begin{cases} \operatorname{inv}_B(v) + \operatorname{sgn}(v(1)) + \operatorname{sgn}(v(2)) + 1, & \text{if } -v(1) < v(2), \\ \operatorname{inv}_B(v) + \operatorname{sgn}(v(1)) + \operatorname{sgn}(v(2)) - 1, & \text{if } -v(1) > v(2) \end{cases}$$

and, hence, we conclude from equation (8.17) that

$$\operatorname{inv}_D(v\, s_0^D) = \begin{cases} \operatorname{inv}_D(v) + 1, & \text{if } v(1) + v(2) > 0, \\ \operatorname{inv}_D(v) - 1, & \text{if } v(1) + v(2) < 0. \end{cases} \tag{8.22}$$

Since $\operatorname{inv}_D(e) = \ell_D(e) = 0$, equations (8.21) and (8.22) imply inequality (8.20), as claimed.

We now prove equation (8.19) by induction on $\operatorname{inv}_D(v)$. If $\operatorname{inv}_D(v) = 0$, then, by equations (8.17) and (8.1), $\operatorname{inv}(v(1), \ldots, v(n)) = 0$ and $\operatorname{nsp}(v(1), \ldots, v(n)) = 0$. Therefore, $v(1) < \cdots < v(n)$ and $\operatorname{neg}(v(1), \ldots, v(n)) \le 1$ and this, since $v \in S_n^D$, implies that $v = e$ and equation (8.19) holds in this case. So, let $t \in \mathbb{N}$ and $v \in S_n^D$ be such that $\operatorname{inv}_D(v) = t+1$. Then, $v \ne e$ and we claim that there exists $s \in S$ such that $\operatorname{inv}_D(v\, s) = t$. In fact, if $\operatorname{inv}_D(v\, s) = t+2$ for all $s \in S$, then equations (8.21) and (8.22) would imply that $v(1) < v(2) < \cdots < v(n)$ and $v(1) + v(2) > 0$. However, this implies that $v(1) > 0$ and hence that $v = e$, which proves our claim. So, let $s \in S$ be such that $\operatorname{inv}_D(v\, s) = t$. Then, by our induction hypothesis, $\ell_D(v\, s) = t$ and hence $\ell_D(v) \le t+1$. Therefore, $\ell_D(v) \le \operatorname{inv}_D(v)$ and this, by inequality (8.20), concludes the induction step and, hence, the proof. $\square$

The preceding result allows a simple description of the descent set of an element of S_n^D.

Proposition 8.2.2 *Let* $v \in S_n^D$. *Then,*

$$D_R(v) = \{ s_i \in S : v(i) > v(i+1) \}, \tag{8.23}$$

where $v(0) \overset{\text{def}}{=} -v(2)$ *and* $v(n+1) \overset{\text{def}}{=} 0$.

Proof. By Proposition 8.2.1, we have that

$$D_R(v) = \{ s \in S : \operatorname{inv}_D(v\, s) < \operatorname{inv}_D(v) \},$$

so the result follows from equations (8.21) and (8.22). $\square$

For example, if $v = [-2, 1, 5, -3, -4, -6]$ then $D_R(v) = \{s_0, s_3, s_4, s_5\}$.

As in the previous section, the combinatorial description of S_n^D allows us to give a simple combinatorial proof of the fact that it is indeed a Coxeter group of type D_n. (Here, the definition of D_n, $n \geq 4$, from Appendix A1 is extended to include also $D_2 = A_1 \times A_1$ and $D_3 = A_3$.)

Proposition 8.2.3 (S_n^D, S_D) *is a Coxeter system of type* D_n.

Proof. We follow the proof of Proposition 1.5.4. Let $i, i_1, \ldots, i_p \in [0, n-1]$ be such that

$$\ell_D(s_{i_1} \ldots s_{i_p} s_i) < \ell_D(s_{i_1} \ldots s_{i_p})$$

and $w \stackrel{\text{def}}{=} s_{i_1} \ldots s_{i_p}$.

If $i \in [n-1]$, then, by Proposition 8.2.2, $b > a$, where $a \stackrel{\text{def}}{=} w(i+1)$ and $b \stackrel{\text{def}}{=} w(i)$. The reasoning in this case then goes exactly as in the proof of Proposition 8.1.3, except that now it is possible that $i_j = 0$. However, if $i_j = 0$, then either $s_{i_1} \ldots s_{i_{j-1}}(-2) = a$ and $s_{i_1} \ldots s_{i_{j-1}}(1) = b$, or $s_{i_1} \ldots s_{i_{j-1}}(-1) = a$ and $s_{i_1} \ldots s_{i_{j-1}}(2) = b$, and in either case it still follows that the complete notations of $s_{i_1} \ldots s_{i_{j-1}}$ and $s_{i_1} \ldots s_{i_j}$ are equal except that a and b as well as $-a$ and $-b$ are interchanged, so the reasoning goes through in the same way.

If $i = 0$, then, by Proposition 8.2.2, $b > a$, where $a \stackrel{\text{def}}{=} w(1)$ and $b \stackrel{\text{def}}{=} w(-2)$. Hence, there exists $j \in [p]$ such that a is to the left of b in the complete notation of $s_{i_1} \ldots s_{i_{j-1}}$ but a is to the right of b in the complete notation of $s_{i_1} \ldots s_{i_j}$. Hence, the complete notations of $s_{i_1} \ldots s_{i_{j-1}}$ and $s_{i_1} \ldots s_{i_j}$ are equal except that a and b as well as $-a$ and $-b$ are interchanged (this is obvious if $i_j \in [n-1]$, and it follows from the reasoning done above in the case $i_j = 0$) and we are done as before since multiplying w on the right by s_0 has exactly the effect of interchanging a and b as well as $-a$ and $-b$ in the complete notation of w. $\square$

For a different proof of this proposition, see Exercise 7.

As done in the previous section, we now describe the parabolic subgroups and the quotients of S_n^D (again, for simplicity, we assume that $|J| = |S|-1$). The proof of the next result is clear.

Proposition 8.2.4 *Let* $i \in [0, n-1]$ *and* $J \stackrel{\text{def}}{=} S \setminus \{s_i\}$. *Then,*

$$(S_n^D)_J = \begin{cases} \text{Stab}\,([i+1, n]), & \text{if } i \neq 1, \\ \text{Stab}\,(\{-1, 2, 3, \ldots, n\}), & \text{if } i = 1 \end{cases}$$

and

$$(S_n^D)^J = \{v \in S_n^D : v(-2) < v(1) < \cdots < v(i), \ v(i+1) < \cdots < v(n)\}.$$

$\square$

We illustrate the general case of the preceding proposition with some examples. Let $n = 9$, and $J = \{s_0, s_1, s_2, s_4, s_5, s_7, s_8\}$. Then $(S_n^D)^J = \{v \in$

S_9^D : $v(-2) < v(1) < v(2) < v(3)$, $v(4) < v(5) < v(6)$, $v(7) < v(8) <$ $v(9)$}. If $J = \{s_2, s_3, s_5, s_6, s_8\}$, then $(S_n^D)^J = \{v \in S_9^D : v(2) < v(3) <$ $v(4)$, $v(5) < v(6) < v(7)$, $v(8) < v(9)\}$.

Again, Proposition 8.2.4 implies a simple combinatorial rule for computing u^J given $u \in S_n^D$ and $J \subseteq S$. As usual, we state it only in the case that $|J| = |S| - 1$. Namely, if $J = S \setminus \{s_i\}$ and $i \neq 1$, the window notation of u^J is obtained by rearranging $\{|u(1)|, \ldots, |u(i)|\}$ and $\{u(i+1), \ldots, u(n)\}$ in increasing order and then (possibly) changing the sign of the first entry so that the resulting sequence has an even number of negative entries. On the other hand, if $i = 1$, the window notation of u^J is obtained from that of u by rearranging the elements $\{-u(1), u(2), \ldots, u(n)\}$ in increasing order, and then changing the sign to the smallest one.

We illustrate this (in the general case) with some examples. Let $n = 9$, $J = \{s_0, s_1, s_2, s_4, s_5, s_7, s_8\}$, and $u = [-9, 2, -7, -6, 1, 5, 4, 3, -8]$. Then the window notation of u^J is obtained by rearranging

$$\{|2|, |-7|, |-9|\}, \quad \{-6, 1, 5\}, \quad \text{and} \quad \{4, 3, -8\}$$

in increasing order and, hence, $u^J = [2, 7, 9, -6, 1, 5, -8, 3, 4]$. On the other hand, if $J = \{s_0, s_2, s_3, s_4, s_6, s_7, s_8\}$, then the window notation of u^J is obtained from that of u by rearranging the elements $\{9, 2, -7, -6, 1\}$ in increasing order, then changing the sign to the smallest one (namely -7) and then rearranging the elements $\{5, 4, 3, -8\}$ in increasing order; hence, $u^J = [7, -6, 1, 2, 9, -8, 3, 4, 5]$.

We now describe the set of reflections of S_n^D.

Proposition 8.2.5 *The set of reflections of S_n^D is*

$$\{(i, j)(-i, -j) : 1 \leq |i| < j \leq n\}.$$

In particular, S_n^D has $n^2 - n$ reflections.

Proof. Let $w \in S_n^D$. Then,

$$w s_0^D w^{-1} = (w(1), -w(2))(-w(1), w(2)).$$

On the other hand, we have already computed $w s_i w^{-1}$, for $i \in [n-1]$, in equation (8.9). Since w is an arbitrary element of S_n^D, there follows that $w(i)$ and $w(i+1)$ can be any two elements of $[\pm n]$ such that $w(i) \neq \pm w(i+1)$, and the result follows. $\square$

The preceding result yields the following description of the Bruhat graph of S_n^D.

Proposition 8.2.6 *Let $u, v \in S_n^D$. Then, the following are equivalent:*

(i) $u \rightarrow v$.

(ii) *There exist $i, j \in [\pm n]$, $|i| < j$, such that $u(i) < u(j)$ and $v = u(i, j)(-i, -j)$.*

Proof. By definition of the Bruhat graph and Proposition 8.2.5, we only have to check that, given i, j, u, and v as in (ii), $\mathrm{inv}_D(u) < \mathrm{inv}_D(v)$ if $u(i) < u(j)$.

So, suppose that $u(i) < u(j)$. Then, from Proposition 8.1.6 we know that $\mathrm{inv}_B(v) > \mathrm{inv}_B(u)$. However, $\mathrm{neg}\,(v(1), \ldots, v(n)) = \mathrm{neg}\,(u(1), \ldots, u(n))$ unless $i < 0 < j$ and $u(i) < 0 < u(j)$. But, if $i < 0 < j$ and $u(i) < 0 < u(j)$, then $\mathrm{neg}\,(v(1), \ldots, v(n)) = \mathrm{neg}\,(u(1), \ldots, u(n)) + 2$ and $u \to u(i, -i) \to u(i, -i)(-j, i)(-i, j) \to v$ in S_n^B. Hence, $\mathrm{inv}_B(v) \geq \mathrm{inv}_B(u) + 3$, and the result follows from equation (8.17) also in this case. $\square$

Proposition 8.2.6 shows that the Bruhat graph of S_n^D is the directed subgraph induced by that of S_n^B on S_n^D. For example, the Bruhat graph of S_2^D is shown in Figure 8.5 (cf. Figure 8.3). The Bruhat order of S_3^D is isomorphic to that shown in Figure 2.4.

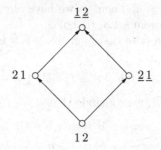

Figure 8.5. The Bruhat graph of S_2^D.

Proposition 8.2.6 also implies that if $u, v \in S_n^D$ and $u \leq v$ in S_n^D, then $u \leq v$ in S_n^B. The converse is, however, not true. For example, $[2, 1] \leq [-2, -1]$ in S_2^B (see Figure 8.3), but $[2, 1]$ and $[-2, -1]$ are incomparable in S_2^D (see Figure 8.5). Therefore, Theorem 8.1.8 is not enough to describe the Bruhat order on S_n^D. However, a slight variation of it does work.

Let $v \in S_n^B$. Given $a, b \in [n]$, we say that $[-a, a] \times [-b, b]$ is an *empty rectangle* (centered at the origin) for v if

$$\{i \in [\pm a] : |v(i)| \leq b\} = \emptyset.$$

The following simple observation is the crucial property of empty rectangles. Its verification is left to the reader (the number $u[i, j]$ is defined in (8.10)).

Lemma 8.2.7 *Let $u \in S_n^B$ and $[-a, a] \times [-b, b]$ be an empty rectangle for u. Then,*

$$u[a, b + 1] - u[-a - 1, b + 1] = a,$$

$$u[-a - 1, -b] - u[-a - 1, b + 1] = b,$$

and

$$u[a, -b] - u[-a - 1, b + 1] = a + b + 1.$$

$\square$

We can now prove the analog of Theorem 8.1.8 for the Bruhat order on S_n^D.

Theorem 8.2.8 *Let $v, u \in S_n^D$. Then, $v \leq u$ if and only if the following hold:*

(i) *$v[i, j] \leq u[i, j]$, for all $i, j \in [-n, n]$.*

(ii) *For all $a, b \in [n]$, if $[-a, a] \times [-b, b]$ is an empty rectangle for both v and u and $u[-a - 1, b + 1] = v[-a - 1, b + 1]$, then $u[-1, b + 1] \equiv v[-1, b + 1] \pmod{2}$.*

Proof. Suppose that $v \leq u$. Then, as we have already observed, $v \leq u$ in S_n^B and, hence, by Theorem 8.1.8, (i) holds.

We now prove (ii). Suppose that $[-a, a] \times [-b, b]$ is empty for both u and v and that

$$u[-a - 1, b + 1] = v[-a - 1, b + 1]. \tag{8.24}$$

Then, from Lemma 8.2.7, we conclude that

$$\begin{aligned} v[a, b + 1] &= u[a, b + 1], \\ v[-a - 1, -b] &= u[-a - 1, -b], \\ v[a, -b] &= u[a, -b]. \end{aligned} \tag{8.25}$$

Assume first that $v \to u$. Then, $v = u(i, j)(-i, -j)$ for some $i, j \in [\pm n]$, $|i| < j$, such that $u(i) > u(j)$ and it is is easy to verify that $v[-1, b + 1] = u[-1, b + 1]$ unless $-a < i \leq -1$, $u(i) \geq b$, $1 \leq j \leq a$, and $u(j) < -b$, in which case clearly $v[-1, b + 1] = u[-1, b + 1] - 2$. This proves (ii) if $v \to u$.

Suppose now that $v \leq u$. Then, we have from the definition of Bruhat order that there exist $u_0, u_1, \ldots, u_k$ such that $v = u_0 \to u_1 \to \cdots \to u_k = u$. Hence, by (i),

$$v[i, j] \leq u_1[i, j] \leq \cdots \leq u_{k-1}[i, j] \leq u[i, j]$$

for all $i, j \in [-n, n]$. Therefore, by equations (8.25), (8.24), and (2.5), $[-a, a] \times [-b, b]$ is empty for u_r for all $r = 0, \ldots, k$, and (ii) follows from the $v \to u$ case.

Conversely, suppose that (i) and (ii) hold. Let, for brevity,

$$M(i, j) \stackrel{\text{def}}{=} u[i, j] - v[i, j]$$

for $i, j \in [-n, n]$. We may assume that $M(i, j) > 0$ for some i, j. Call, for brevity, a rectangle $[a, b - 1] \times [c + 1, d] \subseteq [-n, n] \times [-n, n]$ *allowable* if $u(a) = d$, $u(b) = c$, and

$$M([a, b - 1] \times [c + 1, d]) > 0.$$

Reasoning as in the proof of Theorem 2.1.5, we then conclude that there is at least one allowable rectangle. If there is an allowable rectangle, say $[a_1, a_0 - 1] \times [b_0 + 1, b_1]$, that does not contain the origin, then its "mirror image" $[-a_0, -a_1 - 1] \times [-b_1 + 1, -b_0]$ is also an allowable rectangle (cf. the proof of Theorem 8.1.8) and is disjoint from the first one. In this case, $u' \overset{\text{def}}{=} u(a_1, a_0)(-a_1, -a_0) \to u$ in S_n^D and, as the reader can check, (i) and (ii) still hold with u' in place of u. So, the result follows by induction as in the proof of Theorem 8.1.8.

Therefore, we only have to consider the case when *all* allowable rectangles contain the origin. Call, for convenience, a point $(i, j) \in [\pm n]^2$ a *free point* of u (respectively, v) if $u(i) = j \neq v(i)$ (respectively, $v(i) = j \neq u(i)$). Reasoning as in the proof of Theorem 2.1.5, it is easy to see that every free point of u is either the top left, or bottom right, corner of an allowable rectangle. Since *all* allowable rectangles contain the origin, this implies that if $a_1 < \cdots < a_t < 0 < -a_t < \cdots < -a_1$ are the x-coordinates of the free points of u (and hence of v), then $0 < u(a_1) < \cdots < u(a_t)$.

Notice that this implies that if $a \in [n]$, then $u^{-1}(a) \leq v^{-1}(a)$ (or else $M(v^{-1}(a), a) < 0$). Hence,

$$M(i, j) \geq M(i, j+1) \tag{8.26}$$

for all $i, j \in [\pm n]$, with $j \geq 0$, and $M(0,0) > 0$. Therefore, since $u, v \in S_n^D$, $M(0,0) \geq 2$. Hence, there exists an $r \in [t-1]$ such that $v(a_{r+1}) \neq u(a_r)$ (or else $M(0,0) \leq 1$). Choose r maximal with this property. So, $v(a_{r+1}) \neq u(a_r)$, but $v(a_{j+1}) = u(a_j)$ for $j = r+1, \ldots, t-1$. Note that this implies that $M(-1, 1 + u(a_r)) = 1$ and that, if $a_{r+1} \leq j < -a_{r+1}$, then

$$M(j, u(a_r)) = 2. \tag{8.27}$$

Let $w \overset{\text{def}}{=} u(a_r, -a_{r+1})(a_{r+1}, -a_r)$. Then, by Proposition 8.2.6, $w \to u$ in S_n^D, and we leave it to the reader to verify, using relations (8.12), (8.26), and (8.27), that (i) and (ii) still hold with w in place of u. This, by induction, completes the proof of the theorem. $\square$

We illustrate the preceding theorem with some examples. Let $n = 4$, $u = [4, -3, -2, 1]$, and $v = [4, 3, 2, 1]$. Then, one verifies easily that $v[i,j] \leq u[i,j]$ for all $i, j \in [-4, 4]$. However, $[-2, 2] \times [-2, 2]$ is an empty rectangle for both u and v, and $v[-3, 3] = 0 = u[-3, 3]$, but $v[-1, 3] - u[-1, 3] = 0 - 1 = -1$, so u and v are incomparable in the Bruhat order of S_4^D. On the other hand, if $u = [1, -3, -2, 4]$ and $v = [1, 3, 2, 4]$, then $v[i, j] \leq u[i, j]$ for all $i, j \in [-4, 4]$, and there is no rectangle that is empty for v, so $v \leq u$ in the Bruhat order of S_4^D.

For a different reformulation of Theorem 8.2.8, similar to the tableau criteria for types A and B, see Exercise 11.

8.3 Type $\widetilde{A}$

Let $\widetilde{S}_n$, $n \geq 2$, be the group of all bijections u of $\mathbb{Z}$ in itself such that

$$u(x + n) = u(x) + n \tag{8.28}$$

for all $x \in \mathbb{Z}$ and

$$\sum_{x=1}^{n} u(x) = \binom{n+1}{2}, \tag{8.29}$$

with composition as group operation. Clearly, such a u is uniquely determined by its values on $[n]$, and we write $u = [a_1, \ldots, a_n]$ to mean that $u(i) = a_i$ for $i = 1, \ldots, n$. We call this the *window* notation of u. For example, if $n = 5$, $u = [2, 1, -2, 0, 14]$, and $v = [15, -3, -2, 4, 1]$, then $uv = [24, -4, -7, 0, 2]$. The elements of $\widetilde{S}_n$ are sometimes called *affine permutations* . Note that if $v \in S_n$, then there exists a unique $\widetilde{v} \in \widetilde{S}_n$ such that $\widetilde{v}(i) = v(i)$ for $i \in [n]$. This map $S_n \to \widetilde{S}_n$ is an injective group homomorphism, and for this reason, we sometimes identify a permutation v with the affine permutation $\widetilde{v}$ and consider S_n as a subgroup of $\widetilde{S}_n$.

As a set of generators for $\widetilde{S}_n$ we take $\widetilde{S}_A = \{\widetilde{s}_1^A, \widetilde{s}_2^A, \ldots, \widetilde{s}_n^A\}$, where

$$\widetilde{s}_i^A \stackrel{\text{def}}{=} [1, 2, \ldots, i - 1, i + 1, i, i + 2, \ldots, n]$$

for $i = 1, \ldots, n - 1$ and

$$\widetilde{s}_n^A \stackrel{\text{def}}{=} [0, 2, 3, \ldots, n - 1, n + 1].$$

The action of $\widetilde{s}_5^A$ on $\widetilde{S}_5$ is illustrated in Figure 8.6. Note that multiplying an element $u \in \widetilde{S}_n$ on the right by $\widetilde{s}_i^A$ ($i \in [n]$) causes the interchange of the entries of the complete notation of u in positions $i + kn$ and $i + 1 + kn$ for all $k \in \mathbb{Z}$. Hence,

$$u\widetilde{s}_i^A = \begin{cases} [u(1), \ldots, u(i - 1), u(i + 1), u(i), u(i + 2), \ldots, u(n)], \\ \qquad\qquad\qquad\qquad\qquad\qquad\qquad\quad \text{if } i \in [n - 1], \\ [u(0), u(2), \ldots, u(n - 1), u(n + 1)], \qquad\quad \text{if } i = n. \end{cases}$$

This makes it clear that $\widetilde{s}_1^A, \ldots, \widetilde{s}_n^A$ generate $\widetilde{S}_n$ and shows the symmetry of $\widetilde{s}_1^A, \ldots, \widetilde{s}_n^A$, which is not immediately apparent from their window notations. For the rest of this section, if there is no danger of confusion, we write simply "S" instead of "$\widetilde{S}_A$" and "s_i" instead of "$\widetilde{s}_i^A$," for $i = 1, \ldots, n$.

Note that for all $v \in \widetilde{S}_n$ and $i, j \in \mathbb{Z}$,

$$v(i) \not\equiv v(j) \pmod{n}$$

if and only if $i \not\equiv j \pmod{n}$.

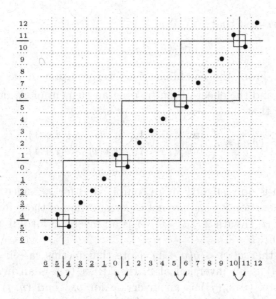

Figure 8.6. Diagram of $\tilde{s}_5^A \in \tilde{S}_5$ and its action on $\tilde{S}_5$.

We now obtain a combinatorial description of the length function $\ell_{\tilde{A}}$ of $\tilde{S}_n$ with respect to S. Given $v \in \tilde{S}_n$ let

$$\mathrm{inv}_{\tilde{A}}(v) \overset{\mathrm{def}}{=} \mathrm{inv}(v(1), \ldots, v(n)) + \sum_{1 \le i < j \le n} \left\lfloor \frac{|v(j) - v(i)|}{n} \right\rfloor$$

$$= \sum_{1 \le i < j \le n} \left\lfloor \left| \frac{v(j) - v(i)}{n} \right| \right\rfloor.$$

For example, if $v = [15, -3, -2, 4, 1] \in \tilde{S}_5$ then $\mathrm{inv}_{\tilde{A}}(v) = 5 + 12 = 17$. Note that $\mathrm{inv}_{\tilde{A}}(v)$ counts certain inversions in the complete notation of v. More precisely,

$$\mathrm{inv}_{\tilde{A}}(v) = |\{(i, j) \in [n] \times \mathbb{P} : i < j, \ v(i) > v(j)\}|. \tag{8.30}$$

In fact,

$$\left\lfloor \frac{|v(j) - v(i)|}{n} \right\rfloor = |\{k \in \mathbb{P} : v(i) > v(j+kn)\}| + |\{k \in \mathbb{P} : v(j) > v(i+kn)\}| \tag{8.31}$$

for all $1 \le i < j \le n$. For this reason we call the inversions counted by the right-hand side of equation (8.30) $\tilde{A}$-*inversions* (sometimes also called *affine inversions*).

Proposition 8.3.1 *We have that*

$$\ell_{\tilde{A}}(v) = \mathrm{inv}_{\tilde{A}}(v) \tag{8.32}$$

for all $v \in \tilde{S}_n$.

Proof. We prove first that

$$\operatorname{inv}_{\widetilde{A}}(v) \le \ell_{\widetilde{A}}(v) \tag{8.33}$$

for all $v \in \widetilde{S}_n$. From the combinatorial interpretation of $\operatorname{inv}_{\widetilde{A}}$ given in equation (8.30) and the effect of multiplying an element of $\widetilde{S}_n$ on the right by s_i, it is clear that

$$\operatorname{inv}_{\widetilde{A}}(vs_i) = \begin{cases} \operatorname{inv}_{\widetilde{A}}(v) + 1, & \text{if } v(i) < v(i+1), \\ \operatorname{inv}_{\widetilde{A}}(v) - 1, & \text{if } v(i) > v(i+1), \end{cases} \tag{8.34}$$

for $i \in [n-1]$.

The situation is only slightly more complicated for s_n, since the values in the window notation for vs_n are not a permutation of those for v. Notice first that a pair (i, j) with $2 \le i \le n - 1$ is an inversion for v if and only if $(i, s_n(j))$ is for vs_n.

Assume now that $v(n) < v(n+1)$. Note that in this case if $j > n + 1$ and a pair (n, j) is an inversion of v, then $(n, s_n(j))$ is an inversion for vs_n. Furthermore, $(n, s_n(j))$ is an inversion for vs_n and (n, j) is not an inversion of v if and only if $v(n+1) \ge v(j) \ge v(n)$. Similarly, if $j > 1$ and a pair $(1, j)$ is an inversion for vs_n, then $(1, s_n(j))$ is an inversion for v, and, furthermore, $(1, s_n(j))$ is an inversion for v and $(1, j)$ is not an inversion for vs_n if and only if $v(1) \ge v(s_n(j)) \ge v(0)$. However, by equation (8.28), the cardinalities of these two sets are equal. Since $(n, n+1)$ is an inversion of vs_n but not of v, we conclude that

$$\operatorname{inv}_{\widetilde{A}}(vs_n) = \operatorname{inv}_{\widetilde{A}}(v) + 1. \tag{8.35}$$

Similarly,

$$\operatorname{inv}_{\widetilde{A}}(vs_n) = \operatorname{inv}_{\widetilde{A}}(v) - 1 \tag{8.36}$$

if $v(n) > v(n+1)$. Since $\operatorname{inv}_{\widetilde{A}}(e) = \ell_{\widetilde{A}}(e) = 0$, equations (8.34), (8.35), and (8.36) prove inequality (8.33), as claimed.

We now prove equation (8.32) by induction on $\operatorname{inv}_{\widetilde{A}}(v)$. If $\operatorname{inv}_{\widetilde{A}}(v) = 0$, then $\left\lfloor \frac{v(j)-v(i)}{n} \right\rfloor = 0$ for all $1 \le i < j \le n$ and, hence, $v(1) < v(2) < \cdots < v(n)$ and $v(n) - v(1) < n$. This implies that $v(i) = v(1)+i-1$ for $i = 1, \ldots, n$ and, therefore, by equation (8.29), that $v = e$, so that equation (8.32) holds. Now, let $t \in \mathbb{N}$ and $v \in \widetilde{S}_n$ be such that $\operatorname{inv}_{\widetilde{A}}(v) = t + 1$. Then, $v \ne e$ and hence there exists $s \in S$ such that $\operatorname{inv}_{\widetilde{A}}(vs) = t$ (otherwise equations (8.34), (8.35) and (8.36) would imply that $v(1) < v(2) < \cdots < v(n) < v(1) + n$ and, hence, that $v = e$, as noted above). This, by the induction hypothesis, implies that $\ell_{\widetilde{A}}(vs) = t$ and hence that $\ell_{\widetilde{A}}(v) \le t + 1$. Therefore, $\ell_{\widetilde{A}}(v) \le \operatorname{inv}_{\widetilde{A}}(v)$ and this, by inequality (8.33), concludes the induction step and hence the proof. $\square$

As a consequence of Proposition 8.3.1, we obtain the following simple description of the descent set of an element of $\widetilde{S}_n$.

Proposition 8.3.2 *Let* $v \in \tilde{S}_n$. *Then,*

$$D_R(v) = \{s_i \in S : v(i) > v(i+1)\}.$$

Proof. By Proposition 8.3.1, we have that

$$D_R(v) = \{s_i \in S : \mathrm{inv}_{\tilde{A}}(v\, s_i) < \mathrm{inv}_{\tilde{A}}(v)\},$$

and the result follows from equations (8.34), (8.35), and (8.36). $\square$

The proof of the following result is essentially identical to that of Proposition 1.5.4 and is therefore omitted.

Proposition 8.3.3 $(\tilde{S}_n, \tilde{S}_A)$ *is a Coxeter system of type* $\tilde{A}_{n-1}$. $\square$

Next, we describe combinatorially the maximal parabolic subgroups and their quotients in the group $\tilde{S}_n$. Again, the proof is clear.

Proposition 8.3.4 *Let* $i \in [n]$, *and* $J \stackrel{\mathrm{def}}{=} S \setminus \{s_i\}$. *Then,*

$$(\tilde{S}_n)_J = \mathrm{Stab}\,([i+1, n+i])$$

and

$$(\tilde{S}_n)^J = \{v \in \tilde{S}_n : v(1) < \cdots < v(i),\ v(i+1) < \cdots < v(n+1)\}.$$

$\square$

Proposition 8.3.4 makes it very easy to describe explicitly the minimal coset representatives of $(\tilde{S}_n)^J$. Consider for notational simplicity only the case $J = S \setminus \{s_i\}$. Then, the complete notation of u^J is obtained from that of u by rearranging the entries $\{u(kn+i+1), \ldots, u(kn+n+i)\}$ in increasing order, for all $k \in \mathbb{Z}$. For example, if $n = 5$, $u = [-3, 6, 3, -5, 14]$ and $J = \{s_1, s_2, s_4, s_5\}$, then $u^J = [3, 6, 9, -5, 2]$, whereas if $J = \{s_1, s_3, s_5\}$, then $u^J = [6, 9, -5, 3, 2]$.

We now describe the set of reflections of $\tilde{S}_n$. It is convenient to introduce the following notation. For $a, b \in \mathbb{Z}$, $a \not\equiv b \pmod{n}$, let

$$t_{a,b} \stackrel{\mathrm{def}}{=} \prod_{r \in \mathbb{Z}} (a + rn, b + rn). \tag{8.37}$$

Thus, $s_i = t_{i,i+1}$ for $i \in [n]$. Note that $t_{a,b} = t_{b,a} = t_{a+kn, b+kn}$ for all $k \in \mathbb{Z}$.

Proposition 8.3.5 *The set of reflections of* $\tilde{S}_n$ *is*

$$\{t_{i,j+kn} : 1 \le i < j \le n, k \in \mathbb{Z}\}.$$

Proof. Let $u \in \tilde{S}_n$, and $i \in [n]$. Then, we have that

$$u s_i u^{-1} = \prod_{r \in \mathbb{Z}} (u(i) + rn, u(i+1) + rn).$$

Since u is any element of $\tilde{S}_n$ we deduce that $u(i)$, $u(i+1)$ can be any two elements of $\mathbb{Z}$ that are not congruent modulo n, and the result follows. $\square$

For example, if $u = [15, -3, -2, 4, 1]$, then $us_1u^{-1} = [1, 20, 3, 4, -13]$, $us_2u^{-1} = [1, 3, 2, 4, 5]$, $us_3u^{-1} = [1, 2, 9, -2, 5]$, $us_4u^{-1} = [4, 2, 3, 1, 5]$, and $us_5u^{-1} = [20, 2, 3, 4, -14]$.

The preceding result enables us to obtain a description of the Bruhat graph of $\widetilde{S}_n$.

Proposition 8.3.6 *Let $u, v \in \widetilde{S}_n$. Then, the following are equivalent:*

(i) $u \to v$.

(ii) *There exist $i, j \in \mathbb{Z}$, $i < j$, $i \not\equiv j \pmod{n}$, such that $u(i) < u(j)$ and $v = ut_{i,j}$.*

Proof. By Proposition 8.3.5 and the definition of the Bruhat graph, the equivalence of (i) and (ii) reduces to showing that if the complete notation of v is obtained from that of u by interchanging $u(i+rn)$ and $u(j+rn)$ for all $r \in \mathbb{Z}$, then $\mathrm{inv}_{\widetilde{A}}(v) > \mathrm{inv}_{\widetilde{A}}(u)$ if $u(i) < u(j)$. This can be established in a way analogous to that used to prove equation (8.35). $\square$

The description of the Bruhat graph of S_n in Section 2.1 and Proposition 8.3.6 show that if $u, v \in S_n$ (and we identify S_n with the subgroup $\mathrm{Stab}([n])$ of $\widetilde{S}_n$ as mentioned at the beginning of this section), then $u \to v$ in S_n if and only if $u \to v$ in $\widetilde{S}_n$. Thus, the Bruhat graph of S_n is the directed subgraph induced by that of $\widetilde{S}_n$ on S_n.

We now obtain a combinatorial characterization of the Bruhat order on $\widetilde{S}_n$. Given $v \in \widetilde{S}_n$, we let

$$v[i, j] \overset{\mathrm{def}}{=} |\{a \le i : v(a) \ge j\}|$$

for all $i, j \in \mathbb{Z}$. For an illustration, see Figure 8.7. Note that $v[i, j] < +\infty$ and

$$v[i, j] = v[i + kn, j + kn] \tag{8.38}$$

for all $i, j, k \in \mathbb{Z}$. Also, note that for any $a \le c$ and $b \ge d$, we have that

$$|\{x \in [a+1, c] : v(x) \in [d, b-1]\}| = v[c, d] - v[c, b] - v[a, d] + v[a, b].$$

Finally, observe that if $v \in \widetilde{S}_n$ and $a < b$ is such that $v(a) < v(b)$, then

$$v(a, b)[i, j] = \begin{cases} v[i, j] + 1, & \text{if } a \le i < b, v(a) < j \le v(b), \\ v[i, j], & \text{otherwise,} \end{cases} \tag{8.39}$$

for all $i, j \in \mathbb{Z}$ (note that $v(a, b)[i, j]$ is well defined even though $v(a, b) \notin \widetilde{S}_n$).

Theorem 8.3.7 *Let $u, v \in \widetilde{S}_n$. Then, the following are equivalent:*

(i) $v \le u$.

(ii) $v[i, j] \le u[i, j]$, *for all $i, j \in \mathbb{Z}$.*

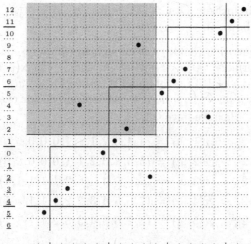

Figure 8.7. $v[4, 2] = 3$ if $v = [1, 2, 9, -2, 5]$.

Proof. Suppose that (i) holds. We may assume that $v \to u$ in $\widetilde{S}_n$. This, by Proposition 8.3.6, implies that there exist $i, j \in \mathbb{Z}$, $i < j$, $i \not\equiv j \pmod{n}$, such that $u = vt_{i,j}$ and $u(i) < u(j)$. Then, $i + rn < j + rn$ and $u(i + rn) < u(j + rn)$ for all $r \in \mathbb{Z}$ and (ii) follows from definition (8.37) and equation (8.39).

Assume now that (ii) holds. Let, for brevity,

$$M(i, j) \stackrel{\text{def}}{=} u[i, j] - v[i, j]$$

for $i, j \in \mathbb{Z}$. Then, reasoning exactly as in the proof of Theorem 2.1.5, we deduce that there exist $a_1, b_1, a_0, b_0 \in \mathbb{Z}$ such that $a_1 < a_0$, $u(a_1) = b_1 > b_0 = u(a_0)$, and

$$M([a_1, a_0 - 1] \times [b_0 + 1, b_1]) > 0. \tag{8.40}$$

If the rectangles

$$\{[a_1 + nk, a_0 - 1 + nk] \times [b_0 + 1 + nk, b_1 + nk]\}_{k \in \mathbb{Z}} \tag{8.41}$$

are all disjoint, let $u' = ut_{a_1, a_0}$. Then, by Proposition 8.3.6, $u' \to u$ in $\widetilde{S}_n$ and $u'[i, j] \geq v[i, j]$ for all $i, j \in \mathbb{Z}$ by relations (8.39) and (8.40), so the result follows by induction.

If the rectangles in (8.41) do intersect, then $b_0 + n < b_1$ and $a_1 + n < a_0$. If so, let $(\alpha_1, \beta_1) \stackrel{\text{def}}{=} (a_0 + n, b_0 + n)$. Then, $u(\alpha_1) = \beta_1$ and $u[i, j] > v[i, j]$ if $(i, j) \in [a_1, \alpha_1 - 1] \times [\beta_1 + 1, b_1]$. If the rectangles $\{[a_1 + kn, \alpha_1 - 1 + kn] \times [\beta_1 + 1 + kn, b_1 + nk]\}_{k \in \mathbb{Z}}$ are all disjoint, then $u_2 \stackrel{\text{def}}{=} ut_{a_1, \alpha_1}$ yields an element u_2 such that $u_2 \to u$ in $\widetilde{S}_n$ and $u_2[i, j] \geq v[i, j]$ for all $i, j \in \mathbb{Z}$, and the result again follows by induction. Otherwise, define $(\alpha_2, \beta_2) = (\alpha_1 + n, \beta_1 + n)$ and continue as above. □

Note that, by equation (8.38), it is enough to compute $v[i,j]$ for $i \in [n]$ and $j \in \mathbb{Z}$. Furthermore, $v[i,j] = 0$ if $i \in [n]$ and $j > \max(v(1), \ldots, v(n))$, and $v[i, j-1] = v[i,j] + 1$ if $i \in [n]$ and $j \leq \min(v(1), \ldots, v(n))$. Therefore, only finitely many values $v[i,j]$ need to be computed. For example, if $v = [1, 2, 9, -2, 5] \in \tilde{S}_5$, then we have that

$$
\begin{aligned}
(v[1, -2], \ldots, v[1, 9]) &= (4, 4, 3, 2, 1, 1, 1, 0, 0, 0, 0, 0), \\
(v[2, -2], \ldots, v[2, 9]) &= (5, 5, 4, 3, 2, 1, 1, 0, 0, 0, 0, 0), \\
(v[3, -2], \ldots, v[3, 9]) &= (6, 6, 5, 4, 3, 2, 2, 1, 1, 1, 1, 1), \\
(v[4, -2], \ldots, v[4, 9]) &= (7, 6, 5, 4, 3, 2, 2, 1, 1, 1, 1, 1), \\
(v[5, -2], \ldots, v[5, 9]) &= (8, 7, 6, 5, 4, 3, 3, 2, 1, 1, 1, 1),
\end{aligned}
$$

and all the other values can be readily computed from these. In particular, this shows that only a finite number of comparisons are needed to check condition (ii) in Theorem 8.3.7.

We illustrate the preceding theorem with an example. Let v be as above and $u = [5, 4, 3, 2, 1] \in \tilde{S}_5$. Then, $v[1, 1] = 2 > 1 = u[1, 1]$, but $v[1, 5] = 0 < 1 = u[1, 5]$, so v and u are incomparable in Bruhat order.

Note that Theorems 2.1.5 and 8.3.7 imply that if $u, v \in S_n$, then $u \leq v$ in S_n if and only if $u \leq v$ in $\tilde{S}_n$. The Hasse diagram of the Bruhat order on the elements of $\tilde{S}_3$ of rank ≤ 3 is shown in Figure 8.8.

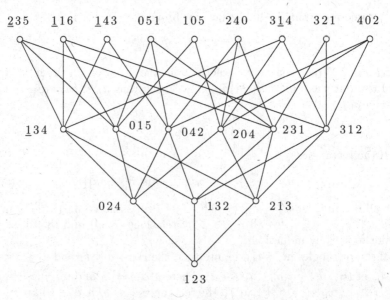

Figure 8.8. The Bruhat order on $\tilde{S}_3$ for elements of length ≤ 3.

8.4 Type $\tilde{C}$

Let $n \geq 2$ and $\tilde{S}_n^C$ be the group of all bijections u of $\mathbb{Z}$ in itself such that

$$u(x + 2n + 1) = u(x) + 2n + 1 \qquad (8.42)$$

and

$$u(-x) = -u(x), \qquad (8.43)$$

for all $x \in \mathbb{Z}$, with composition as group operation. Note that this implies that

$$u(k(2n + 1)) = k(2n + 1)$$

for all $k \in \mathbb{Z}$. From now on, and for the rest of this chapter, we set $N \stackrel{\text{def}}{=} 2n + 1$.

Clearly, an element $u \in \tilde{S}_n^C$ is uniquely determined by its values on $[n]$, and we write $u = [a_1, \ldots, a_n]$ to mean that $u(i) = a_i$ for $i \in [n]$. We call this the *window* notation of u. For example, if $n = 5$, $u = [4, -3, 16, 2, -1]$ and $v = [2, -4, 12, -5, 3]$, then $uv = [-3, -2, 15, 1, 16]$. Note that it follows easily from equations (8.42) and (8.43) that

$$\sum_{i=1}^{N} u(i) = Nn + N = \binom{N+1}{2}$$

for all $u \in \tilde{S}_n^C$. Hence, $\tilde{S}_n^C \subseteq \tilde{S}_{2n+1}$. Also, note that if $v \in S_n^B$, then there is a unique element $\tilde{v} \in \tilde{S}_n^C$ such that $\tilde{v}(i) = v(i)$ for $i = 1, \ldots, n$. This is an injective group homomorphism $S_n^B \to \tilde{S}_n^C$ and, for this reason, we often identify v with $\tilde{v}$ and consider S_n^B as a subgroup of $\tilde{S}_n^C$.

As a set of generators for $\tilde{S}_n^C$ we take $\tilde{S}_C = \{\tilde{s}_0^C, \tilde{s}_1^C, \ldots, \tilde{s}_n^C\}$, where

$$\tilde{s}_i^C \stackrel{\text{def}}{=} \prod_{r \in \mathbb{Z}} (i + rN, i + 1 + rN)(-i + rN, -i - 1 + rN)$$

for $i = 1, \ldots, n - 1$,

$$\tilde{s}_n^C \stackrel{\text{def}}{=} \prod_{r \in \mathbb{Z}} (n + rN, n + 1 + rN),$$

and

$$\tilde{s}_0^C \stackrel{\text{def}}{=} \prod_{r \in \mathbb{Z}} (1 + rN, -1 + rN).$$

The action of $\tilde{s}_0^C$, $\tilde{s}_3^C$, and $\tilde{s}_4^C$ on $\tilde{S}_4^C$ is illustrated in Figures 8.9, 8.10, and 8.11, respectively. Thus,

$$
w\tilde{s}_i^C =
\begin{cases}
[w(1),\dots,w(i-1),w(i+1),w(i),w(i+2),\dots,w(n)], \\
\qquad\qquad\qquad\qquad\qquad\qquad\quad \text{if } i \in [n-1], \\
[w(1),\dots,w(n-1),w(n+1)], \qquad \text{if } i = n, \\
[w(-1),w(2),\dots,w(n)], \qquad\quad\ \text{if } i = 0,
\end{cases}
$$

for all $w \in \tilde{S}_n^C$. Note that, by equations (8.43) and (8.42), $w(-1) = -w(1)$ and $w(n+1) = 2n+1-w(n)$, so that the window notation of $w\tilde{s}_i^C$ can be computed directly from that of w. For the rest of this section, if there is no danger of confusion, we write simply "S" instead of "$\tilde{S}_C$" and "s_i" instead of "$\tilde{s}_i^C$," for $i = 0,\dots,n$. Note that $u(i) \equiv u(j) \pmod{N}$ if and only if $i \equiv j \pmod{N}$, for all $u \in \tilde{S}_n^C$ and $i,j \in \mathbb{Z}$.

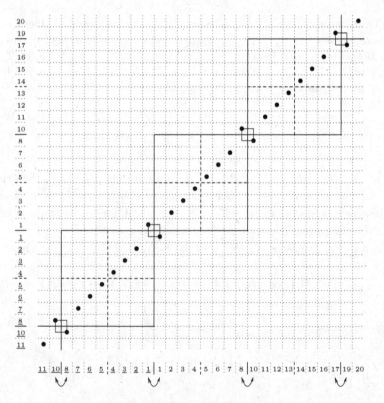

Figure 8.9. Diagram of $\tilde{s}_0^C \in \tilde{S}_4^C$ and its action on $\tilde{S}_4^C$.

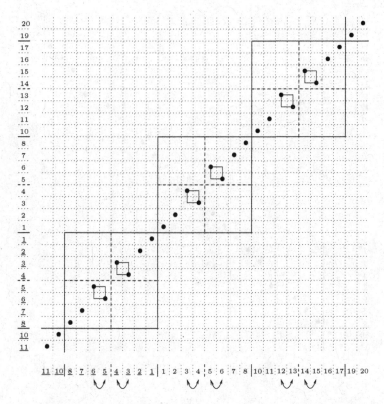

Figure 8.10. Diagram of $\tilde{s}_3^C \in \tilde{S}_4^C$ and its action on $\tilde{S}_4^C$.

The above remarks make it clear that S generates $\tilde{S}_n^C$, and our first goal is that of determining the length function $\ell_{\tilde{C}}$ of $\tilde{S}_n^C$ with respect to S.

Given $v \in \tilde{S}_n^C$, we let

$$\mathrm{inv}_{\tilde{C}}(v) \stackrel{\text{def}}{=} \mathrm{inv}_B(v(1), \ldots, v(n))$$
$$+ \sum_{1 \leq i \leq j \leq n} \left(\left\lfloor \frac{|v(i) - v(j)|}{N} \right\rfloor + \left\lfloor \frac{|v(i) + v(j)|}{N} \right\rfloor \right). \quad (8.44)$$

For example, $\mathrm{inv}_{\tilde{C}}([4, -3, 16, 2, -1]) = 10 + 10 = 20$.

Note that, by equations (8.31) and (8.2), $\mathrm{inv}_{\tilde{C}}(v)$ counts certain inversions in the complete notation of v.

Proposition 8.4.1 Let $v \in \tilde{S}_n^C$. Then,

$$\ell_{\tilde{C}}(v) = \mathrm{inv}_{\tilde{C}}(v). \quad (8.45)$$

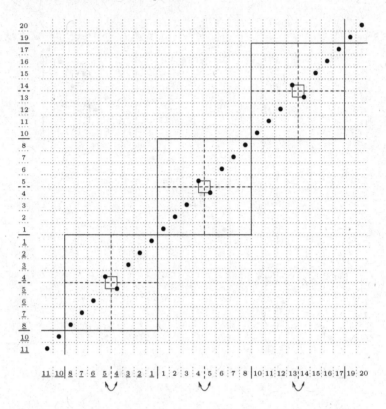

Figure 8.11. Diagram of $\widetilde{s}_4^C \in \widetilde{S}_4^C$ and its action on $\widetilde{S}_4^C$.

Proof. We prove first that

$$\mathrm{inv}_{\widetilde{C}}(v) \leq \ell_{\widetilde{C}}(v) \tag{8.46}$$

for all $v \in \widetilde{S}_n^C$. It is clear from definition (8.44) that

$$\mathrm{inv}_{\widetilde{C}}(vs_i) - \mathrm{inv}_{\widetilde{C}}(v) = \mathrm{sgn}(v(i+1) - v(i)) \tag{8.47}$$

for $i \in [n-1]$. Furthermore, we have from equations (8.44) and (8.6) that

$$\mathrm{inv}_{\widetilde{C}}(vs_0) - \mathrm{inv}_{\widetilde{C}}(v) = \mathrm{sgn}(v(1)). \tag{8.48}$$

Finally, noting that

$$\left\lfloor \frac{|a-k|}{k} \right\rfloor - \left\lfloor \frac{|a|}{k} \right\rfloor = \begin{cases} -1, & \text{if } a \geq k, \\ 0, & \text{if } 0 < a < k, \\ 1, & \text{if } a \leq 0, \end{cases} \tag{8.49}$$

for $a \in \mathbb{Z}$ and $k \in \mathbb{P}$, we conclude from equations (8.44) and (8.1) that

$$
\begin{aligned}
\mathrm{inv}_{\widetilde{C}}(vs_n) - \mathrm{inv}_{\widetilde{C}}(v) = {}& \mathrm{inv}_B(v(1), \dots, v(n-1), N - v(n)) \\
& - \mathrm{inv}_B(v(1), \dots, v(n)) \\
& + |\{i \in [n-1] : v(i) + v(n) \leq 0\}| \\
& - |\{i \in [n-1] : v(i) + v(n) \geq N\}| \\
& + |\{i \in [n-1] : v(n) - v(i) \leq 0\}| \\
& - |\{i \in [n-1] : v(n) - v(i) \geq N\}| \\
& + \left\lfloor \frac{|2N - 2v(n)|}{N} \right\rfloor - \left\lfloor \frac{|2v(n)|}{N} \right\rfloor \\
= {}& \left| \left\lfloor \frac{2N - 2v(n)}{N} \right\rfloor \right| - \left| \left\lfloor \frac{2v(n)}{N} \right\rfloor \right| \\
= {}& \mathrm{sgn}(N - 2v(n)). \qquad\qquad (8.50)
\end{aligned}
$$

Since $\mathrm{inv}_{\widetilde{C}}(e) = 0 = \ell_{\widetilde{C}}(e)$, equations (8.47), (8.48), and (8.50) prove (8.46), as claimed.

We now prove equation (8.45) by induction on $\mathrm{inv}_{\widetilde{C}}(v)$. If $\mathrm{inv}_{\widetilde{C}}(v) = 0$, then $0 < v(1) < v(2) < \cdots < v(n) < n + 1$ and, hence, $v = e$ and equation (8.45) clearly holds. So, let $t \in \mathbb{N}$ and $v \in \widetilde{S}_n^C$ be such that $\mathrm{inv}_{\widetilde{C}}(v) = t + 1$. Then, $v \neq e$ and, hence, there exists $s \in S$ such that $\mathrm{inv}_{\widetilde{C}}(vs) = t$ (otherwise equations (8.47), (8.48), and (8.50) would imply that $0 < v(1) < v(2) < \cdots < v(n) < n + 1$ and hence that $v = e$). This, by the induction hypothesis, implies that $\ell_{\widetilde{C}}(vs) = t$, and the result follows as in the proof of Proposition 8.1.1. $\square$

The proof of the preceding result allows us to obtain a simple description of the descent set of an element of $\widetilde{S}_n^C$.

Proposition 8.4.2 *Let $v \in \widetilde{S}_n^C$. Then,*

$$
D_R(v) = \{s_i \in S : v(i) > v(i+1)\}.
$$

Proof. This follows immediately from equations (8.47), (8.48), and (8.50), and the observation that $v(0) = 0$ and $v(n+1) = N - v(n)$. $\square$

As in the previous sections, we can now give a simple combinatorial proof of the fact that $\widetilde{S}_n^C$ is a Coxeter group of type $\tilde{C}_n$.

Proposition 8.4.3 *$(\widetilde{S}_n^C, \widetilde{S}_C)$ is a Coxeter system of type $\tilde{C}_n$.*

Proof. We proceed as in the proof of Proposition 1.5.4. Let $i, i_1, \dots, i_p \in [0, n]$ be such that

$$
\ell_{\widetilde{C}}(s_{i_1} \dots s_{i_p} s_i) < \ell_{\widetilde{C}}(s_{i_1} \dots s_{i_p})
$$

and $w \stackrel{\mathrm{def}}{=} s_{i_1} \dots s_{i_p}$. If $i \in [n-1]$, then, by Proposition 8.4.2, $b > a$, where $a \stackrel{\mathrm{def}}{=} w(i+1)$ and $b \stackrel{\mathrm{def}}{=} w(i)$, and the reasoning goes through as in Proposition 8.1.3, except that $i_j \neq 0, n$ since $a \not\equiv \pm b \pmod{N}$.

If $i = 0$, then, by Proposition 8.4.2, $0 > a$, where $a \overset{\text{def}}{=} w(1)$. Hence, a appears to the left of 0 in the complete notation of the identity, but to the right of it in that of w. Hence, there exists a $j \in [p]$ such that a is to the left of 0 in the complete notation of $s_{i_1} \ldots s_{i_{j-1}}$ but to the right of it in that of $s_{i_1} \ldots s_{i_j}$. Hence, $a = s_{i_1} \ldots s_{i_{j-1}}(-1) = s_{i_1} \ldots s_{i_j}(1)$ and $i_j = 0$, and, therefore, the complete notations of $s_{i_1} \ldots s_{i_{j-1}}$ and $s_{i_1} \ldots s_{i_j}$ are equal except that $kN + a$ and $kN - a$ are interchanged for each $k \in \mathbb{Z}$. Hence, the same is true for the complete notations of $s_{i_1} \ldots \widehat{s_{i_j}} \ldots s_{i_p}$ and $s_{i_1} \ldots s_{i_p}$ and, therefore,

$$s_{i_1} \ldots s_{i_p} s_i = s_{i_1} \ldots \widehat{s_{i_j}} \ldots s_{i_p}, \tag{8.51}$$

since $i = 0$ and the effect of s_0 on w is exactly that of interchanging $kN + a$ and $kN - a$, for each $k \in \mathbb{Z}$.

If $i = n$, then, by Proposition 8.4.2, $a > N - a$, where $a \overset{\text{def}}{=} w(n)$. Hence, a is to the right of $N - a$ in the complete notation of the identity but to the left of it in that of w. Reasoning as above, we conclude that there exists $j \in [p]$ such that $i_j = n$ and the complete notations of $s_{i_1} \ldots s_{i_{j-1}}$ and $s_{i_1} \ldots s_{i_j}$ are equal, except that $kN + a$ and $(k+1)N - a$ are interchanged for each $k \in \mathbb{Z}$. Thus, the same is true for the complete notations of $s_{i_1} \ldots \widehat{s_{i_j}} \ldots s_{i_p}$ and $s_{i_1} \ldots s_{i_p}$, and this proves equation (8.51) since $i = n$ and the effect of s_n on w is exactly that of interchanging $kN + a$ and $(k+1)N - a$, for each $k \in \mathbb{Z}$. $\square$

We now describe combinatorially the parabolic subgroups and quotients of $\widetilde{S}_n^C$. As in the previous sections, we describe for notational convenience only the case $|S \setminus J| = 1$. The reader should be able to see the validity of the following result "by inspection."

Proposition 8.4.4 Let $i \in [0, n]$, and $J \overset{\text{def}}{=} S \setminus \{s_i\}$. Then,

$$(\widetilde{S}_n^C)_J = \mathrm{Stab}\,([-i, i]) \cap \mathrm{Stab}\,([i+1, 2n-i])$$

and

$$(\widetilde{S}_n^C)^J = \{v \in \widetilde{S}_n^C : v(0) < v(1) < \cdots < v(i),\ v(i+1) < \cdots < v(n+1)\}.$$

$\square$

The preceding result yields a simple combinatorial rule to compute, given $u \in \widetilde{S}_n^C$ and $J \subseteq S$, the minimal coset representative u^J. Namely, if $J = S \setminus \{s_i\}$, then the complete notation of u^J is obtained from that of u by rearranging in increasing order the elements of $\{u(-i+rN), \ldots, u(i+rN)\}$ and $\{u(i+1+rN), \ldots, u((r+1)N - i - 1)\}$ for all $r \in \mathbb{Z}$. Equivalently, the window notation of u^J is obtained from that of u by first writing the elements of $\{|u(1)|, \ldots, |u(i)|\}$ in increasing order, and then writing the $n - i$ smallest elements of $\{u(i+1), \ldots, u(n), N - u(n), \ldots, N - u(i+1)\}$ in increasing order. For example, if $u = [4, -3, 16, 2, -1] \in \widetilde{S}_5^C$ and

$J = S \setminus \{s_3\}$, then $u^J = [3, 4, 16, -1, 2]$, whereas if $J = S \setminus \{s_0, s_2\}$, then $u^J = [-3, 4, -5, -1, 2]$, and if $J = S \setminus \{s_0, s_5\}$, then $u^J = [-3, -1, 2, 4, 16]$.

Next, we describe the set of reflections of $\tilde{S}_n^C$. For $a, b \in \mathbb{Z}$, $a \not\equiv b$ (mod N), let

$$t_{a,b} \overset{\text{def}}{=} \prod_{r \in \mathbb{Z}} (a + rN, b + rN). \tag{8.52}$$

Thus, $s_i = t_{i,i+1} t_{-i,-i-1}$ for $i \in [n-1]$, $s_0 = t_{1,-1}$, and $t_{n,n+1} = s_n$.

Proposition 8.4.5 *The set of reflections of $\tilde{S}_n^C$ is*

$$\{t_{i,j+kN} t_{-i,-j-kN} : 1 \leq i < |j| \leq n, k \in \mathbb{Z}\} \cup \{t_{i,-i+kN} : i \in [n], \ k \in \mathbb{Z}\}.$$

Proof. Let $u \in \tilde{S}_n^C$. Then, we have that

$$u s_i u^{-1} = \prod_{r \in \mathbb{Z}} (u(i) + rN, u(i+1) + rN)(-u(i) + rN, -u(i+1) + rN), \tag{8.53}$$

if $i \in [n-1]$,

$$u s_n u^{-1} = \prod_{r \in \mathbb{Z}} (u(n) + rN, u(n+1) + rN)$$

(note that $u(n+1) = N - u(n)$), and

$$u s_0 u^{-1} = \prod_{r \in \mathbb{Z}} (u(1) + rN, -u(1) + rN). \tag{8.54}$$

Since u is an arbitrary element of $\tilde{S}_n^C$, we conclude that $u(i)$ and $u(i+1)$ can be any two elements of $\mathbb{Z}$, not congruent to 0 modulo N, such that $\pm u(i) \not\equiv u(i+1)$ (mod N). Similarly, we conclude that $u(1)$ (respectively, $u(n)$) can be any element of $\mathbb{Z}$ not congruent to 0 modulo N. The result follows. $\square$

The preceding proposition makes it easy to describe the Bruhat graph of $\tilde{S}_n^C$.

Proposition 8.4.6 *Let $u, v \in \tilde{S}_n^C$. Then, the following are equivalent:*

(i) $u \to v$.

(ii) *There exist $i, j \in \mathbb{Z}$, $j \not\equiv i$ (mod N), $i, j \not\equiv 0$ (mod N), $i < j$, such that $u(i) < u(j)$ and either $v = u t_{i,j} t_{-i,-j}$ (if $i \not\equiv -j$ (mod N)) or $v = u t_{i,j}$ (if $i \equiv -j$ (mod N)).*

Proof. By Proposition 8.4.5 and the definition of the Bruhat graph, it is enough to show that if v, u, i, and j are as in (ii), then $\text{inv}_{\tilde{C}}(v) > \text{inv}_{\tilde{C}}(u)$ if $u(i) < u(j)$. However, this can be verified, using relations (8.44) and (8.1), in a way similar to the proof of 8.50. $\square$

Note that Propositions 8.1.6 and 8.4.6 imply that if $u, v \in S_n^B$, then $u \to v$ in S_n^B if and only if $u \to v$ in $\tilde{S}_n^C$.

We now give a combinatorial characterization of the Bruhat order on $\tilde{S}_n^C$. For $v \in \tilde{S}_n^C$, let

$$v[i,j] \overset{\text{def}}{=} |\{a \in \mathbb{Z} : a \leq i,\, v(a) \geq j\}|, \qquad (8.55)$$

for $i, j \in \mathbb{Z}$. Note that $v[i,j] < +\infty$ and that $v[i,j] = v[i+kN, j+kN]$ for all $i, j, k \in \mathbb{Z}$. The proof of the following result is identical to that of Proposition 8.1.7 and is therefore omitted.

Proposition 8.4.7 *Let $v \in \tilde{S}_n^C$. Then,*

$$v[-i-1, -j+1] - v[i,j] = j - i - 1,$$

for all $i, j \in \mathbb{Z}$. $\square$

We then have the following characterization of Bruhat order on $\tilde{S}_n^C$.

Theorem 8.4.8 *Let $u, v \in \tilde{S}_n^C$. Then, the following are equivalent:*

(i) $v \leq u$.

(ii) $v[i,j] \leq u[i,j]$, *for all $i, j \in \mathbb{Z}$.*

Proof. Suppose that (i) holds. Then, by Propositions 8.3.6 and 8.4.6, $v \leq u$ in the Bruhat order of $\tilde{S}_N$ and, therefore, (ii) holds by Theorem 8.3.7.

Conversely, assume that (ii) holds. Let, as in the previous sections,

$$M(i,j) \overset{\text{def}}{=} u[i,j] - v[i,j] \qquad (8.56)$$

for $i, j \in \mathbb{Z}$. Reasoning as in the proof of Theorem 8.3.7 (recall that $v, u \in \tilde{S}_N$), we then conclude that there exist $a_1, b_1, a_0, b_0 \in \mathbb{Z}$ such that $a_1 < a_0$, $u(a_1) = b_1 > b_0 = u(a_0)$,

$$M([a_1, a_0 - 1] \times [b_0 + 1, b_1]) > 0, \qquad (8.57)$$

and the rectangles

$$\{[a_1 + kN, a_0 - 1 + kN] \times [b_0 + 1 + kN, b_1 + kN]\}_{k \in \mathbb{Z}} \qquad (8.58)$$

are all disjoint. Then, from definition (8.56) and Proposition 8.4.7, we deduce that

$$M(i,j) = M(-i-1, -j+1)$$

for all $i, j \in \mathbb{Z}$. Hence, by inequality (8.57),

$$M([-a_0, -a_1 - 1] \times [-b_1 + 1, -b_0]) > 0. \qquad (8.59)$$

Also, since $u \in \tilde{S}_n^C$, we have that $u(-a_0) = -b_0 > -b_1 = u(-a_1)$. If the rectangles

$$\{[-a_0 + kN, -a_1 - 1 + kN] \times [-b_1 + 1 + kN, -b_0 + kN]\}_{k \in \mathbb{Z}} \qquad (8.60)$$

(which are themselves pairwise disjoint) are disjoint with the ones in (8.58), then $u' \overset{\text{def}}{=} ut_{a_1, a_0} t_{-a_1, -a_0}$ is such that $u' \to u$ (by Proposition 8.4.6) and

$u'[i,j] \geq v[i,j]$ (by relations (8.57), (8.59), and (8.39)), and (i) follows by induction.

If the rectangles in (8.58) and (8.60) intersect, then there exists $\ell \in \mathbb{Z}$ such that the rectangles $[a_1, a_0 - 1] \times [b_0 + 1, b_1]$ and $[-a_0 + \ell N, -a_1 - 1 + \ell N] \times [-b_1 + 1 + \ell N, -b_0 + \ell N]$ have nonempty intersection. Hence, $\ell N - a_0 < a_0$, $a_1 < \ell N - a_1$, $b_0 < \ell N - b_0$, and $\ell N - b_1 < b_1$. Note that this implies that $(\ell N - a_0, a_0)$ and $(a_1, \ell N - a_1)$ are both inversions of u and, hence, that $a_0, a_1 \not\equiv 0 \pmod{N}$. Let $a \overset{\text{def}}{=} \max(a_1, \ell N - a_0)$ and

$$ r \overset{\text{def}}{=} \left\lfloor \frac{2u(a)}{N} \right\rfloor - \ell. $$

Then, the rectangles

$$ \{[a + kN, (\ell + r + k)N - a - 1] \times [(\ell + r + k)N - u(a) + 1, u(a) + kN]\}_{k \in \mathbb{Z}} $$

are all disjoint and, by inequalities (8.57) and (8.59),

$$ M([a, (\ell + r)N - a - 1] \times [(\ell + r)N - u(a) + 1, u(a)]) > 0. \qquad (8.61) $$

Therefore, if $u' \overset{\text{def}}{=} ut_{a,(\ell+r)N-a}$, then $u' \to u$ and, by relations (8.61) and (8.39), $u'[i,j] \geq v[i,j]$ for all $i, j \in \mathbb{Z}$, so (i) again follows by induction. $\square$

We illustrate the preceding theorem with an example. Let $v = [4, -3, 16, 2, -1]$ and $u = [-1, -2, -3, -4, -5]$. Then, $v[-4, 4] = 1 < 2 = u[-4, 4]$ and $v[3, 10] = 1 > 0 = u[3, 10]$, so v and u are incomparable in the Bruhat order of $\tilde{S}_5^C$.

Note that, since $\tilde{S}_n^C \subseteq \tilde{S}_{2n+1}$, the comments made after Theorem 8.3.7 apply to $\tilde{S}_n^C$. In particular, only a finite number of comparisons is actually needed to check condition (ii) of Theorem 8.4.8.

The last theorem can be restated in the following way, which is the "affine analog" of Corollary 8.1.9.

Corollary 8.4.9 Let $u, v \in \tilde{S}_n^C$. Then, $v \leq u$ in $\tilde{S}_n^C$ if and only if $v \leq u$ in $\tilde{S}_{2n+1}$. $\square$

Finally, note that Theorems 8.1.8 and 8.4.8 imply that if $u, v \in S_n^B$ (and we identify S_n^B as a subgroup of $\tilde{S}_n^C$ as explained at the beginning of this section), then $v \leq u$ in S_n^B if and only if $v \leq u$ in $\tilde{S}_n^C$.

The Hasse diagram of the Bruhat order on the elements of $\tilde{S}_2^C$ of rank ≤ 3 is shown in Figure 8.12.

8.5 Type $\tilde{B}$

Let $\tilde{S}_n^B$ be the subgroup of $\tilde{S}_n^C$ consisting of all the elements of $\tilde{S}_n^C$ that have, in the complete notation, an even number of entries to the left of position

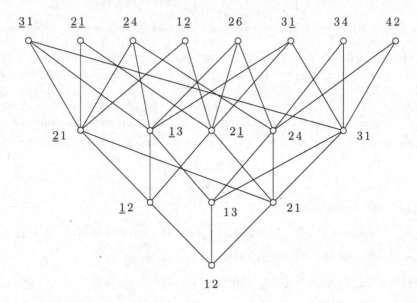

Figure 8.12. The Bruhat order on $\widetilde{S}_2^C$ for the elements of length ≤ 3.

n that are greater than n (note that this number cannot be infinite). More precisely,

$$\widetilde{S}_n^B = \{u \in \widetilde{S}_n^C \; : \; u[n, n+1] \equiv 0 \pmod{2}\}, \qquad (8.62)$$

where $u[n, n+1]$ is defined by (8.55). So, for example, if $u = [4, -3, 16, 2, 1] \in \widetilde{S}_5^C$, then $u[5, 6] = 1$ and, hence, $u \notin \widetilde{S}_5^B$. Thus, $\widetilde{S}_n^B$ is a subgroup of $\widetilde{S}_n^C$ of index 2. In fact, $\widetilde{S}_n^C = \widetilde{S}_n^B \uplus (\widetilde{S}_n^B t_{n,n+1})$ (where $t_{n,n+1}$ has the same meaning as in definition (8.52), so $t_{n,n+1} = \widetilde{s}_n^C$).

Note that if $u \in \widetilde{S}_n^C$, then $u[n, n+1] = 0$ if and only if $u \in S_n^B$ (where we identify S_n^B with a subgroup of $\widetilde{S}_n^C$, as done in the previous section). Hence, in particular, $S_n^B \subseteq \widetilde{S}_n^B$.

As a set of generators for $\widetilde{S}_n^B$ we take $\widetilde{S}_B \overset{\text{def}}{=} \{\widetilde{s}_0^B, \widetilde{s}_1^B, \ldots, \widetilde{s}_{n-1}^B, \widetilde{s}_n^B\}$, where $\widetilde{s}_i^B \overset{\text{def}}{=} \widetilde{s}_i^C$ for $i = 0, \ldots, n-1$,

$$\widetilde{s}_n^B = t_{n-1,n+1} t_{-n+1,-n-1}, \qquad (8.63)$$

and $t_{a,b}$ is defined by (8.52). Figure 8.13 illustrates the action of $\widetilde{s}_4^B$ on $\widetilde{S}_4^B$. From equation (8.63), we deduce that

$$w\widetilde{s}_n^B = [w(1), w(2), \ldots, w(n-2), w(n+1), w(n+2)]$$
$$= [w(1), w(2), \ldots, w(n-2), N - w(n), N - w(n-1)]$$

for all $w \in \widetilde{S}_n^C$. For the rest of this section, if there is no danger of confusion, we write simply "S" instead of "$\widetilde{S}_B$" and "s_i" instead of "$\widetilde{s}_i^B$," for $i = 0, \ldots, n$.

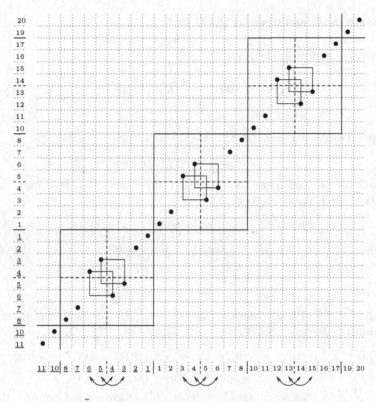

Figure 8.13. Diagram of $\widetilde{s}_4^B \in \widetilde{S}_4^B$ and its action on $\widetilde{S}_4^B$.

It is, of course, useful to recognize, from the window notation, which elements of $\widetilde{S}_n^C$ are in $\widetilde{S}_n^B$. To this end, note that if $i \in [n]$ and $u \in \widetilde{S}_n^C$ then

$$|\{k \in \mathbb{N}: u(i - kN) > n\}| + |\{k \in \mathbb{N}: u(-i - kN) > n\}| = \left\lfloor \frac{|u(i)| + n}{N} \right\rfloor$$

and, hence,

$$\sum_{i=1}^{n} \left\lfloor \frac{|u(i)| + n}{N} \right\rfloor = u[n, n+1]. \tag{8.64}$$

As in the previous sections, we first obtain a combinatorial description of the length function $\ell_{\widetilde{B}}$ of $\widetilde{S}_n^B$ with respect to S. Given $v \in \widetilde{S}_n^B$, let

$$\mathrm{inv}_{\widetilde{B}}(v) = \mathrm{inv}_{\widetilde{C}}(v) - v[n, n+1]. \tag{8.65}$$

For example, if $v = [-1, -2, 6, 4, 14] \in \widetilde{S}_5^B$, then $\mathrm{inv}_{\widetilde{B}}(v) = 14 - 2 = 12$. Note that, since $\left\lfloor \frac{|a|}{N} \right\rfloor = \left\lfloor \frac{|2a|}{N} \right\rfloor - \left\lfloor \frac{|a|+n}{N} \right\rfloor$ for all $a \in \mathbb{Z}$, we have from definition (8.44) and equation (8.64) that

$$
\operatorname{inv}_{\widetilde{B}}(v) = \operatorname{inv}_B(v(1), \ldots, v(n)) + \sum_{i=1}^{n} \left\lfloor \frac{|v(i)|}{N} \right\rfloor
$$

$$
+ \sum_{1 \leq i < j \leq n} \left(\left\lfloor \frac{|v(i) - v(j)|}{N} \right\rfloor + \left\lfloor \frac{|v(i) + v(j)|}{N} \right\rfloor \right). \tag{8.66}
$$

Proposition 8.5.1 *Let* $v \in \widetilde{S}_n^B$. *Then,*

$$
\ell_{\widetilde{B}}(v) = \operatorname{inv}_{\widetilde{B}}(v). \tag{8.67}
$$

Proof. We proceed as in the proof of Proposition 8.4.1. It is clear from equations (8.47) and (8.65) that

$$
\operatorname{inv}_{\widetilde{B}}(vs_i) - \operatorname{inv}_{\widetilde{B}}(v) = \operatorname{sgn}(v(i+1) - v(i)), \tag{8.68}
$$

for all $v \in \widetilde{S}_n^B$ and $i \in [n-1]$. Similarly, from equations (8.48) and (8.65), we obtain that

$$
\operatorname{inv}_{\widetilde{B}}(vs_0) - \operatorname{inv}_{\widetilde{B}}(v) = \operatorname{sgn}(v(1)), \tag{8.69}
$$

for all $v \in \widetilde{S}_n^B$. Finally, noting that $\widetilde{s}_n^B = \widetilde{s}_n^C \widetilde{s}_{n-1}^C \widetilde{s}_n^C$, we conclude from equations (8.50) and (8.47) that

$$
\begin{aligned}
\operatorname{inv}_{\widetilde{C}}(v\widetilde{s}_n^B) - \operatorname{inv}_{\widetilde{C}}(v) &= \operatorname{sgn}(N - 2(v\widetilde{s}_n^C \widetilde{s}_{n-1}^C)(n)) + \operatorname{sgn}(N - 2v(n)) \\
&\quad + \operatorname{sgn}((v\widetilde{s}_n^C)(n) - (v\widetilde{s}_n^C)(n-1)) \\
&= \operatorname{sgn}(N - 2v(n-1)) + \operatorname{sgn}(v(n+1) - v(n-1)) \\
&\quad + \operatorname{sgn}(N - 2v(n)). \tag{8.70}
\end{aligned}
$$

Therefore, we have from equation (8.65) that

$$
\begin{aligned}
\operatorname{inv}_{\widetilde{B}}(v\widetilde{s}_n^B) - \operatorname{inv}_{\widetilde{B}}(v) &= \operatorname{sgn}(v(n+1) - v(n-1)) \\
&= \operatorname{sgn}(N - v(n) - v(n-1)), \tag{8.71}
\end{aligned}
$$

for all $v \in \widetilde{S}_n^B$. Since $\operatorname{inv}_{\widetilde{B}}(e) = 0$, equations (8.68), (8.69), and (8.71) imply that

$$
\operatorname{inv}_{\widetilde{B}}(v) \leq \ell_{\widetilde{B}}(v)
$$

for all $v \in \widetilde{S}_n^B$.

We now prove equation (8.67) by induction on $\operatorname{inv}_{\widetilde{B}}(v)$, in a way exactly analog to the one used in the proof of Proposition 8.4.1. This works since if $v \in \widetilde{S}_n^B$ and $\operatorname{inv}_{\widetilde{B}}(v) = 0$, then $0 < v(1) < v(2) < \cdots < v(n)$ and $v(n-1) + v(n) < N$. Hence, $v(n) < n+1$ (for if $v(n) = n+1$, then $v(n-1) = n-1$ and, hence, $v(i) = i$ for $i = 1, \ldots, n-1$, which is a contradiction since $v \in \widetilde{S}_n^B$), which implies that $v = e$. Similarly, if $v \in \widetilde{S}_n^B$ is such that $\operatorname{inv}_{\widetilde{B}}(vs) > \operatorname{inv}_{\widetilde{B}}(v)$ for all $s \in S$, then equations (8.68), (8.69), and (8.71) would imply that $0 < v(1) < v(2) < \cdots < v(n)$ and $v(n) + v(n-1) < N$. This, as just observed, implies that $v = e$. $\square$

As done in the previous sections, we can now easily obtain a description of the descent set of an element of $\widetilde{S}_n^B$ and a combinatorial proof of the fact that $\widetilde{S}_n^B$ is a Coxeter group of type $\tilde{B}_n$.

Proposition 8.5.2 Let $v \in \widetilde{S}_n^B$. Then,

$$D_R(v) = \{s_i \in S : i \in D(v(0), v(1), \ldots, v(n), v(n+2))\}.$$

Proof. This follows immediately from equations (8.68), (8.69), and (8.71) and the fact that $v(0) = 0$ and $v(n+2) = N - v(n-1)$. $\square$

Proposition 8.5.3 $(\widetilde{S}_n^B, \widetilde{S}_B)$ is a Coxeter system of type $\tilde{B}_n$.

Proof. We prove that the exchange condition holds, as in the proofs of the corresponding results in the previous sections. Let $i, i_1, \ldots, i_p \in [0, n]$ be such that

$$\ell_{\tilde{B}}(s_{i_1} \ldots s_{i_p} s_i) < \ell_{\tilde{B}}(s_{i_1} \ldots s_{i_p})$$

and $w \overset{\text{def}}{=} s_{i_1} \ldots s_{i_p}$.

If $i \in [n-1]$, then, by Proposition 8.5.2, $b > a$, where $a \overset{\text{def}}{=} w(i+1)$, and $b \overset{\text{def}}{=} w(i)$. Hence, there exists $j \in [p]$ such that a is to the left of b in the complete notation of $s_{i_1} \ldots s_{i_{j-1}}$ but is to the right of b in that of $s_{i_1} \ldots s_{i_j}$. Since $|a| \not\equiv |b| \pmod{N}$, this implies that $i_j \in [n]$. If $i_j \in [n-1]$, then the reasoning goes through as in the proof of Proposition 8.4.3. If $i_j = n$, then reasoning as in the proof of the case $i \in [n-1]$, $i_j = 0$ of Proposition 8.2.3, we conclude that the complete notations of $s_{i_1} \ldots s_{i_p}$ and $s_{i_1} \ldots \widehat{s}_{i_j} \ldots s_{i_p}$ are equal, except that $kN + a$ and $kN + b$ as well as $kN - a$ and $kN - b$ are interchanged for each $k \in \mathbb{Z}$. This implies that $s_{i_1} \ldots s_{i_p} s_i = s_{i_1} \ldots \widehat{s}_{i_j} \ldots s_{i_p}$ by the definitions of w, a, and b, since $i \in [n-1]$.

If $i = 0$, then the reasoning goes through exactly as in the proof of the case $i = 0$ of Proposition 8.4.3.

If $i = n$, then, by Proposition 8.5.2, $b > a$, where $b \overset{\text{def}}{=} w(n-1)$ and $a \overset{\text{def}}{=} w(n+1)$, and the reasoning goes through analogously to the case $i = 0$ of the proof of Proposition 8.2.3, since $i_j \neq 0$. $\square$

We now describe combinatorially the (maximal, for notational simplicity) parabolic subgroups and quotients of $\widetilde{S}_n^B$. The reader should be able to verify the following result "by inspection."

Proposition 8.5.4 Let $i \in [0, n]$, and $J \overset{\text{def}}{=} S \setminus \{s_i\}$. Then,

$$(\widetilde{S}_n^B)_J = \begin{cases} \text{Stab}\,([-i, i]) \cap \text{Stab}\,([i+1, 2n-i]), & \text{if } i \neq n-1, \\ \text{Stab}\,([-n-1, n+1] \setminus \{n, -n\}), & \text{if } i = n-1 \end{cases}$$

and

$$(\widetilde{S}_n^B)^J = \{v \in \widetilde{S}_n^B : v(0) < \cdots < v(i), \; v(i+1) < \cdots < v(n) < v(n+2)\}.$$

□

The preceding proposition yields a simple combinatorial rule for computing the minimal coset representatives of $\widetilde{S}_n^B$. Namely, if $u \in \widetilde{S}_n^B$ and $J = S \setminus \{s_i\}$, with $i \neq n-1$, then the complete notation of u^J is obtained from that of u by rearranging $\{u(rN-i), \ldots, u(rN), \ldots, u(rN+i)\}$ and $\{u(rN+i+1), \ldots, u(rN+N-i-1)\}$ in increasing order, for all $r \in \mathbb{Z}$, and then (possibly) switching the elements in positions $rN+n$ and $rN+n+1$, for all $r \in \mathbb{Z}$, so that the resulting element is in $\widetilde{S}_n^B$ (see equation (8.62)). Equivalently, the window notation of u^J is obtained from that of u by first writing $\{|u(1)|, \ldots, |u(i)|\}$ in increasing order, then writing the $n-i$ smallest elements of $\{u(i+1), \ldots, u(n), N-u(n), \ldots, N-u(i+1)\}$ in increasing order, and then (possibly) changing a to $N-a$, where a is the rightmost element written down. On the other hand, if $i = n-1$, then the complete notation of u^J is obtained from that of u by rearranging the elements $\{u(rN-n-1), u(rN-n+1), \ldots, u(rN-1), u(rN), u(rN+1), \ldots, u(rN+n-1), u(rN+n+1)\}$ in increasing order, for all $r \in \mathbb{Z}$. Equivalently, the window notation of u^J is obtained from that of u by writing down the elements of $\{|u(1)|, \ldots, |u(n-1)|, |N-u(n)|\}$ in increasing order and then changing a to $N-a$, where a is the rightmost element written down. For example, if $v = [-1, -2, 6, 4, 14] \in \widetilde{S}_5^B$ and $J = S \setminus \{s_2\}$, then $v^J = [1, 2, -3, 4, 5]$, whereas if $J = S \setminus \{s_0\}$, then $v^J = [-3, -2, -1, 4, 5]$, and if $J = S \setminus \{s_4\}$, then $v^J = [1, 2, 3, 4, 5]$ (as was to be expected).

We now describe the set of reflections of $\widetilde{S}_n^B$.

Proposition 8.5.5 *The set of reflections of $\widetilde{S}_n^B$ is*

$$\{t_{i,j+kN}t_{-i,-j-kN} : 1 \leq i < |j| \leq n, \; k \in \mathbb{Z}\} \cup \{t_{i,2kN-i} : i \in [n], \; k \in \mathbb{Z}\}.$$

Proof. Let $w \in \widetilde{S}_n^B$. We have already computed $w s_i w^{-1}$ for $i = 0, 1, \ldots,$ $n-1$ in equations (8.53) and (8.54). Furthermore,

$$u s_n u^{-1} = \prod_{r \in \mathbb{Z}} (rN+u(n), rN+N-u(n-1))(rN+u(n-1), rN+N-u(n)).$$

(8.72)

Since u is an arbitrary element of $\widetilde{S}_n^B$, we conclude that $u(i)$ and $u(i+1)$ can be any two elements of $\mathbb{Z}$, not congruent to 0 modulo N, such that $u(i) \not\equiv \pm u(i+1) \pmod{N}$. Similarly, $u(1)$ can be any element of $\mathbb{Z}$ such that $u(1) \not\equiv 0 \pmod{N}$. The result follows from equations (8.53), (8.54), and (8.72). □

The proof of the following result is similar to that of Proposition 8.2.6 and is left to the reader.

Proposition 8.5.6 *Let $u, v \in \widetilde{S}_n^B$. Then, $u \to v$ in $\widetilde{S}_n^B$ if and only if $u \to v$ in $\widetilde{S}_n^C$.* □

Thus, the Bruhat graph of $\tilde{S}_n^B$ is the directed subgraph of that of $\tilde{S}_n^C$ induced by $\tilde{S}_n^B$, and hence a combinatorial description of it is given by Proposition 8.4.6.

The preceding proposition implies, in particular, that if $u, v \in \tilde{S}_n^B$ and $u \leq v$ in $\tilde{S}_n^B$, then $u \leq v$ in $\tilde{S}_n^C$. The converse, however, is false. For example, if $v = [4, 3, 2, 1]$ and $u = [4, 3, 7, 8]$, then $v, u \in \tilde{S}_4^B$ and $v \leq u$ in $\tilde{S}_4^C$ (by Proposition 8.4.6, since $u = vt_{4,5}t_{3,6}$), but $v \not\leq u$ in $\tilde{S}_4^B$ since $\operatorname{inv}_{\tilde{B}}(v) = 6 = \operatorname{inv}_{\tilde{B}}(u)$.

A combinatorial characterization of Bruhat order in $\tilde{S}_n^B$, similar to that of Theorem 8.2.8, is not known at present.

The Hasse diagram of the Bruhat order on the elements of $\tilde{S}_2^B$ of rank ≤ 3 is shown in Figure 8.14 (the reader should compare this with Figure 8.12).

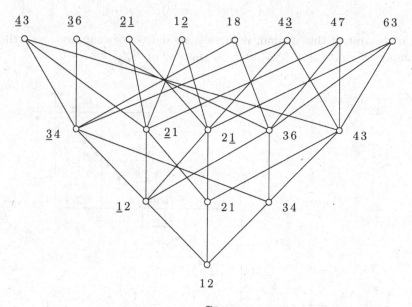

Figure 8.14. The Bruhat order on $\tilde{S}_2^B$ for the elements of length ≤ 3.

8.6 Type $\tilde{D}$

Let $\tilde{S}_n^D$ be the subgroup of $\tilde{S}_n^B$ consisting of all the elements of $\tilde{S}_n^B$ that have, in their complete notation, an even number of negative entries to the right of 0 (note that this number cannot be infinite). More precisely,

$$\tilde{S}_n^D = \left\{ u \in \tilde{S}_n^B : u[0, 1] \equiv 0 \pmod{2} \right\}, \tag{8.73}$$

where $u[a, b]$ is defined by (8.55). So, for example, if $u = [5, -3, 4, 2, 1] \in \widetilde{S}_5^B$, then $u[0, 1] = 1$ and, hence, $u \notin \widetilde{S}_5^D$. Thus, $\widetilde{S}_n^D$ is a subgroup of $\widetilde{S}_n^B$ of index 2. In fact $\widetilde{S}_n^B = \widetilde{S}_n^D \uplus (\widetilde{S}_n^D t_{-1,1})$ (where $t_{-1,1}$ has the same meaning as in definition (8.52), so $t_{-1,1} = \widetilde{s}_0^C$). Note that we may identify S_n^D as a subgroup of $\widetilde{S}_n^D$ in a natural way.

As a set of generators for $\widetilde{S}_n^D$ we take $\widetilde{S}_D \overset{\text{def}}{=} \{\widetilde{s}_0^D, \widetilde{s}_1^D, \ldots, \widetilde{s}_{n-1}^D, \widetilde{s}_n^D\}$, where $\widetilde{s}_i^D \overset{\text{def}}{=} \widetilde{s}_i^B$ for $i = 1, \ldots, n$, and

$$\widetilde{s}_0^D = t_{1,-2} t_{-1,2}. \tag{8.74}$$

The action of $\widetilde{s}_0^D$ on $\widetilde{S}_4^D$ is illustrated in Figure 8.15. From this, we deduce that if $w \in \widetilde{S}_n^C$, then

$$w \widetilde{s}_0^D = [w(-2), w(-1), w(3), \ldots, w(n)].$$

For the rest of this section, if there is no danger of confusion, we write simply "S" instead of "$\widetilde{S}_D$" and "s_i" instead of "$\widetilde{s}_i^D$," for $i = 0, \ldots, n$.

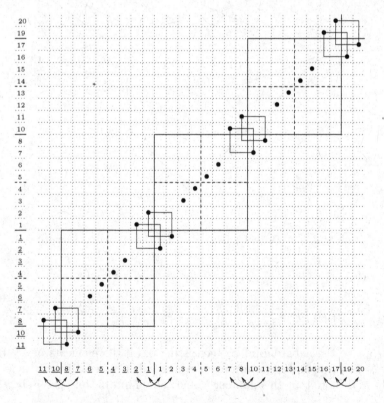

Figure 8.15. Diagram of $\widetilde{s}_0^D \in \widetilde{S}_4^D$ and its action on $\widetilde{S}_4^D$.

It is, of course, useful to be able to recognize, from the window notation, which elements of $\tilde{S}_n^B$ are in $\tilde{S}_n^D$. To this end, note that if $i \in [n]$ and $u \in \tilde{S}_n^C$, then

$$|\{k \in \mathbb{P} : u(i - kN) > 0\}| + |\{k \in \mathbb{N} : u(-i - kN) > 0\}| = \left|\left\lfloor \frac{u(i)}{N} \right\rfloor\right|,$$

and, hence,

$$u[0, 1] = \sum_{i=1}^{n} \left|\left\lfloor \frac{u(i)}{N} \right\rfloor\right| = \sum_{i=1}^{n} \left\lfloor \frac{|u(i)|}{N} \right\rfloor + \mathrm{neg}\,(u(1), \dots, u(n)). \qquad (8.75)$$

As usual, we begin by obtaining a combinatorial description of the length function $\ell_{\tilde{D}}$ of $\tilde{S}_n^D$ with respect to S. Given $v \in \tilde{S}_n^D$, we let

$$\mathrm{inv}_{\tilde{D}}(v) \stackrel{\mathrm{def}}{=} \mathrm{inv}_{\tilde{B}}(v) - v[0, 1]. \qquad (8.76)$$

For example, if $v = [-1, -2, 6, 4, -14] \in \tilde{S}_5^D$, then $\mathrm{inv}_{\tilde{D}}(v) = 21 - 4 = 17$. Note that, from relations (8.75), (8.66), and (8.1), we deduce that

$$\mathrm{inv}_{\tilde{D}}(v) = \mathrm{inv}(v(1), \dots, v(n)) + \mathrm{nsp}\,(v(1), \dots, v(n))$$
$$+ \sum_{1 \le i < j \le n} \left(\left\lfloor \frac{|v(i) - v(j)|}{N} \right\rfloor + \left\lfloor \frac{|v(i) + v(j)|}{N} \right\rfloor \right). \qquad (8.77)$$

Proposition 8.6.1 Let $v \in \tilde{S}_n^D$. Then,

$$\ell_{\tilde{D}}(v) = \mathrm{inv}_{\tilde{D}}(v). \qquad (8.78)$$

Proof. We prove first that

$$\mathrm{inv}_{\tilde{D}}(v) \le \ell_{\tilde{D}}(v), \qquad (8.79)$$

for all $v \in \tilde{S}_n^D$. It is clear from definition (8.76) and equation (8.68) that

$$\mathrm{inv}_{\tilde{D}}(vs_i) = \mathrm{inv}_{\tilde{D}}(v) + \mathrm{sgn}(v(i+1) - v(i)), \qquad (8.80)$$

for all $v \in \tilde{S}_n^D$ and $i \in [n-1]$. Similarly, we conclude from relations (8.76) and (8.71) that

$$\mathrm{inv}_{\tilde{D}}(vs_n) - \mathrm{inv}_{\tilde{D}}(v) = \mathrm{sgn}(N - v(n) - v(n-1)), \qquad (8.81)$$

for all $v \in \tilde{S}_n^D$. Now, using definition (8.76) and proceeding analogously to the proof of equation (8.71) in Proposition 8.5.1 we conclude that

$$\mathrm{inv}_{\tilde{D}}(v\tilde{s}_0^D) - \mathrm{inv}_{\tilde{D}}(v) = \mathrm{sgn}(v(1) + v(2)), \qquad (8.82)$$

for all $v \in \tilde{S}_n^D$. Since $\mathrm{inv}_{\tilde{D}}(e) = \ell_{\tilde{D}}(e) = 0$, equations (8.80), (8.81), and (8.82) prove inequality (8.79), as desired. To prove equation (8.78), we now proceed by induction on $\mathrm{inv}_{\tilde{D}}(v)$ in a way analogous to the proof of Proposition 8.5.1. This works because if $v(1) < \cdots < v(n)$, $0 < v(1) + v(2)$, and $v(n) + v(n-1) < N$, then $0 < v(2) < \cdots < v(n) < N$ and $|v(1)| < N$, which implies that $v[0, 1] = \text{neg}\,(v(1), \ldots, v(n))$ and, therefore, that $v(1) > 0$ since $v \in \widetilde{S}_n^D$. $\square$

As done in the previous sections, we deduce from the proof of the preceding result the following description of the descent set of an element of $\widetilde{S}_n^D$.

Proposition 8.6.2 *Let* $v \in \widetilde{S}_n^D$. *Then,*

$$D_R(v) = \{s_i \in S : i \in D(v(-2), v(1), \ldots, v(n), v(n+2))\}.$$

$\square$

At this point in the chapter, the reader should have no trouble proving the following result by herself.

Proposition 8.6.3 $(\widetilde{S}_n^D, \widetilde{S}_D)$ *is a Coxeter system of type* $\widetilde{D}_n$. $\square$

The next result describes combinatorially the (maximal, for notational simplicity) parabolic subgroups and quotients of $\widetilde{S}_n^D$. Its verification is left to the reader.

Proposition 8.6.4 *Let* $i \in [0, n]$, *and* $J \stackrel{\text{def}}{=} S \setminus \{s_i\}$. *Then,*

$$(\widetilde{S}_n^D)_J = \begin{cases} \text{Stab}\,([-i, i]) \cap \text{Stab}\,([i+1, 2n-i]), & \textit{if } i \neq 1, n-1, \\ \text{Stab}\,([-1, N+1] \setminus \{1, N-1\}), & \textit{if } i = 1, \\ \text{Stab}\,([-n-1, n+1] \setminus \{-n, n\}), & \textit{if } i = n-1, \end{cases}$$

and

$$(\widetilde{S}_n^D)^J = \{v \in \widetilde{S}_n^D : v(-2) < v(1) < \ldots < v(i),$$
$$v(i+1) < \ldots < v(n) < v(n+2)\}.$$

$\square$

As in the preceding sections, Proposition 8.6.4 can be used to describe combinatorially the minimal coset representatives of $\widetilde{S}_n^D$. Namely, if $u \in \widetilde{S}_n^D$ and $J = S \setminus \{s_i\}$ with $i \neq 1, n-1$, then the complete notation of u^J is obtained from that of u by rearranging the elements of $\{u(rN - i), \ldots, u(rN), \ldots, u(rN + i)\}$ and $\{u(rN + i + 1), \ldots, u(rN + N - i - 1)\}$ in increasing order and then (possibly) switching the elements in positions $rN + n$ and $rN + n + 1$ for all $r \in \mathbb{Z}$, and (possibly) those in positions $rN + 1$ and $rN - 1$ for all $r \in \mathbb{Z}$, so that the resulting element is in $\widetilde{S}_n^D$. Equivalently, the window notation of u^J is obtained from that of u by first writing $\{|u(1)|, \ldots, |u(i)|\}$ in increasing order, then writing the $n - i$ smallest elements of $\{u(i+1), \ldots, u(n), N - u(n), \ldots, N - u(i+1)\}$ in increasing order, and then (possibly) changing the sign of the leftmost element written down and (possibly) changing a to $N - a$, where a is the rightmost element written down so that the resulting window notation represents an element of $\widetilde{S}_n^D$.

On the other hand, if $i = n - 1$, then the complete notation of u^J is obtained from that of u by rearranging the elements of $\{u(rN - n - 1), u(rN-n+1), \ldots, u(rN-1), u(rN+1), \ldots, u(rN+n-1), u(rN+n+1)\}$ in increasing order, for all $r \in \mathbb{Z}$, and then (possibly) switching the elements in positions $rN + 1$ and $rN - 1$, for all $r \in \mathbb{Z}$, so that the resulting element is in $\tilde{S}_n^D$. Equivalently, the window notation of u^J is obtained from that of u by writing down the elements of $\{|u(1)|, \ldots, |u(n - 1)|, |N - u(n)|\}$ in increasing order, then changing a to $N-a$, where a is the rightmost element written down, and then (possibly) changing the sign of the leftmost element written down so that the resulting window notation represents an element of $\tilde{S}_n^D$.

Finally, if $i = 1$, then the complete notation of u^J is obtained from that of u by rearranging the elements of $\{u(rN - 1), u(rN + 2), \ldots, u(rN + N - 2), u(rN + N + 1)\}$ in increasing order, for all $r \in \mathbb{Z}$, and then (possibly) switching the elements in positions $rN + n$ and $rN + n + 1$, for all $r \in \mathbb{Z}$, so that the resulting element is in $\tilde{S}_n^D$. Equivalently, the window notation of u^J is obtained from that of u by writing down the n smallest elements of $\{-u(1), u(2), \ldots, u(n), N - u(n), \ldots, N - u(2), N + u(1)\}$ in increasing order, then changing the sign of the leftmost element written down, and then (possibly) changing a to $N - a$, where a is the rightmost element written down, so that the resulting window notation represents an element of $\tilde{S}_n^D$.

For example, if $v = [5, -3, -15, 2, 12] \in \tilde{S}_5^D$ and $J = S \setminus \{s_3\}$, then $v^J = [3, 5, 15, -1, 9]$, whereas if $J = S \setminus \{s_4\}$, then $v^J = [-1, 2, 3, 5, -4]$, and if $J = S \setminus \{s_1\}$, then $v^J = [15, -5, -3, -1, 9]$.

We now describe the set of reflections of $\tilde{S}_n^D$.

Proposition 8.6.5 *The set of reflections of $\tilde{S}_n^D$ is*

$$\{t_{i,j+kN}t_{-i,-j-kN} : 1 \leq i < |j| \leq n, \ k \in \mathbb{Z}\}.$$

Proof. The computations for $s_1, \ldots, s_n$ are the same as in equations (8.53) and (8.72). For s_0, we now obtain that

$$us_0u^{-1} = \prod_{r \in \mathbb{Z}} (rN + u(1), rN - u(2))(rN - u(1), rN + u(2))$$

for all $u \in \tilde{S}_n^D$, and the result follows as in the proof of Proposition 8.5.5. $\square$

The proof of the following result is similar to that of Proposition 8.2.6 and is left to the reader.

Proposition 8.6.6 *Let $u, v \in \tilde{S}_n^D$. Then, $u \rightarrow v$ in $\tilde{S}_n^D$ if and only if $u \rightarrow v$ in $\tilde{S}_n^B$.* $\square$

Thus, the Bruhat graph of $\tilde{S}_n^D$ is the directed subgraph induced on $\tilde{S}_n^D$ by the Bruhat graph of $\tilde{S}_n^B$. Hence, a combinatorial criterion for deciding if

$u \to v$ in $\widetilde{S}_n^D$ is given by Proposition 8.5.6 (and hence by Proposition 8.4.6). Note, however, that when Proposition 8.4.6 is applied to two elements $u, v \in \widetilde{S}_n^D$, then in part (ii) necessarily $i \not\equiv -j \pmod{N}$.

From Proposition 8.6.6, there follows that if $u, v \in \widetilde{S}_n^D$ and $u \leq v$ in $\widetilde{S}_n^D$, then $u \leq v$ in $\widetilde{S}_n^B$. The converse, however, is false; for example, if $u = [5, 3, -15, 2, -12]$ and $v = [-5, -3, -15, 2, -12]$, then $u, v \in \widetilde{S}_5^D$ and $u \leq v$ in $\widetilde{S}_5^B$ (by Proposition 8.5.6, since $v = ut_{-2,2}t_{-1,1}$), but $u \not\leq v$ in $\widetilde{S}_5^D$ since $\mathrm{inv}_{\widetilde{D}}(v) = 25 = \mathrm{inv}_{\widetilde{D}}(u)$.

A result analogous to Theorem 8.2.8, allowing a direct comparison of any two elements of $\widetilde{S}_n^D$ under Bruhat order is not known at present.

Exercises

1. For $n \geq 2$, let $(B_n, \{s_0, s_1, \ldots, s_{n-1}\})$ be a Coxeter system of type B_n (see Appendix A1), so $B_2 \subseteq B_3 \subseteq \cdots$.

   (a) Show that there is a unique group homomorphism $g : B_n \to S_n^B$ such that $g(s_i) = s_i^B$, for $i = 0, \ldots, n-1$, and that g is surjective.

   (b) For $x \in B_n \setminus B_{n-1}$, let $p \overset{\text{def}}{=} \min\{\ell(y) : y \in B_{n-1}x\}$ and $s_{i_1} \ldots s_{i_p} \in B_{n-1}x$. Show that

   $$(i_1, \ldots, i_p) = \begin{cases} (n-1, n-2, \ldots, n-p), & \text{if } p \leq n, \\ (n-1, n-2, \ldots, 1, 0, 1, \ldots, p-n), & \text{if } p \geq n. \end{cases}$$

   (c) Deduce from (b) that there are at most $2n$ right cosets of B_{n-1} in B_n and hence, by induction, that g is a bijection.

2. Prove that the length of $\sigma \in S_n^B$ is given by

   $$\ell_B(\sigma) = \frac{\mathrm{inv}_{\pm}(\sigma) + \mathrm{neg}(\sigma)}{2},$$

   where $\mathrm{inv}_{\pm}(\sigma)$ is the length of σ in the symmetric group $S([\pm n])$; that is,

   $$\mathrm{inv}_{\pm}(\sigma) = |\{(i, j) \in [\pm n]^2 : i < j, \ \sigma(i) > \sigma(j)\}|.$$

   [For example, for $\sigma = [-3, -2, 1] \in S_3^B$, we have $\mathrm{inv}_{\pm}(\sigma) = 8$ and $\mathrm{neg}(\sigma) = 2$, thus $\ell_B(\sigma) = 5$.]

3. Let $w = [-9, 3, 1, -5, -6, 8, 2, -4, 7] \in S_9^B$, and $J = S \setminus \{s_0^B, s_2^B, s_7^B\}$. Compute w^J and $^J w$.

4. Let $k \in [0, n-1]$, and $J \overset{\text{def}}{=} S_B \setminus \{s_k^B\}$.

   (a) For $u, v \in (S_n^B)^J$ show that the following are equivalent:
      (i) $v \leq u$.
      (ii) $v(j) \geq u(j)$, for $j = k+1, \ldots, n$.

(b) Prove (a) without using Theorem 8.1.8.

(c) Deduce Theorem 8.1.8 from part (b) and Theorem 2.6.1.

5. A poset $M(n)$ is defined as follows. The elements of $M(n)$ are the subsets of $[n]$, and the partial order is so defined: If $A, B \subseteq [n]$, with $A = \{a_1, \ldots, a_j\}_<$ and $B = \{b_1, \ldots, b_k\}_<$, then $A \leq B$ if and only if $j \leq k$ and $a_{j-i} \leq b_{k-i}$ for $i = 0, \ldots, j-1$. The diagram of $M(4)$ is shown in Figure 8.16. Show the following:

(a) $M(n)$ is isomorphic to Bruhat order on the quotient $(S_n^B)^J$, where $J = S \setminus \{s_0^B\}$.

(b) $M(n)$ is a distributive lattice.

[*Hint:* Part (a) of Exercise 4 is useful.]

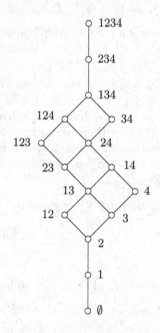

Figure 8.16. The poset $M(4)$.

6. Given $w \in S_n^B$, define an array $A(w) \overset{\text{def}}{=} (A(w)_{i,j})_{1 \leq i \leq n, 1 \leq j \leq n+1-i}$ by letting

$$\{A(w)_{i,1}, \ldots, A(w)_{i,n+1-i}\}_< \overset{\text{def}}{=} \{k \in [\pm n] : w(k) \geq i\}_<,$$

for $i = 1, \ldots, n$. For example, if $w = [-7, 2, 6, -1, -4, 5, -3]$, then

$$
A(w) = \begin{pmatrix}
-7 & -5 & -4 & -1 & 2 & 3 & 6 \\
-7 & -5 & -1 & 2 & 3 & 6 \\
-7 & -5 & -1 & 3 & 6 \\
-5 & -1 & 3 & 6 \\
-1 & 3 & 6 \\
-1 & 3 \\
-1
\end{pmatrix}.
$$

It is easy to verify that $A(w)$ is always weakly increasing down each column. Show that for $u, v \in S_n^B$, the following are equivalent:

(i) $u \leq v$.
(ii) $A(u)_{i,j} \geq A(v)_{i,j}$, for all $i \in [n]$, $j \in [n+1-i]$.

7. For $n \geq 3$, let $(D_n, \{s_0, s_1, \ldots s_{n-1}\})$ be a Coxeter system of type D_n (see Appendix A1), so $A_3 \cong D_3 \subseteq D_4 \subseteq \cdots$.

   (a) Show that there is a unique group homomorphism $h : D_n \to S_n^D$ such that $h(s_i) = s_i^D$, for $i = 0, \ldots n-1$, and that h is surjective.

   (b) For $x \in D_n \setminus D_{n-1}$ let $p \overset{\text{def}}{=} \min\{\ell(y) : y \in D_{n-1}x\}$, and $s_{i_1} \ldots s_{i_p} \in D_{n-1}x$. Show that either $(i_1, \ldots, i_p) = (n - 1, n - 2, \ldots, 3, 2, 0)$ or $(i_1, \ldots, i_p)$ is the initial segment of length p of the sequence $(n - 1, n - 2, \ldots, 2, 1, 0, 2, 3, \ldots, n - 2, n - 1)$.

   (c) Deduce from (b) that there are at most $2n$ right cosets of D_{n-1} in D_n and, hence, by induction, that h is a bijection.

   (d) Show that $h : D_n \to S_n^D$ is an isomorphism directly as a consequence of Exercise 1.

8. Prove that the length of $\sigma \in S_n^D$ is given by

$$
\ell_D(\sigma) = \frac{\mathrm{inv}_\pm(\sigma) - \mathrm{neg}(\sigma)}{2}
$$

(cf. Exercise 2).
[For example, for $\sigma = [-3, -2, 1] \in S_3^D$, we have $\ell_D(\sigma) = 3$.]

9. Let $w = [9, 2, -7, -6, -1, 5, 4, 3, -8] \in S_9^D$. Compute w^J and $^J w$ when $J = \{s_3^D, s_6^D, s_7^D\}$, $\{s_1^D, s_3^D, s_6^D, s_7^D\}$, $\{s_0^D, s_3^D, s_6^D, s_7^D\}$, and $\{s_0^D, s_1^D, s_3^D, s_6^D, s_7^D\}$.

10. Let $u = [-6, 2, 1, -4, 9, -7, 3, 8, -5]$.

    (a) Compute u^J and $^J u$ in S_9^D, when $J = \{s_2^D, s_3^D, s_6^D\}$ and $\{s_1^D, s_2^D, s_3^D, s_6^D\}$.
    (b) Do part (a) considering u as an element of S_9^B.

11. Say that two vectors $(a_1, \ldots, a_k), (b_1, \ldots, b_k) \in \mathbb{Z}^k$ are D-compatible if the following condition is satisfied:

If $\{|a_i|, \ldots, |a_j|\} = \{|b_i|, \ldots, |b_j|\} = [j - i + 1]$ for some $1 \leq i \leq j \leq k$, then $\mathrm{neg}\,(a_i, \ldots, a_j) \equiv \mathrm{neg}\,(b_i, \ldots, b_j)$ (mod 2).

For $u, v \in S_n^D$, let $A(u)_{i,j}$ and $A(v)_{i,j}$ ($i \in [n]$, $j \in [n + 1 - i]$) have the same meaning as in Exercise 6. Show that the following are equivalent:

(i) $u \leq v$.

(ii) $A(u)_{i,j} \geq A(v)_{i,j}$, for all $i \in [n]$ and $j \in [n + 1 - i]$, and the two vectors $(A(u)_{i,1}, \ldots, A(u)_{i,n+1-i})$ and $(A(v)_{i,1}, \ldots, A(v)_{i,n+1-i})$ are D-compatible for all $i \in [n]$.

12. (a) Prove that the one-way infinite path in Example 1.2.6 is indeed the Coxeter graph of S_∞.

(b) Describe as a permutation group the Coxeter group given by the two-way infinite path

(c) Describe as permutation groups the Coxeter groups given by the one-way infinite graphs

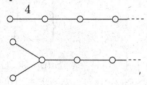

13. Let $w = [-5, 4, 2, 8, 6] \in \widetilde{S}_5$. Compute w^J and $^J w$ when $J = \{\widetilde{s}_1^A, \widetilde{s}_2^A, \widetilde{s}_4^A, \widetilde{s}_5^A\}$.

14. Let $u, v \in \widetilde{S}_n$. Show that the following are equivalent:

(i) $u \to v$.

(ii) There exist $i, j \in [n]$, $i \neq j$, and $k \in \mathbb{N}$ such that

$$\begin{cases} v(a) = u(a), & \text{if } a \in [n] \setminus \{i, j\}, \\ v(i) = u(j) + kn, \\ v(j) = u(i) - kn, \end{cases}$$

and either $|v(i) - v(j)| > |u(i) - u(j)|$, or $|v(i) - v(j)| = |u(i) - u(j)|$ and $(u(i) - u(j))(i - j) > 0$.

15. Given $u \in \widetilde{S}_n$ and $i \in [n]$, let

$$\mathrm{Inv}_i(u) \stackrel{\mathrm{def}}{=} |\{j \in \mathbb{P} : i < j, \ u(i) > u(j)\}|$$

and

$$\mathrm{Inv}(u) \stackrel{\mathrm{def}}{=} (\mathrm{Inv}_1(u), \ldots, \mathrm{Inv}_n(u)).$$

For example, if $u = [5, 3, -2] \in \widetilde{S}_3$, then $\mathrm{Inv}(u) = (4, 2, 0)$. $\mathrm{Inv}(u)$ is called the *affine inversion table* of u. Show that the map $\mathrm{Inv} \colon \widetilde{S}_n \to \mathbb{N}^n \setminus \mathbb{P}^n$ is a bijection.

16. Let $(a_1, \ldots, a_n) \in \mathbb{Z}^n$. Show that the following conditions are equivalent:

    (i) There exists $u \in \tilde{S}_n^C$ such that $(u(1), \ldots, u(n)) = (a_1, \ldots, a_n)$.
    (ii) If $i, j \in [0, n]$ and $i \neq j$, then $\pm a_i \not\equiv a_j \pmod{N}$, (where $a_0 \stackrel{\text{def}}{=} 0$).

17. Let $w = [-5, 7, 2, -1, 8] \in \tilde{S}_5^C$ and $J = \{\tilde{s}_0^C, \tilde{s}_1^C, \tilde{s}_3^C, \tilde{s}_5^C\}$. Compute w^J and $^J w$.

18. Let $w = [-3, 1, 2, 7, 6] \in \tilde{S}_5^B$. Compute w^J and $^J w$ when J equals $\{\tilde{s}_1^B, \tilde{s}_2^B, \tilde{s}_3^B, \tilde{s}_5^B\}$ and $\{\tilde{s}_0^B, \tilde{s}_2^B, \tilde{s}_3^B, \tilde{s}_4^B\}$.

19. Let $w = [2, -7, -1, 3, 6] \in \tilde{S}_5^D$. Compute w^J and $^J w$ when J equals $\{\tilde{s}_0^D, \tilde{s}_5^D\}$, $\{\tilde{s}_1^D, \tilde{s}_3^D, \tilde{s}_5^D\}$, $\{\tilde{s}_0^D, \tilde{s}_2^D, \tilde{s}_3^D, \tilde{s}_4^D\}$, and $\{\tilde{s}_0^D, \tilde{s}_1^D, \tilde{s}_3^D, \tilde{s}_4^D\}$.

20. Let $e_1, \ldots, e_6 \in \mathbb{P}$. Define

$$\mathcal{E}_6 \stackrel{\text{def}}{=} \{e_i : i \in [6]\} \cup \left\{ \frac{\Sigma}{3} - e_i - e_j : \{i, j\} \in \binom{[6]}{2} \right\}$$
$$\cup \left\{ e_i - \frac{\Sigma}{3} : i \in [6] \right\},$$

where $\Sigma \stackrel{\text{def}}{=} e_1 + \cdots + e_6$. Suppose that $|\mathcal{E}_6| \stackrel{\cdot}{=} 27$ (this is easily achieved; take, e.g., $(e_1, \ldots, e_6) = (3, 6, 11, 13, 19, 20)$). Let S_6^E be the subgroup of $S(\mathcal{E}_6)$ consisting of all the $v \in S(\mathcal{E}_6)$ such that the following hold:

    (i) $v\left(\frac{1}{3}\Sigma - e_i - e_j\right) = \frac{1}{3} \sum_{r=1}^{6} v(e_r) - v(e_i) - v(e_j)$, for all $\{i, j\} \in \binom{[6]}{2}$,
    (ii) $v\left(e_i - \frac{1}{3}\Sigma\right) = v(e_i) - \frac{1}{3} \sum_{r=1}^{6} v(e_r)$, for all $i \in [6]$.

It is clear that such a $v \in S_6^E$ is uniquely determined by its values on $\{e_1, \ldots, e_6\}$. We therefore write $v = [a_1, \ldots, a_6]$ to mean that $v(e_i) = a_i$ for $i = 1, \ldots, 6$. So, for example, $e = [e_1, e_2, \ldots, e_6]$. Let

$$s_i = [e_1, \ldots, e_{i-1}, e_{i+1}, e_i, e_{i+2}, \ldots, e_6]$$

for $i = 1, \ldots, 5$, and

$$s_0 = \left[\frac{1}{3}\Sigma - e_2 - e_3, \frac{1}{3}\Sigma - e_1 - e_3, \frac{1}{3}\Sigma - e_1 - e_2, e_4, e_5, e_6 \right].$$

    (a) Show that $S \stackrel{\text{def}}{=} \{s_0, \ldots, s_5\}$ generates S_6^E.
    (b) Show that (S_6^E, S) is a Coxeter system of type E_6.

21. Let $e_1, \ldots, e_7 \in \mathbb{P}$. Define

$$\mathcal{E}_7 \stackrel{\text{def}}{=} \{\pm e_i : i \in [7]\} \cup \left\{ \pm \left(\frac{1}{3}\Sigma - e_i - e_j \right) : \{i, j\} \in \binom{[7]}{2} \right\},$$

where $\Sigma \overset{\text{def}}{=} e_1 + \cdots + e_7$. Suppose that $|\mathcal{E}_7| = 56$. Let S_7^E be the subgroup of $S(\mathcal{E}_7)$ consisting of all the $v \in S(\mathcal{E}_7)$ such that the following hold:

(i) $v(-a) = -v(a)$, for all $a \in \mathcal{E}_7$,

(ii) $v\left(\frac{1}{3}\Sigma - e_i - e_j\right) = \frac{1}{3}\sum_{r=1}^{7} v(e_r) - v(e_i) - v(e_j)$,
   for all $\{i, j\} \in \binom{[7]}{2}$.

It is clear that such a $v \in S_7^E$ is uniquely determined by its values on $\{e_1, \ldots, e_7\}$. We therefore write $v = [a_1, \ldots, a_7]$ to mean that $v(e_i) = a_i$ for $i = 1, \ldots, 7$. Let

$$s_i = [e_1, \ldots, e_{i-1}, e_{i+1}, e_i, e_{i+2}, \ldots, e_7]$$

for $i = 1, \ldots, 6$, and

$$s_0 = \left[\frac{1}{3}\Sigma - e_2 - e_3, \frac{1}{3}\Sigma - e_1 - e_3, \frac{1}{3}\Sigma - e_1 - e_2, e_4, \ldots, e_7\right].$$

(a) Show that $S = \{s_0, \ldots, s_6\}$ generates S_7^E.

(b) Show that (S_7^E, S) is a Coxeter system of type E_7.

22. Let $e_1, \ldots, e_8 \in \mathbb{P}$. Define

$$\mathcal{E}_8 \overset{\text{def}}{=} \{\pm e_i : i \in [8]\} \cup \left\{ \pm \left(\frac{1}{3}\Sigma - e_i - e_j\right) : \{i, j\} \in \binom{[8]}{2} \right\}$$

$$\cup \left\{ \pm \left(e_i + e_j + e_k - \frac{1}{3}\Sigma\right) : \{i, j, k\} \in \binom{[8]}{3} \right\}$$

$$\cup \left\{ \pm(e_i - e_j) : \{i, j\} \in \binom{[8]}{2} \right\},$$

where $\Sigma \overset{\text{def}}{=} e_1 + \ldots + e_8$. Suppose that $|\mathcal{E}_8| = 240$. Let S_8^E be the subgroup of $S(\mathcal{E}_8)$ consisting of all the $v \in S(\mathcal{E}_8)$ such that the following hold:

(i) $v(-a) = -v(a)$, for all $a \in \mathcal{E}_8$,

(ii) $v\left(\frac{1}{3}\Sigma - e_i - e_j\right) = \frac{1}{3}\sum_{r=1}^{8} v(e_r) - v(e_i) - v(e_j)$,
   for all $\{i, j\} \in \binom{[8]}{2}$,

(iii) $v\left(\frac{1}{3}\Sigma - e_i - e_j - e_k\right) = \frac{1}{3}\sum_{r=1}^{8} v(e_r) - v(e_i) - v(e_j) - v(e_k)$,
   for all $\{i, j, k\} \in \binom{[8]}{3}$,

(iv) $v(e_i - e_j) = v(e_i) - v(e_j)$, for all $1 \leq i < j \leq 8$.

It is clear that such a $v \in S_8^E$ is uniquely determined by its values on $\{e_1, \ldots, e_8\}$, so we write $v = [a_1, \ldots, a_8]$ to mean that $v(e_i) = a_i$ for $i = 1, \ldots, 8$. Let

$$s_i = [e_1, \ldots, e_{i-1}, e_{i+1}, e_i, e_{i+2}, \ldots, e_8]$$

for $i = 1, \ldots, 7$, and

$$s_0 = \left[\tfrac{1}{3}\Sigma - e_2 - e_3, \tfrac{1}{3}\Sigma - e_1 - e_3, \tfrac{1}{3}\Sigma - e_1 - e_2, e_4, \ldots, e_8\right].$$

(a) Show that $S = \{s_0, \ldots, s_7\}$ generates S_8^E.
(b) Show that (S_8^E, S) is a Coxeter system of type E_8.

23. Let $e_1, \ldots, e_5 \in \mathbb{P}$. Define

$$\mathcal{F}_4 \stackrel{\text{def}}{=} \{e_i - e_j : i, j \in [5], i \neq j\} \cup \{\pm(e_1 + e_2 - e_j) : j = 3, 4, 5\}$$
$$\cup \{\pm e_i : i \in [5]\} \cup \{\pm(2e_i - e_j) : 1 \leq i < 3 \leq j \leq 5\}.$$

Suppose that $|\mathcal{F}_4| = 48$ (take, e.g., $(e_1, \ldots, e_5) = (7, 8, 9, 10, 11)$). Let S_4^F be the subgroup of $S(\mathcal{F}_4)$ consisting of all the $v \in S(\mathcal{F}_4)$ such that the following hold:

(i) $v(e_i - e_j) = v(e_i) - v(e_j)$, for $i, j \in [5]$, $i \neq j$,
(ii) $v(-a) = -v(a)$, for all $a \in \mathcal{F}_4$,
(iii) $v(e_1 + e_2 - e_j) = v(e_1) + v(e_2) - v(e_j)$, for $j = 3, 4, 5$,
(iv) $v(2e_i - e_j) = 2v(e_i) - v(e_j)$, for $1 \leq i < 3 \leq j \leq 5$.

Such a $v \in S_4^F$ is uniquely determined by its values on $\{e_1, \ldots, e_5\}$ and we write $v = [a_1, \ldots, a_5]$ to mean that $v(e_i) = a_i$ for $i = 1, \ldots, 5$. Let $s_1 = [e_2, e_1, e_3, e_4, e_5]$, $s_3 = [e_1, e_2, e_4, e_3, e_5]$, $s_4 = [e_1, e_2, e_3, e_5, e_4]$, and $s_2 = [e_1 + e_2 - e_3, e_2, 2e_2 - e_3, e_4, e_5]$.

(a) Show that $S = \{s_1, s_2, s_3, s_4\}$ generates S_4^F.
(b) Show that (S_4^F, S) is a Coxeter system of type F_4.

24. Let α be the golden ratio, namely $\alpha = \frac{1}{2}(1 + \sqrt{5})$, and $e_1, e_2, e_3 \in \mathbb{P}$. Define

$$\mathcal{H}_3 \stackrel{\text{def}}{=} \{\pm e_i : i \in [3]\} \cup \{\pm(\alpha e_i - (\alpha - 1)\Sigma) : i \in [3]\},$$

where $\Sigma \stackrel{\text{def}}{=} e_1 + e_2 + e_3$. Suppose that $|\mathcal{H}_3| = 12$ (this is easily achieved; take, e.g., $(e_1, e_2, e_3) = (2, 1 + \alpha, 2\alpha)$). Let S_3^H be the subgroup of $S(\mathcal{H}_3)$ consisting of all the $v \in S(\mathcal{H}_3)$ such that the following hold:

(i) $v(-a) = -v(a)$, for all $a \in \mathcal{H}_3$,
(ii) $v(\alpha e_i - (\alpha - 1)\Sigma) = \alpha v(e_i) - (\alpha - 1)(v(e_1) + v(e_2) + v(e_3))$, for $i = 1, 2, 3$.

It is clear that such a $v \in S_3^H$ is uniquely determined by its values on $\{e_1, e_2, e_3\}$. We therefore write $v = [a_1, a_2, a_3]$ to mean that $v(e_i) = a_i$ for $i = 1, 2, 3$. So, for example, $e = [e_1, e_2, e_3]$. Let $s_1 = [e_2, e_1, e_3]$, $s_2 = [e_1, e_3, e_2]$, and $s_3 = [-e_1 + (\alpha - 1)(e_2 + e_3), e_2, e_3]$.

(a) Show that $S = \{s_0, s_1, s_2\}$ generates S_3^H.
(b) Show that (S_3^H, S) is a Coxeter system of type H_3.

25. Let α be the golden ratio, namely $\alpha = \frac{1}{2}(1+\sqrt{5})$, and $e_1, e_2, e_3, e_4 \in \mathbb{P}$. Define

$$\mathcal{H}_4 \overset{\text{def}}{=} \{\pm e_i : i \in [4]\}$$
$$\cup \{\pm((\alpha + 1)e_i - (\alpha - 1)\Sigma) : i \in [4]\}$$
$$\cup \{\pm(\alpha e_i - (\alpha - 1)\Sigma) : i \in [4]\}$$
$$\cup \{\pm(\alpha e_i - \Sigma) : i \in [4]\}$$
$$\cup \{\pm(e_i + (\alpha + 1)e_j - \Sigma) : i, j \in [4], i \neq j\}$$
$$\cup \{\pm(\alpha e_i + (\alpha + 1)e_j - \Sigma) : i, j \in [4], i \neq j\}$$
$$\cup \{e_i - e_j : i, j \in [4], i \neq j\}$$
$$\cup \{\pm(e_i - \alpha e_j) : i, j \in [4], i \neq j\},$$

where $\Sigma = e_1 + \cdots + e_4$. Suppose that $|\mathcal{H}_4| = 116$ (take, e.g., $(e_1, e_2, e_3, e_4) = (3\alpha, 5, 2 + 2\alpha, 7 - \alpha)$). Let S_4^H be the subgroup of $S(\mathcal{H}_4)$ consisting of all the $v \in S(\mathcal{H}_4)$ such that the following hold:

(i) $v(-a) = -v(a)$, for all $a \in \mathcal{H}_4$,
(ii) $v(\alpha e_i - (\alpha - 1)\Sigma) = \alpha v(e_i) - (\alpha - 1)v(\Sigma)$,
    $v((\alpha + 1)e_i - (\alpha - 1)\Sigma) = (\alpha + 1)v(e_i) - (\alpha - 1)v(\Sigma)$,
    $v(\alpha e_i - \Sigma) = \alpha v(e_i) - v(\Sigma)$, for all $i \in [4]$,
(iii) $v(e_i + (\alpha + 1)e_j - \Sigma) = v(e_i) + (\alpha + 1)v(e_j) - v(\Sigma)$,
    $v(\alpha e_i + (\alpha + 1)e_j - \Sigma) = \alpha v(e_i) + (\alpha + 1)v(e_j) - v(\Sigma)$,
    $v(e_i - \alpha e_j) = v(e_i) - \alpha v(e_j)$,
    $v(e_i - e_j) = v(e_i) - v(e_j)$, for all $i, j \in [4], i \neq j$,

where $v(\Sigma) \overset{\text{def}}{=} v(e_1) + v(e_2) + v(e_3) + v(e_4)$. It is clear that such a $v \in S_4^H$ is uniquely determined by its values on $\{e_1, e_2, e_3, e_4\}$, and we therefore write $v = [a_1, \ldots, a_4]$ to mean that $v(e_i) = a_i$ for $i = 1, \ldots, 4$. Let $s_1 = [e_2, e_1, e_3, e_4]$, $s_2 = [e_1, e_3, e_2, e_4]$, $s_3 = [e_1, e_2, e_4, e_3]$, and $s_0 = [-e_1 + (\alpha - 1)(e_2 + e_3 + e_4), e_2, e_3, e_4]$.

(a) Show that $S = \{s_0, s_1, s_2, s_3\}$ generates S_4^H.
(b) Show that (S_4^H, S) is a Coxeter system of type H_4.

Notes

The fact that the groups S_n^B and S_n^D of "signed permutations" and "even signed permutations" are Coxeter systems of types B_n and D_n (with respect to the generating sets S_B and S_D) is part of the folklore of the subject. That the group $\widetilde{S}_n$ of affine permutations gives a realization of the Coxeter system of type $\widetilde{A}_{n-1}$ was first explicitly mentioned by Lusztig [365]. This realization was then further studied by Shi [455] and Björner and Brenti [62]. Combinatorial descriptions of the groups of type $\widetilde{C}_n$ appear in Bédard [19] and Shi [467].

A unified and comprehensive study of combinatorial descriptions of a large class of Coxeter groups, which includes the affine Weyl groups of types $\widetilde{A}_n$, $\widetilde{C}_n$, $\widetilde{B}_n$, and $\widetilde{D}_n$, is given in the thesis of H. Eriksson [223] (see also H. and K. Eriksson [224]). The combinatorial descriptions given in Sections 8.4, 8.5, and 8.6 are essentially equivalent to those given in [223].

Exercises 2 and 8 are due to Incitti [311].
Exercise 5. See Stanley [491].
Exercise 11. See Proctor [423].
Exercise 15. See Björner and Brenti [62].
Exercises 20, 21, 22, 23, 24, and 25. See Eriksson [223].

Appendix A1

Classification of finite and affine Coxeter groups

Table I. The finite irreducible Coxeter systems

Name	Diagram	Order	$|\mathbf{T}|$	Exponents
A_n $(n \geq 1)$		$(n+1)!$	$\binom{n+1}{2}$	$1, 2, \ldots, n$
B_n $(n \geq 2)$		$2^n n!$	n^2	$1, 3, \ldots, 2n-1$
D_n $(n \geq 4)$		$2^{n-1} n!$	$n^2 - n$	$1, 3, \ldots, 2n-3, n-1$

	Diagram	Order		Exponents
E_6		$2^7\,3^4\,5$	36	$1, 4, 5, 7, 8, 11$
E_7		$2^{10}\,3^4\,5\,7$	63	$1, 5, 7, 9, 11, 13, 17$
E_8		$2^{14}\,3^5\,5^2\,7$	120	$1, 7, 11, 13, 17, 19, 23, 29$
F_4		1152	24	$1, 5, 7, 11$
G_2		12	6	$1, 5$
H_3		120	15	$1, 5, 9$
H_4		14400	60	$1, 11, 19, 29$
$I_2(m)$ $(m \geq 3)$		$2m$	m	$1, m-1$

The underlying groups are pairwise nonisomorphic, except that $I_2(3) = A_2$, $I_2(4) = B_2$ and $I_2(6) = G_2$.

Table II. The affine irreducible Coxeter systems

Name	Diagram
$\widetilde{A}_1$ $= I_2(\infty)$	(two nodes joined by ∞)
$\widetilde{A}_{n-1}$ ($n \geq 3$)	(cycle: $1, 2, 3, \dots, n-2, n-1$ with apex n)
$\widetilde{C}_n$ ($n \geq 2$)	($0 \overset{4}{-} 1 - 2 - \cdots - n-2 - n-1 \overset{4}{-} n$)
$\widetilde{B}_n$ ($n \geq 3$)	($0 \overset{4}{-} 1 - 2 - \cdots - n-2 - n-1$, branch to n)
$\widetilde{D}_n$ ($n \geq 4$)	($0, 1$ branch to $2 - 3 - \cdots - n-2 - n-1$, branch to n)

Name	Diagram
$\widetilde{E}_6$	(E6 affine diagram)
$\widetilde{E}_7$	(E7 affine diagram)
$\widetilde{E}_8$	(E8 affine diagram)
$\widetilde{F}_4$	(four-node chain with 4)
$\widetilde{G}_2$	(three-node chain with 6)

Appendix A2

Graphs, posets, and complexes

Graphs, posets, and simplicial complexes are, together with permutations and tableaux, the basic combinatorial notions used. They play an important role throughout the book.

In this appendix, we define and recall some terminology, notation, and results. More details, proofs and references for the first two sections can be found, for example, in [497], and for the last three, for example, in [64, Section 4.7].

A2.1 Graphs and Digraphs

By a *graph* we mean a pair $G = (V, E)$, where V is a set and $E \subseteq \binom{V}{2}$. We call V the set of *nodes* or *vertices* of G, and E the set of *edges* of G. A *path* in G is a sequence $\Gamma = (x_0, \ldots, x_k) \in V^{k+1}$ such that $\{x_i, x_{i+1}\} \in E$ for all $i = 0, \ldots, k - 1$. If $x_0 = x_k$ and $k \geq 1$, then we call Γ a *cycle*. We also say that the path Γ *connects* x_0 and x_k. A graph is *connected* if for all $x, y \in V$ there is a path Γ that connects x and y.

A *rooted graph* is a pair (G, x), where G is a graph and x is a vertex of G, called the *root*. A *tree* is a connected graph with no cycles. A vertex v of a tree is a *leaf* if $|\{x \in V : \{x, v\} \in E\}| = 1$. Note that if (G, x) is a rooted tree, then for every $v \in V$ there is a unique path $\Gamma(v)$ connecting x and v. Given two vertices $u, v \in V$, we then say that u is a *descendant* of v if $v \in \Gamma(u)$.

By a *directed graph* (or *digraph*, for short) we mean a pair $D = (V, A)$, where V is a set and $A \subseteq V^2$. We call V the set of *nodes* or *vertices* of D and A the set of *directed edges* of D. We write $x \to y$ to mean that $(x, y) \in A$. An edge $x \to x$ is called a *loop*. A *directed path* (or, *path*, for short) in D is a sequence $\Gamma = (x_0, x_1, \ldots, x_k) \in V^{k+1}$ such that $x_0 \to x_1 \to \cdots \to x_k$. We say that Γ goes *from* x_0 *to* x_k, and we call k the *length* of Γ. If $x_0 = x_k$, we call Γ a *directed cycle*. The edges of a directed cycle of length $k = 2$ are sometimes referred to as a pair of *antiparallel edges*. If $S \subseteq V$, then $(S, A \cap S^2)$ is also a digraph called the *induced directed subgraph*, induced by D on S.

It is sometimes convenient to allow multiple edges in graphs and digraphs. This means that E is a multiset of elements from $\binom{V}{2}$ (resp., A is a multiset of elements from V^2). Such graphs with multiple edges appear a few times in the book. We do not distinguish this more general case terminologically or notationally.

A2.2 Posets

The word *poset* is an abbreviation of *partially ordered set*. Thus, a poset $P = (P, \leq)$ consists of a set P together with a partial order relation $\leq$. The relation is suppressed from the notation when it is clear from context. If $Q \subseteq P$, we may refer to Q also as a poset, having in mind the *induced subposet* $(Q, \leq)$, whose order relation is the restriction of P's. Two elements $x, y \in P$ are said to be *comparable* if either $x \leq y$ or $y \leq x$, and *incomparable* otherwise.

A sequence $(x_0, x_1, \ldots, x_h)$ of elements of P is called a *chain* (respectively *multichain*) if $x_0 < x_1 < \cdots < x_h$ (respectively, $x_0 \leq x_1 \leq \cdots \leq x_h$). We then also say that the chain (respectively, multichain) goes from x_0 to x_h. The integer h is called the *length* of the chain (respectively, multichain). The supremum of this number over all chains of P is the *rank* (or *length*) of P. A chain is *maximal* if its elements are not a proper subset of those of any other chain. If all maximal chains are of the same finite length, then P is *pure*. An element $x \in P$ is *maximal* if there is no element $y \in P$ such that $x < y$.

Suppose that P is pure of length k. Define the *rank* $r(x)$ of $x \in P$ to be the length of the subposet $\{y \in P : y \leq x\}$. The rank function $r : P \to [0, k]$ restricts to a bijection on each maximal chain, and decomposes P into *rank levels* $P_i = \{x \in P : r(x) = i\}$, $i \in [0, k]$.

If $x \leq y$ in P, we define the *closed interval* (or *interval*, for short) $[x, y] = \{z \in P : x \leq z \leq y\}$, the *open interval* $(x, y) = \{z \in P : x < z < y\}$, and the *half-open interval* $(x, y] = \{z \in P : x < z \leq y\}$. A *bottom element* $\widehat{0}$ (resp. a *top element* $\widehat{1}$) is an element satisfying $\widehat{0} \leq x$ (resp. $x \leq \widehat{1}$) for all $x \in P$. If P has a bottom element $\widehat{0}$ and every interval $[\widehat{0}, x]$ is pure, then

P is *graded*. The *rank function* $r : P \to \mathbb{N}$ is defined for a graded poset P as for a pure one. If $|P_i| < \infty$ for all $i \geq 0$, then we call the formal power series $\sum_{i \geq 0} |P_i| q^i$ the *rank generating function* of P.

Suppose from now on that all intervals in P are finite (only such posets appear in this book). A pair (x, y) such that $x < y$ and no $z \in P$ satisfies $x < z < y$ is called a *covering* and is denoted by $x \lessdot y$ (or $y \gtrdot x$). Let $\mathrm{Cov}(P)$ be the set of all coverings in P. This set of ordered pairs implies all other order relations by transitivity, and $\mathrm{Cov}(P)$ is clearly minimal with this property. A chain is *saturated* if all successive relations are coverings: $x_0 \lessdot x_1 \lessdot \cdots \lessdot x_h$. If P has a $\hat{0}$ (respectively, a $\hat{1}$), then an element $x \in P$ is an *atom* (respectively, *coatom*) of P if $\hat{0} \lessdot x$ (respectively, $x \lessdot \hat{1}$).

The standard way of depicting a poset P is to draw a digraph with the elements of P as nodes and the elements of $\mathrm{Cov}(P)$ as upward-directed edges. This graph is called the *diagram* of P (sometimes the *Hasse diagram*). For instance, Figure 2.10 depicts a graded poset of length 3, with the rank levels P_1 and P_2 both of cardinality k.

A map $f : P \to Q$ of posets is *order-preserving* if $x \leq y$ implies $f(x) \leq f(y)$, for all $x, y \in P$. If, instead, $x \leq y$ implies $f(x) \geq f(y)$, the map is *order-reversing*. Two posets P and Q are *isomorphic* if there exists an order-preserving bijection $f : P \to Q$ such that f^{-1} is also order-preserving. A poset P is a *Boolean algebra* if there is a set X such that P is isomorphic to the set of all subsets of X, partially ordered by inclusion. A bijection $f : P \to P$ is an *automorphism* if f and f^{-1} are order-preserving, and an *antiautomorphism* if f and f^{-1} are order-reversing.

A poset P is a *lattice* if for all $x, y \in P$, the subposet $\{z \in P : z \leq x, z \leq y\}$ has a top element, the *meet* $x \wedge y$, and — dually — the subposet $\{z \in P : z \geq x, z \geq y\}$ has a bottom element, the *join* $x \vee y$. If only the meet $x \wedge y$ is guaranteed to exist, P is a *meet-semilattice*. See Section 3.2 for a few more definitions pertaining to (semi)lattices.

The *Möbius function* of P assigns to each ordered pair $x \leq y$ an integer $\mu(x, y)$ according to the following recursion:

$$\mu(x, y) = \begin{cases} 1, & \text{if } x = y, \\ -\sum_{x \leq z < y} \mu(x, z), & \text{if } x < y. \end{cases} \tag{A2.1}$$

Let $\mathrm{Int}(P) \overset{\text{def}}{=} \{(x, y) \in P^2 : x \leq y\}$. Given a commutative ring R, the *incidence algebra* of P with coefficients in R, denoted $I(P; R)$, is the set of all functions $f : \mathrm{Int}(P) \to R$ with sum and product defined by

$$(f + g)(x, y) \overset{\text{def}}{=} f(x, y) + g(x, y)$$

and

$$(fg)(x, y) \overset{\text{def}}{=} \sum_{x \leq z \leq y} f(x, z)\, g(z, y), \tag{A2.2}$$

for all $f, g \in I(P; R)$ and $(x, y) \in \mathrm{Int}(P)$. For instance, $\mu \in I(P; \mathbb{Z})$.

It is well known (see, e.g., [497, Section 3.6]) that $I(P; R)$ is an associative algebra having δ as identity element (where $\delta(x, y) \overset{\text{def}}{=} 1$ if $x = y$, and $\overset{\text{def}}{=} 0$ otherwise) and that an element $f \in I(P; R)$ is invertible if and only if $f(x, x)$ is invertible for all $x \in P$. If f is invertible, then we denote by f^{-1} its (two-sided) inverse.

See [497, Chapter 3] for more about posets and the Möbius function.

A2.3   Simplicial complexes

By an *(abstract simplicial) complex* on vertex set V is meant a nonempty collection Δ of finite subsets of V, called *faces*, which is closed under containment: $F \subseteq F' \in \Delta$ implies $F \in \Delta$. Since we assume that $\Delta \neq \emptyset$, it follows that $\emptyset \in \Delta$. If $F \subseteq F' \in \Delta$, let $[F, F'] = \{G \in \Delta : F \subseteq G \subseteq F'\}$. The *dimension* of a face is defined by $\dim F = |F| - 1$, and the dimension of Δ by $\dim\Delta = \sup_{F \in \Delta} \dim F$ (which can be equal to ∞). So, for example, $\dim\{\emptyset\} = -1$.

A complex Δ is *pure d-dimensional* if every face is contained in some d-dimensional face. Almost all complexes dealt with in this book are pure. In this case, the d-dimensional faces are called *facets* and the $(d - 1)$-dimensional faces are called *panels*. The collection of all facets is denoted by $\mathcal{F}(\Delta)$. Two facets C and C' are *adjacent* if $\dim(C \cap C') = d - 1$.

Let $\Delta \subseteq \Delta'$ be complexes and assume that x is a vertex of Δ' but not of Δ. Then, Δ' is said to be a *cone* over Δ with *cone point* x if

$$C \in \mathcal{F}(\Delta) \quad \Leftrightarrow \quad C \cup \{x\} \in \mathcal{F}(\Delta').$$

For any complex Δ, we let $\|\Delta\|$ denote its topological space, or *geometric realization*. For this construction, as well as such notions as Euler characteristic, simplicial homology and homotopy type, and their connections, see any textbook on algebraic topology (e.g. [401]).

Let Δ be a finite d-dimensional simplicial complex, and let f_i be the number of i-dimensional faces of Δ. The sequence $f = (f_0, \ldots, f_d)$ is called the *f-vector* of Δ. We put $f_{-1} = 1$. The *h-vector* $h = (h_0, \ldots, h_{d+1})$ of Δ is defined by the equation

$$\sum_{i=0}^{d+1} f_{i-1} x^{d+1-i} = \sum_{i=0}^{d+1} h_i (x + 1)^{d+1-i}. \tag{A2.3}$$

Note that $h_0 = 1$, $h_1 = n - d - 1$, and

$$h_{d+1} = f_d - f_{d-1} + \cdots + (-1)^d f_0 + (-1)^{d+1} = (-1)^d \widetilde{\chi}(\Delta),$$

where $\widetilde{\chi}(\Delta)$ is the reduced Euler characteristic of Δ. In particular,

$$h_{d+1} = \begin{cases} 1, & \text{if } \|\Delta\| \text{ is homeomorphic to a sphere,} \\ 0, & \text{if } \|\Delta\| \text{ is homeomorphic to a ball.} \end{cases}$$

An important way in which complexes arise in combinatorics is from posets. If P is a poset, let $\Delta(P)$ be the collection of all finite chains $x_0 < x_1 < \cdots < x_k$ in P. A subset of a chain is again a chain, so this is a simplicial complex, called the *order complex* of P.

We make use of the following two facts. Let $x < y$ in P. Then, the Möbius function $\mu(x, y)$ is equal to the reduced Euler characteristic of the order complex of the open interval (x, y):

Fact A2.3.1 $\mu(x, y) = \widetilde{\chi}(\Delta((x, y)))$.

See [497, Proposition 3.8.6] for a proof.

Fact A2.3.2 *Let* $f : P \to P$ *be an order-preserving map such that* $x \geq f(x) = f^2(x)$ *for all* $x \in P$. *Then, the order complexes of* P *and of* $f(P)$ *are homotopy equivalent.*

It is not hard to give a direct proof that $\Delta(f(P))$ is a strong deformation retract of $\Delta(P)$ in this situation. For another simple proof, see [60, Corollary 10.12].

A2.4   Shellability

Throughout this section, let Δ be a pure d-dimensional complex of at most countable cardinality. We will be considering linear orderings $C_1, C_2, C_3,$ ... of $\mathcal{F}(\Delta)$. Given such an ordering, let $\Delta_k = [\emptyset, C_1] \cup [\emptyset, C_2] \cup \cdots \cup [\emptyset, C_k]$, for $k \geq 1$. Thus, Δ_k is the subcomplex generated by the k first facets.

Definition A2.4.1 *The complex* Δ *is said to be* shellable *if its facets can be arranged in linear order* $C_1, C_2, C_3, \ldots$ *in such a way that* $\Delta_{k-1} \cap [\emptyset, C_k]$ *is pure* $(d-1)$-*dimensional, for* $k = 2, 3, \ldots$. *Such an ordering of* $\mathcal{F}(\Delta)$ *is called a* shelling.

In other words, a linear order $C_1, C_2, C_3, \ldots$ is a shelling if and only if whenever $i < k$, there exists some $j < k$ such that $C_i \cap C_k \subseteq C_j \cap C_k$, and $|C_j \cap C_k| = |C_k| - 1$. Note that

$$\text{if } \Delta' \text{ is a cone over } \Delta, \text{ then } \Delta' \text{ is shellable if and only if } \Delta \text{ is.} \quad \text{(A2.4)}$$

Given a shelling, define the *restriction* of facet C_k by

$$\mathcal{R}(C_k) = \{x \in C_k : C_k \setminus \{x\} \in \Delta_{k-1}\}. \quad \text{(A2.5)}$$

(Here, and whenever else needed, we let $\Delta_0 = \emptyset$.) Shellings and their restriction maps have several useful characterizations, of which we mention the following.

Fact A2.4.2 *Given an ordering* $C_1, C_2, C_3, \ldots$ *of* $\mathcal{F}(\Delta)$ *and a map* $\mathcal{R} : \mathcal{F}(\Delta) \to \Delta$, *the following are equivalent:*

(i) $C_1, C_2, C_3, \ldots$ is a shelling and $\mathcal{R}$ its restriction map.

(ii) $\Delta_k = \biguplus_{i=1}^{k} [\mathcal{R}(C_i), C_i]$, for all $k \geq 1$.

A pure complex Δ is said to be *thin* if every panel is contained in exactly two facets. It is called *subthin* if every panel is contained in at most two facets and it is not thin. It is *locally finite* if every vertex is contained in only finitely many facets.

Let $\mathbb{B}^d$ and $\mathbb{S}^d$ denote the standard PL d-ball and d-sphere (i.e., a geometric d-simplex and the boundary of a geometric $(d+1)$-simplex, respectively). Using some simple facts from PL (piecewise linear) topology, one derives the following.

Fact A2.4.3 *Let Δ be a shellable pure d-dimensional simplicial complex.*

(i) If Δ is finite and subthin, then $\|\Delta\|$ is PL homeomorphic to $\mathbb{B}^d$.

(ii) If Δ is finite and thin, then $\|\Delta\|$ is PL homeomorphic to $\mathbb{S}^d$.

(iii) If Δ is infinite and thin, then $\|\Delta\|$ is contractible.

(iv) If Δ is infinite, thin, and locally finite, then $\|\Delta\|$ is PL homeomorphic to $\mathbb{R}^d$.

If Δ is finite and shellable, then the h-vector has the following interpretation in terms of the restriction map:

$$h_i = \operatorname{card}\{C \in \mathcal{F}(\Delta) : |\mathcal{R}(C)| = i\}. \tag{A2.6}$$

The definition of the h-vector can be extended to infinite shellable complexes via equation (A2.6).

Fact A2.4.4 *Let Δ be a shellable pure d-dimensional complex and let $h \overset{\text{def}}{=} h_{d+1}$. Then, $\|\Delta\|$ has the homotopy type of a wedge of h copies of the d-sphere. Consequently,*

$$\widetilde{H}_i(\Delta; \mathbb{Z}) = \begin{cases} \mathbb{Z}^h, & \text{if } i = d, \\ 0, & \text{if } i \neq d. \end{cases} \tag{A2.7}$$

Here, $\widetilde{H}_i(\Delta; \mathbb{Z})$ denotes reduced simplicial homology with integer coefficients.

The *link* of a face $F \in \Delta$ (including $F = \emptyset$) is the subcomplex $lk_\Delta(F) \overset{\text{def}}{=} \{G \in \Delta : G \cup F \in \Delta \text{ and } G \cap F = \emptyset\}$. The complex Δ is said to be *Cohen-Macaulay* if

$$\widetilde{H}_i(lk_\Delta(F); \mathbb{Z}) = 0, \text{ for all } F \in \Delta \text{ and } i < \dim lk_\Delta(F). \tag{A2.8}$$

By a theorem of Reisner, this property is (in the finite case) equivalent to the Cohen-Macaulayness (in the sense of commutative algebra) of a certain ring $\mathbf{k}[\Delta]$, for every coefficient field $\mathbf{k}$. The ring $\mathbf{k}[\Delta]$ is the quotient of the polynomial ring $\mathbf{k}[x_1, \ldots, x_n]$, whose indeterminates are the vertices x_i of Δ, modulo the ideal generated by the square-free monomials $x_{i_1} x_{i_2} \cdots x_{i_k}$

corresponding to nonfaces $\{x_{i_1}, x_{i_2}, \ldots, x_{i_k}\} \notin \Delta$. See [496] for a detailed discussion of this connection to ring theory.

It is easy to see that shellability of Δ is inherited by all links $lk_\Delta(F)$. Hence, from equation (A2.7) we get the following:

Fact A2.4.5 *If Δ is shellable, then Δ is Cohen-Macaulay.*

A *colored* complex Δ, on vertex set V and with color set S, is by definition a pure d-dimensional complex Δ with a partition $V = \biguplus_{s \in S} V_s$ such that $|C \cap V_s| = 1$ for all $C \in \mathcal{F}(\Delta)$ and all $s \in S$. It is convenient to think of S as a set of colors, the condition being that every facet has exactly one vertex of each color. Clearly, $|S| = d+1$. Examples of colored complexes are provided by order complexes of pure posets P of length d, where $S = [0, d]$ and the color classes V_s are the rank levels P_i.

Let Δ be a colored complex as above. Define the *type* of a face $F \in \Delta$ as its set of colors: $\tau(F) = \{s \in S : F \cap V_s \neq \emptyset\}$. Then, for $E \subseteq S$, let $\Delta_E = \{F \in \Delta : \tau(F) \subseteq E\}$. The *type-selected subcomplex* Δ_E is pure $(|E| - 1)$-dimensional.

Fact A2.4.6 *Suppose that Δ is colored and shellable. Fix $E \subseteq S$. Then, Δ_E is shellable and*

$$h_{|E|}(\Delta_E) = \mathrm{card}\,\{C \in \mathcal{F}(\Delta) : \tau(\mathcal{R}(C)) = E\}.$$

A2.5   Regular CW complexes

By a *ball* in a topological space T we mean a subspace $\sigma \subseteq T$ that is homeomorphic to the ball $\mathbb{B}^d$, for some $d \geq 0$. The (relative) interior $\overset{\circ}{\sigma}$ and boundary $\partial\sigma = \sigma \setminus \overset{\circ}{\sigma}$ are defined via transfer from $\mathbb{B}^d$. If $\dim \sigma = 0$, then $\overset{\circ}{\sigma} = \sigma = \{\text{point}\}$.

Definition A2.5.1 *A regular CW complex Γ is a collection of balls in a Hausdorff space $\|\Gamma\| = \cup_{\sigma \in \Gamma} \sigma$ such that the following hold:*

(i) *The interiors $\overset{\circ}{\sigma}$ partition $\|\Gamma\|$.*

(ii) *The boundary $\partial\sigma$ is a union of some members of Γ, for all $\sigma \in \Gamma$ of positive dimension.*

This definition of a regular CW complex is not the standard one in the topological literature, where an approach via attaching maps (applicable also to general "non-regular" CW complexes) is more common. In that setting, regularity means that the attaching map of each cell should be a homeomorphism on *the whole* cell that is being attached, not only on its interior. For detailed topological treatments of regular cell complexes, see [158] or [363]. For a discussion of regular CW complexes from a combinatorial point of view, including motivation for the equality of the two definitions, see [64, Section 4.7].

The balls $\sigma \in \Gamma$ are the *closed cells* of Γ; their interiors $\overset{\circ}{\sigma}$ are the *open cells*. If $\|\Gamma\| \cong T$, then Γ is said to provide (via the homeomorphism) a *regular CW decomposition* of the space T. The geometric realizations of abstract simplicial complexes are examples of regular CW complexes whose cells are the geometric simplices representing the abstract faces.

The *cell poset* $\mathcal{C}(\Gamma)$ is the set of closed cells of Γ ordered by containment. Now, it turns out that the order complex of $\mathcal{C}(\Gamma)$ is homeomorphic to $\|\Gamma\|$, which has the following consequence.

Fact A2.5.2 *The cell poset determines the topology of $\|\Gamma\|$ and its cellular structure up to cell-preserving homeomorphism.*

For any CW complex Γ, there exists an algebraic chain complex, the *cellular chain complex*,

$$\cdots \longrightarrow C_{i+1} \overset{d_{i+1}}{\longrightarrow} C_i \overset{d_i}{\longrightarrow} C_{i-1} \longrightarrow \cdots$$

with the following properties:

(i) C_i is a free Abelian group with a basis indexed by the i-dimensional cells of Γ.

(ii) $H_i(\|\Gamma\|; \mathbb{Z}) \cong \operatorname{Ker} d_i \,/\, \operatorname{Im} d_{i+1}$.

Furthermore, if Γ is regular, there exists a mapping from pairs of cells (σ, τ) such that $\sigma \subset \tau$ and $\dim \sigma + 1 = \dim \tau$ (or, in other words, from coverings $\sigma \lhd \tau$ in $\mathcal{C}(\Gamma)$) to numbers $[\sigma : \tau] \in \{+1, -1\}$ (called *incidence numbers*) such that the boundary maps are given by

$$d_i(\tau) = \sum_{\sigma \lhd \tau} [\sigma : \tau]\, \sigma, \tag{A2.9}$$

where we identify cells with the corresponding basis elements.

Appendix A3

Permutations and tableaux

Permutations play a central role throughout the book. They have close connections with the combinatorics of tableaux, which is of importance in Chapters 6 and 7.

Here, we first review the basic definitions and establish notation for permutations and tableaux. Then, in Sections A3.3 – A3.9 we summarize those properties of the Robinson-Schensted correspondence that are needed in Chapter 6. The final section concerns properties of dual equivalence of skew tableaux, needed in Chapter 7.

For a detailed treatment with proofs of this material, see [450]. Much of the material can also be found in [248], [328], and [498].

A3.1 Permutations

Fix a set E, finite or infinite. Bijections $x : E \to E$ are called *permutations* of E. They form a group under composition that we denote by $S(E)$. Subgroups of $S(E)$ are called *permutation groups*. A *permutation representation* of a group W is a homomorphism $f : W \to S(E)$ for some set E.

If $G \subseteq S(E)$ is a permutation group and $A \subseteq E$, let

$$\operatorname{Stab}(A) \stackrel{\text{def}}{=} \{x \in G : x(A) = A\}.$$

The notation means that x maps A onto A *as a set* (not necessarily fixing each of its elements). This defines a subgroup of G called the *stabilizer* of A.

The finite groups $S_n \overset{\text{def}}{=} S([n])$ are called the *symmetric groups*. Suppose that E is a finite subset of $\mathbb{Z}$, such as $[n]$ or $[\pm n]$. Then, permutations $x \in S(E)$ can be denoted by listing all the values $x(i)$ left to right in order of increasing argument i. For instance, 74185236 denotes the permutation $1 \mapsto 7$, $2 \mapsto 4$, $3 \mapsto 1$, etc., an element of S_8. We call this the *complete notation*[1] for x. To keep notation simple, we omit commas in the complete notation wherever, as in the given example, confusion cannot arise. So, writing $x = x_1 x_2 \ldots x_n$ for a permutation x of a finite set $E \subset \mathbb{Z}$, this means that $x_i = x(e_i)$ for all i, where e_i is the i-th element of E in increasing order.

At times, we also write permutations in *disjoint cycle form*, omitting to write the 1-cycles. For instance, we have that

$$74185236 = (1, 7, 3)(2, 4, 8, 6),$$

where the left-hand side uses complete notation and the right hand side disjoint cycle form. Permutations of the form (i, j) are called *transpositions*.

Our convention for multiplying permutations is to read the product right to left as composition of mappings. For instance, with permutations of [5] expressed in complete notation, we have

$$31524 \cdot 15243 = 34125.$$

This has the consequence for S_n that multiplying $x = x_1 x_2 \ldots x_n$ (complete notation) on the right by a transposition $t_{i,j} = (i, j)$ has the effect of transposing the values in *positions* i and j, whereas multiplying on the left transposes the *values* i and j. For example,

$$24531 \cdot t_{2,5} = 21534 \quad \text{and} \quad t_{2,5} \cdot 24531 = 54231.$$

Given a sequence $(x_1, x_2, \ldots, x_n) \in \mathbb{Z}^n$, define

$$\text{inv}\,(x_1, x_2, \ldots, x_n) \overset{\text{def}}{=} |\{(i, j) : 1 \le i < j \le n \text{ and } x_i > x_j\}|,$$
$$\text{neg}\,(x_1, x_2, \ldots, x_n) \overset{\text{def}}{=} |\{i \in [n] : x_i < 0\}|,$$
$$\text{nsp}\,(x_1, x_2, \ldots, x_n) \overset{\text{def}}{=} \left|\left\{\{i, j\} \in \binom{[n]}{2} : x_i + x_j < 0\right\}\right|,$$
$$D(x_1, x_2, \ldots, x_n) \overset{\text{def}}{=} \{i \in [n-1] : x_i > x_{i+1}\}.$$

These definitions apply, in particular, to permutations $x = x_1 x_2 \ldots x_n \in S(E)$, where $E \subseteq \mathbb{Z}$, $|E| = n$.

---

[1] Of course, listing only the first *seven* values (i.e., the images of $1, 2, \ldots, 7$, in this order) uniquely identifies a permutation of S_8. Thus, the last entry of the complete notation of a permutation is redundant and could be omitted. Although it would make no sense to use this shorter notation for the elements of the symmetric group, such "window notation" is extremely convenient for other permutation groups, including all those discussed in Chapter 8.

The functions "neg" and "nsp" (short for "number of negative entries" and "number of negative sum pairs") appear only in Chapter 8. For more about the "descent set" $D = D_R$, see Section A3.4.

A pair $(i,j) \in [n]^2$ is an *inversion* of a permutation $x = x_1 x_2 \ldots x_n$ (or of a sequence $(x_1, \ldots, x_n)$) if $i < j$ and $x_i > x_j$. The *inversion table* of x (or of $(x_1, \ldots, x_n)$) is the sequence

$$(I_1(x), \ldots, I_n(x)),$$

where

$$I_i(x) \overset{\text{def}}{=} |\{j \in [n]: \ i < j, \ x_i > x_j\}|.$$

In the rest of this appendix, all permutations will be elements of S_n.

A3.2   Tableaux

A *partition* $\lambda = (\lambda_1, \ldots, \lambda_k)$ of the integer n (written $\lambda \vdash n$ or $|\lambda| = n$) is a weakly decreasing sequence of positive integers $\lambda_1 \geq \lambda_2 \geq \cdots \geq \lambda_k$ such that $\lambda_1 + \cdots + \lambda_k = n$. The integers $\lambda_1, \ldots, \lambda_k$ are called the *parts* of λ, and k is called the *length* of λ. If $k = 0$, then λ is the *empty partition*. A partition is geometrically represented by its *(Ferrers) diagram*, a left-justified arrangement of boxes (also called *cells*) having λ_i boxes in row i. For instance, the following is the diagram of $(5, 3, 2, 2, 1)$:

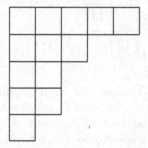

A *partititon* $\lambda = (\lambda_1, \ldots, \lambda_k)$ is a *hook* if $\lambda_2 = \lambda_3 = \cdots = \lambda_k = 1$, a *rectangle* if $\lambda_1 = \lambda_2 = \cdots = \lambda_k$, a *square* if $\lambda_1 = \cdots = \lambda_k = k$, and a *staircase* if $(\lambda_1, \ldots, \lambda_k) = (k, k-1, \ldots, 2, 1)$. We let $\delta_n \overset{\text{def}}{=} (n, n-1, \ldots, 2, 1)$ for all $n \in \mathbb{P}$.

Given two partitions $\mu = (\mu_1, \ldots, \mu_r)$, $\lambda = (\lambda_1, \ldots, \lambda_k)$, we write $\mu \subseteq \lambda$ to mean that $r \leq k$ and $\mu_i \leq \lambda_i$ for $i = 1, \ldots, r$. In this case, we call $\lambda \setminus \mu$ a *skew partition*. Skew partitions are also represented geometrically as diagrams. A skew partition of the form $\lambda \setminus \delta_n$, where λ is a square of length n, is called an *antistaircase*.

Given two skew partitions θ and ρ, we say that θ is an *extension* of (or *extends*) ρ if there exist three partitions λ, μ, and ν, with $\nu \subseteq \mu \subseteq \lambda$ such that $\rho = \mu \setminus \nu$ and $\theta = \lambda \setminus \mu$. We then write $\lambda \setminus \nu = \rho \cup \theta$. We say that

θ is a *final segment* (respectively, *initial segment*) of ρ if there exists three partitions λ, μ, ν with $\nu \subseteq \mu \subseteq \lambda$ such that $\rho = \lambda \setminus \nu$ and $\lambda \setminus \mu = \theta$ (respectively, $\mu \setminus \nu = \theta$). Flipping the diagram of a skew partition along the main diagonal yields the diagram of another skew partition, called its *conjugate*. For example, the conjugate of $(5, 3, 2, 2, 1) \setminus (2, 2, 1)$ is $(5, 4, 2, 1, 1) \setminus (3, 2)$. A skew partition is called *self-conjugate* (or *symmetric*) if it coincides with its conjugate. For example, $(4, 2, 1, 1) \setminus (1)$ is self-conjugate.

By a *connected* skew partition we mean a skew partition whose diagram is rookwise connected. The diagram of a partition (and hence of a skew partition) can be naturally identified with a subset of $\mathbb{N}^2$. Giving $\mathbb{N}^2$ its natural partial order ($(a, b) \leq (c, d)$ if and only if $a \leq c$ and $b \leq d$) then gives a partial order on the cells of the diagram. For this reason, we often identify a diagram with its corresponding poset in this way.

A *tableau* is a filling of the boxes of a diagram by distinct integers so that each row and each column is strictly increasing when read left to right and top to bottom. We call these integers the *entries* of the tableau. A tableau is called *standard* (or a *standard Young tableau*) if its entries are the numbers $1, 2, \ldots, n$, for some n. For instance,

1	2	5	8	11
3	6	10		
4	9			
7	13			
12				

is a standard Young tableau.

If we reflect a tableau T across the main diagonal, then we get another tableau, which we call the *transpose* of T, and denote by T'. Given a tableau T and $i \in \mathbb{Z}$, we let $T_{|+i}$ be the tableau obtained by adding i to each entry of T. If T is standard, we sometimes abuse terminology and call $T_{|+i}$ also a standard tableau. Given a tableau T, we denote by $T_{a,b}$ its b-th entry (from the left) in its a-th row (from the top).

The (possibly skew) partition associated with the diagram of a tableau T is called its *shape* , denoted $\mathrm{sh}(T)$, so, for example, the shape of the preceding tableau is $(5, 3, 2, 2, 1)$. We say that a tableau T has *normal shape* if $\mathrm{sh}((T)$ is a partition. If we wish to emphasize that there are no restrictions on the shape of T, then we say that T is a *skew tableau* (or that it has *skew shape*).

We let SYT_n denote the set of all standard Young tableaux with n boxes, SYT_λ the subset of those having shape λ, and $f^\lambda \stackrel{\text{def}}{=} |SYT_\lambda|$. The word "tableau" in this book means (unless otherwise explicitly stated) standard Young tableau.

A3.3 The Robinson-Schensted correspondence

With each permutation $x \in S_n$ is associated a pair $(P(x), Q(x))$ of tableaux of the same normal shape according to the following rule.

Let $x = x_1 x_2 \ldots x_n$. Starting with the pair of empty tableaux $(\emptyset, \emptyset)$, iterate the following procedure n times. Assume that (P_i, Q_i) has been constructed after i steps. Now, if x_{i+1} is greater than all entries in the first row of P_i, then place it at the end of that row (adding a new box). Otherwise, if, say, $p_{1,j} < x_{i+1} < p_{1,j+1}$, then replace (or "bump") $p_{1,j+1}$ by x_{i+1}. Then, repeat the same operation on the second row with $p_{1,j+1}$ playing the role of x_{i+1}. This bumping process will continue row by row until either a new box is created at the end of some existing row or a new one-box row is created. With this algorithm, a new box is created somewhere and the left tableau P_i grows to P_{i+1}. Let the right tableau Q_i grow to Q_{i+1} by placing $i+1$ in the correspondingly located new box. Due to this formation algorithm, the left tableau $P(x)$ is often called the *insertion tableau* and the right tableau $Q(x)$ is called the *recording tableau*.

The whole procedure is best explained by an example. Let $x = 35214$. The various steps in the formation of P and Q are:

$$P_i \qquad\qquad Q_i$$

Step1 : $\boxed{3}$ $\qquad$ $\boxed{1}$

Step2 : $\begin{array}{|c|c|} \hline 3 & 5 \\ \hline \end{array}$ $\qquad$ $\begin{array}{|c|c|} \hline 1 & 2 \\ \hline \end{array}$

Step3 : $\begin{array}{|c|c|} \hline 2 & 5 \\ \hline 3 \\ \cline{1-1} \end{array}$ $\qquad$ $\begin{array}{|c|c|} \hline 1 & 2 \\ \hline 3 \\ \cline{1-1} \end{array}$

Step4 : $\begin{array}{|c|c|} \hline 1 & 5 \\ \hline 2 \\ \cline{1-1} 3 \\ \cline{1-1} \end{array}$ $\qquad$ $\begin{array}{|c|c|} \hline 1 & 2 \\ \hline 3 \\ \cline{1-1} 4 \\ \cline{1-1} \end{array}$

Step5 : $\begin{array}{|c|c|} \hline 1 & 4 \\ \hline 2 & 5 \\ \hline 3 \\ \cline{1-1} \end{array}$ $\qquad$ $\begin{array}{|c|c|} \hline 1 & 2 \\ \hline 3 & 5 \\ \hline 4 \\ \cline{1-1} \end{array}$

The last pair of tableaux is $(P(x), Q(x))$.

Fact A3.3.1 *The mapping* $x \mapsto (P(x), Q(x))$ *is a bijection between permutations* $x \in S_n$ *and pairs of tableaux* $(P, Q) \in \bigcup_{\lambda \vdash n} SYT_\lambda^2$.

This is called the *Robinson-Schensted correspondence*, and much of this appendix is concerned with summarizing its key properties.

A3.4 Descent sets

The *right* and *left descent sets* of $x \in S_n$ are, by definition,

$$D_R(x) = \{i \in [n-1] : x(i) > x(i+1)\}$$

and $D_L(x) = D_R(x^{-1})$. For example,

$$D_R(41253) = \{1,4\}, \quad D_L(41253) = \{3\}.$$

Note that the left descent set of a permutation $x \in S_n$ consists of those $i \in [n-1]$ such that $i+1$ appears to the left of i in the complete notation of x.

The *descent set* of a tableau T is the set $D(T)$ consisting of those entries i such that $i+1$ appears in a strictly lower row. This is related to the previous definition via the Robinson-Schensted correspondence as follows.

Fact A3.4.1 $D_L(x) = D(P(x))$ *and* $D_R(x) = D(Q(x))$.

For instance, for

$$Q(35214) = \begin{array}{|c|c|}\hline 1 & 2 \\\hline 3 & 5 \\\hline 4 \\\cline{1-1}\end{array}$$

computed earlier, both x and $Q(x)$ have (right) descent set $= \{2,3\}$.

A3.5 Special tableaux

A tableau is *row superstandard* if when reading its rows from left to right and from top to bottom, we get the integers $1, 2, \ldots, n$ in their natural order. For instance, the following is the row superstandard tableau of shape $(5,3,1)$:

$$\begin{array}{|c|c|c|c|c|}\hline 1 & 2 & 3 & 4 & 5 \\\hline 6 & 7 & 8 \\\cline{1-3} 9 \\\cline{1-1}\end{array}$$

By symmetry there is a corresponding notion of *column superstandard* tableaux.

The *reading word* $\rho(T)$ of a tableau T is obtained by reading the rows of T from left to right and from bottom to top. For instance, let T be the row superstandard tableau of shape $(5,3,1)$. Then, $\rho(T) = 967812345$. We will sometimes consider $\rho(T)$ as an element of S_n if T has n boxes (and is standard). For a given partition $\lambda = (\lambda_1, \ldots, \lambda_k)$, there is a unique tableau T_λ having descent set $\{\lambda_k, \lambda_k + \lambda_{k-1}, \ldots, \lambda_k + \cdots + \lambda_2\}$. It is called the *reading tableau* of shape λ, because of the bijection

$$\rho(T) \longleftrightarrow (T, T_\lambda),$$

which holds under Robinson-Schensted for all $T \in SYT_\lambda$. For example, the reading tableau of shape $(5, 3, 1)$ is

1	3	4	8	9
2	6	7		
5				

A3.6 Knuth equivalence

Let $x, y \in S_n$. We write $x \underset{K}{\approx} y$ if there exist $1 < i < n$ such that $x_1 x_2 \ldots x_n$ and $y_1 y_2 \ldots y_n$ differ only on the substrings $x_{i-1} x_i x_{i+1}$ and $y_{i-1} y_i y_{i+1}$, and these substrings are related to each other in either of the following two ways:

$$bca \leftrightarrow bac \quad \text{or} \quad cab \leftrightarrow acb,$$

where $a < b < c$. This is called *elementary Knuth equivalence*. It means that a commutation $ac \leftrightarrow ca$ is allowed if and only if the commuted pair has a neighbor b of intermediate value placed immediately to the left or immediately to the right.

Let $x \underset{K}{\sim} y$ be the equivalence relation (called *Knuth equivalence*) generated by $x \underset{K}{\approx} y$. For instance,

$$215\underline{436} \underset{K}{\approx} 215\underline{463} \underset{K}{\approx} \underline{215}643 \underset{K}{\approx} 2\underline{51}643 \underset{K}{\approx} 256143$$

shows that $215436 \underset{K}{\sim} 256143$.

The equivalence classes with respect to $\underset{K}{\sim}$ are called *Knuth classes*. This turns out to characterize the relation of having the same insertion tableau.

Fact A3.6.1 $P(x) = P(y)$ *if and only if* $x \underset{K}{\sim} y$, *for all* $x, y \in S_n$.

This result has a dual form characterizing equality of recording tableaux. The dual form can be easily deduced using Fact A3.9.1 below. Because of its importance in Chapter 6, we nevertheless give the explicit statement.

For $x, y \in S_n$, we write $x \underset{dK}{\approx} y$ if x and y differ by transposition of two values i and $i+1$, and either $i-1$ or $i+2$ occurs in a position between those of i and $i + 1$. This defines *elementary dual Knuth equivalence*, and *dual Knuth equivalence* (written $x \underset{dK}{\sim} y$) is the transitive closure. For instance,

$$\underline{21}5436 \underset{dK}{\approx} 315\underline{42}6 \underset{dK}{\approx} 415\underline{32}6 \underset{dK}{\approx} 425316$$

shows that $215436 \underset{dK}{\sim} 425316$.

Fact A3.6.2 $Q(x) = Q(y)$ *if and only if* $x \underset{dK}{\sim} y$, *for all* $x, y \in S_n$.

The equivalence classes under "$\underset{dK}{\sim}$" are called *dual Knuth classes*.

A3.7 Jeu de taquin slides

Let T be a skew tableau and $x \in \mathbb{N}^2$ be such that sh(T) extends $\{x\}$. We then define the *backward jeu de taquin slide* (or *backward slide*, for short) of T into x to be the skew tableau, denoted $j^x(T)$, obtained as follows. We first fill cell x by "sliding" into it the smaller (or only one) of the entries of T that occupy the cells immediately to the right and immediately below cell x. This will vacate a new cell x', which we now fill by the same sliding rule, and so on until we have vacated a cell y for which neither the cell directly below it nor the one directly to its right are in sh(T).

For example, if

$$
T = \begin{array}{|c|c|c|c|}
\hline
\bullet & 1 & 4 & 8 \\
\hline
\end{array}
\quad (A3.1)
$$

and x is the cell marked by a $\bullet$, then we obtain

$$
\begin{array}{|c|c|c|c|}
\hline
\bullet & 1 & 4 & 8 \\
\hline
2 & 5 & 6 \\
\cline{1-3}
3 & 7 \\
\cline{1-2}
\end{array}
\rightarrow
\begin{array}{|c|c|c|c|}
\hline
1 & \bullet & 4 & 8 \\
\hline
2 & 5 & 6 \\
\cline{1-3}
3 & 7 \\
\cline{1-2}
\end{array}
\rightarrow
\begin{array}{|c|c|c|c|}
\hline
1 & 4 & \bullet & 8 \\
\hline
2 & 5 & 6 \\
\cline{1-3}
3 & 7 \\
\cline{1-2}
\end{array}
\rightarrow
\begin{array}{|c|c|c|c|}
\hline
1 & 4 & 6 & 8 \\
\hline
2 & 5 & \bullet \\
\cline{1-3}
3 & 7 \\
\cline{1-2}
\end{array}
$$

and so

$$
j^x(T) = \begin{array}{|c|c|c|c|}
\hline
1 & 4 & 6 & 8 \\
\hline
2 & 5 \\
\cline{1-2}
3 & 7 \\
\cline{1-2}
\end{array}
$$

Similarly, if $x \in \mathbb{N}^2$ is such that $\{x\}$ extends sh(T), then we define a *forward jeu de taquin slide* (or *forward slide*, for short) of T into x to be the skew tableau, denoted $j_x(T)$, defined as follows. We first fill cell x by "sliding" into it the largest (or only one) of the entries of T that occupy the cells immediately to the left and immediately above cell x. This vacates a new cell x', which we fill by the same rule until we have vacated a cell y for which neither the cell immediately to its left nor the one immediately above it are in sh(T).

For example, if T is the tableau in (A3.1) and x is the cell marked by a $\square$, then we obtain

$$
\begin{array}{|c|c|c|c|}
\hline
1 & 4 & 8 \\
\hline
2 & 5 & 6 & \square \\
\hline
3 & 7 \\
\cline{1-2}
\end{array}
\rightarrow
\begin{array}{|c|c|c|c|}
\hline
1 & 4 & \square \\
\hline
2 & 5 & 6 & 8 \\
\hline
3 & 7 \\
\cline{1-2}
\end{array}
\rightarrow
\begin{array}{|c|c|c|c|}
\hline
1 & \square & 4 \\
\hline
2 & 5 & 6 & 8 \\
\hline
3 & 7 \\
\cline{1-2}
\end{array}
\rightarrow
\begin{array}{|c|c|c|c|}
\hline
\square & 1 & 4 \\
\hline
2 & 5 & 6 & 8 \\
\hline
3 & 7 \\
\cline{1-2}
\end{array}
$$

so

$$j_x(T) = \begin{array}{|c|c|c|c|} \hline & & 1 & 4 \\ \hline 2 & 5 & 6 & 8 \\ \hline 3 & 7 \\ \hline \end{array}$$

A *slide sequence* for T is a sequence of cells $(x_1, \ldots, x_r)$ such that it is meaningful to form the tableaux $T_r, \ldots, T_1$, where, for each $i = 1, \ldots, r$, either $T_i = j_{x_i}(T_{i-1})$ or $T_i = j^{x_i}(T_{i-1})$ (and where $T_0 \overset{\text{def}}{=} T$).

A3.8   Evacuation and antievacuation

We describe two operations called *evacuation* and *antievacuation* that transform a tableau T to other tableaux $e(T)$, $e^*(T)$ of the same shape. These operations turn out to be involutions.

Let T be a tableau with n cells. Delete entry "n" from T and perform a forward slide into the cell that contained it. This will vacate a cell of T. Now do the same for entry "$n-1$," then for entry "$n-2$," etc., and finally for entry "1." You will now have the empty tableau. Then, the *antievacuation tableau* of T, denoted $e^*(T)$, is the tableau whose entries record the order in which the cells of T have been vacated.

For example, if

$$T = \begin{array}{|c|c|c|} \hline & 1 & 3 & 6 \\ \hline 2 & 4 & 7 \\ \hline 5 \\ \hline \end{array}$$

then we obtain from it the following sequence of tableaux:

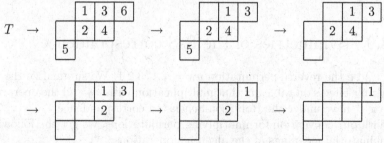

and, therefore,

$$e^*(T) = \begin{array}{|c|c|c|} \hline & 2 & 5 & 7 \\ \hline 1 & 4 & 6 \\ \hline 3 \\ \hline \end{array}$$

Note that $e^*(T)$ has the same shape as T. It is a fact that the mapping e^* is an involution on the set of tableaux of any given shape.

Next, starting from T, we first delete the entry "1" and perform a backward slide on the cell that contained it; then we do the same for the entry "2," etc.... . Then, the tableau that records (in reverse) the order in which the cells of T have been vacated in this process is called the *evacuation* of T and is denoted by $e(T)$.

For example, if

$$
T = \begin{array}{|c|c|c|}
\hline
\multicolumn{1}{c}{} & \multicolumn{1}{c}{1} & \multicolumn{1}{c}{3} \\
\hline
\end{array}
$$

$$
T = \begin{array}{ccc}
& \boxed{1}\ \boxed{3}\ \boxed{4} \\
\boxed{2}\ \boxed{5}\ \boxed{6} \\
\boxed{7}
\end{array}
$$

then we obtain the following sequence of tableaux :

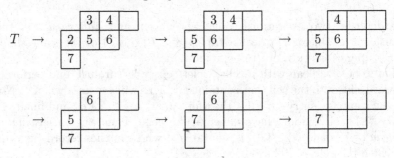

and, hence,

$$
e(T) = \begin{array}{ccc}
& \boxed{2}\ \boxed{5}\ \boxed{7} \\
\boxed{1}\ \boxed{4}\ \boxed{6} \\
\boxed{3}
\end{array}
$$

Again, $e(T)$ has the same shape as T, and the mapping e is an involution.

A3.9   Symmetries of the R-S correspondence

Let w_0 be the reverse permutation $w_0 = n \ldots 3\,2\,1$. We summarize the remarkable effects on tableaux that multiplication with w_0 and the operation $x \mapsto x^{-1}$ have under the Robinson-Schensted correspondence.

With our convention for multiplying permutations, we get the following combinatorial meanings of the algebraic operations:

$$
\begin{array}{rcl}
x^{-1} & \longleftrightarrow & \text{switch places and values,} \\
xw_0 & \longleftrightarrow & \text{reverse the places,} \\
w_0 x & \longleftrightarrow & \text{reverse the values,} \\
w_0 x w_0 & \longleftrightarrow & \text{reverse both.}
\end{array}
$$

For instance, if $x = 24135$, then $x^{-1} = 31425$, $xw_0 = 53142$, $w_0 x = 42531$, and $w_0 x w_0 = 13524$.

Recall that for a tableau P, we let P' denote the transposed tableau (i.e., P mirrored in its main diagonal). This clearly commutes with evacuation: $e(P') = e(P)'$.

Fact A3.9.1 *If $x \leftrightarrow (P,Q)$ are matched under the Robinson-Schensted correspondence, then so are*

$$
\begin{aligned}
x^{-1} &\longleftrightarrow (Q,P), \\
xw_0 &\longleftrightarrow (P', e(Q)'), \\
w_0 x &\longleftrightarrow (e(P)', Q'), \\
w_0 x w_0 &\longleftrightarrow (e(P), e(Q)).
\end{aligned}
$$

Note that the last relation implies that evacuation is an involution on SYT_λ.

We leave it to the reader to exemplify these relations; for instance, starting from the pair

$$
24135 \quad \longleftrightarrow \quad \left(\begin{array}{ccc} \boxed{\begin{array}{ccc} 1 & 3 & 5 \\ \hline 2 & 4 \end{array}} \end{array}, \begin{array}{ccc} \boxed{\begin{array}{ccc} 1 & 2 & 5 \\ \hline 3 & 4 \end{array}} \end{array} \right).
$$

A3.10 Dual equivalence

In this section, we summarize the properties of an equivalence relation for skew tableaux that is closely connected to the dual Knuth equivalence of permutations, as discussed in Section A3.6.

Let P and Q be skew tableaux. We say that P is *dual equivalent* to Q, denoted $P \approx Q$, if whenever a slide sequence can be applied to both P and Q, then the resulting tableaux are of the same shape.

Note that the sequence in the definition can be empty. Thus, two dual equivalent tableaux necessarily have the same shape. The converse, however, is not true in general. For example, if

$$
S = \begin{array}{cc} & \boxed{\begin{array}{cc} 2 & 3 \end{array}} \\ \boxed{1} & \end{array} \quad \text{and} \quad T = \begin{array}{cc} & \boxed{\begin{array}{cc} 1 & 3 \end{array}} \\ \boxed{2} & \end{array}
$$

then S and T are not dual equivalent. In fact, performing a backward slide into the cell $(1,1)$ yields

$$
j^{(1,1)}(S) = \boxed{\begin{array}{ccc} 1 & 2 & 3 \end{array}} \quad \text{and} \quad j^{(1,1)}(T) = \begin{array}{cc} \boxed{\begin{array}{cc} 1 & 3 \end{array}} \\ \boxed{2} \end{array}
$$

which do not have the same shape. We do have, however, the following remarkable result.

Fact A3.10.1 *Let U and V be two tableaux of the same normal shape. Then, $U \approx V$.*

Although the definition of dual equivalence is a global one, this concept can be characterized locally, and this is one of the most important of its many properties.

Fact A3.10.2 *Let $X, S, T,$ and Y be four tableaux such that* $\mathrm{sh}(Y)$ *extends* $\mathrm{sh}(T)$, $\mathrm{sh}(T)$ *extends* $\mathrm{sh}(X)$, *and* $S \approx T$. *Then,* $X \cup S \cup Y \approx X \cup T \cup Y$.

Note that in this lemma we are writing, for simplicity, $X \cup S \cup Y$ instead of $X \cup S_{|+|X|} \cup Y_{|+|X|+|S|}$, etc., thereby tacitly using the convention stated in Section A3.2. We will do this routinely.

A skew partition $\lambda \setminus \mu$ is said to be *miniature* if $|\lambda \setminus \mu| = 3$. A tableau T is said to be *miniature* if $\mathrm{sh}(T)$ is miniature.

Fact A3.10.3 *Each dual equivalence class of miniature tableaux consists of at most two tableaux. Furthermore, a miniature tableau T is in a two-element dual equivalence class if and only if its reading word is either 132, 231, 213, or 312. In each case, the unique tableau S dual equivalent to T is equal to T except that its reading word is the reverse of that of T.*

Two tableaux U and V are *elementary dual equivalent* if there exist four tableaux $X, S, T,$ and Y as in the hypotheses of Fact A3.10.2 such that S and T are miniature, $S \approx T$, $U = X \cup S \cup Y$, and $V = X \cup T \cup Y$. So, for example,

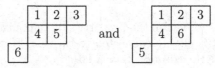

are elementary dual equivalent, but

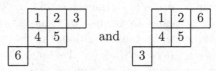

are not.

It is clear from this definition and Fact A3.10.2 that two tableaux that are related by a chain of elementary dual equivalences are dual equivalent. Remarkably, the converse is also true.

Fact A3.10.4 *Let U and V be two tableaux. Then, $U \approx V$ if and only if U can be obtained from V by a sequence of elementary dual equivalences.*

Appendix A4

Enumeration and symmetric functions

In this appendix, we review some results, notation and terminology regarding formal power series and symmetric functions. These are needed in connection with the enumerative theory of Coxeter groups. Further details, including proofs, can be found in [450], [497], [498], and [258].

A4.1 Formal power series

Let $\mathbf{x} = (x_1, x_2, \ldots)$ be a sequence of independent variables and R be a commutative ring with identity. We denote by $R[[\mathbf{x}]]$ the ring of formal power series in $x_1, x_2, \ldots$. Given $(a_1, a_2, \ldots, a_p) \in \mathbb{N}^p$ and $F \in R[[\mathbf{x}]]$, we denote by $[x_1^{a_1} \cdots x_p^{a_p}](F)$ the coefficient of the monomial $x_1^{a_1} \cdots x_p^{a_p}$ in F, and we also write $F(0) \overset{\text{def}}{=} [x_1^0 x_2^0 \cdots](F)$. If $F \in R[[\mathbf{x}]]$ is invertible, we write $G = F^{-1}$ (or $G = 1/F$) to mean that $FG = GF = 1$. An element $F \in R[[\mathbf{x}]]$ is *rational* if there exist polynomials $P, Q \in R[x_1, x_2, \ldots]$ such that $Q(0)$ is invertible in R and

$$F = \frac{P}{Q}.$$

Recall that there is a notion of convergence in $R[[\mathbf{x}]]$. Namely, if $F, F_1, F_2, \ldots \in R[[\mathbf{x}]]$, then we write

$$\lim_{n \to \infty} F_n = F$$

if for each monomial $x_1^{a_1} x_2^{a_2} \cdots$, there is an integer N (depending on $a_1, a_2, \ldots$) such that $[x_1^{a_1} x_2^{a_2} \cdots](F_n) = [x_1^{a_1} x_2^{a_2} \cdots](F)$ for all $n \geq N$. We then say that the sequence $\{F_n\}_{n=1,2,\ldots}$ *converges* to F. In particular, we write

$$\prod_{n \geq 1} F_n = F$$

to mean that

$$\lim_{n \to \infty} \prod_{i=1}^{n} F_i = F.$$

Given $F \in R[[\mathbf{x}]]$ such that $F(0) = 1$, we define

$$\log(F) \overset{\text{def}}{=} \sum_{n \geq 1} (-1)^{n-1} \frac{(F-1)^n}{n}.$$

Let now z, x, and q be three independent variables. The following result is usually known as the q-Binomial Theorem, see, e.g., [258, Appendix II.3].

Fact A4.1.1 *We have that*

$$\sum_{n \geq 0} \prod_{i=0}^{n-1} \left(\frac{1 - zq^i}{1 - q^{i+1}} \right) x^n = \prod_{i \geq 0} \frac{(1 - zxq^i)}{(1 - xq^i)}$$

in $\mathbb{Q}[[z, x, q]]$.

Let D be a directed graph on vertex set $[n]$. The *adjacency matrix* of D is the matrix $Z \overset{\text{def}}{=} (Z_{u,v})_{u,v \in [n]}$ defined by

$$Z_{u,v} = \begin{cases} 1, & \text{if } u \to v, \\ 0, & \text{otherwise.} \end{cases}$$

For $u, v \in [n]$, define a formal power series $F_{u,v}(q) \in \mathbb{Z}[[q]]$ by

$$F_{u,v}(q) \overset{\text{def}}{=} \sum_{n \geq 0} F_{u,v}(n) \, q^n,$$

where $F_{u,v}(n)$ equals the number of paths of length n from u to v (so $F_{u,v}(1) = Z_{u,v}$, $F_{u,v}(0) = \delta_{u,v}$).

The following basic result is sometimes known as the Transfer Matrix Method. See Theorem 4.7.2 in [497] for a detailed discussion

Fact A4.1.2 *Let* $u, v \in [n]$. *Then,*

$$F_{u,v}(q) = \frac{(-1)^{u+v} \det(I - qZ; v, u)}{\det(I - qZ)},$$

where $(I - qZ; v, u)$ *denotes the matrix obtained from* $I - qZ$ *by deleting its v-th row and u-th column, and I is the $n \times n$ identity matrix.*

Corollary A4.1.3 *The series* $F_{u,v}(q)$ *is rational.*

A4.2 Symmetric functions

An element $F \in R[[\mathbf{x}]]$ is said to be *symmetric* if

$$F(x_1, x_2, \ldots) = F(x_{u(1)}, x_{u(2)}, \ldots)$$

for all bijections $u : \mathbb{P} \to \mathbb{P}$, and it is said to be *bounded* if there is a constant M such that all of the monomials appearing in F have degree $\leq M$. F is called a *symmetric function* if it is both symmetric and bounded. For example, $\prod_{i \geq 1}(1 + x_i)$ is symmetric but not a symmetric function.

Let $\lambda = (\lambda_1, \ldots, \lambda_k)$ be a partition. A *column strict plane partition* T of shape λ is a filling of the boxes of the diagram of λ with positive integers so that each row is weakly decreasing when read from left to right and each column is strictly decreasing when read from top to bottom. The *content* of T is the vector

$$m(T) = (m_1(T), m_2(T), \ldots),$$

where $m_i(T)$ equals the number of entries of T that are equal to i ($i \in \mathbb{P}$).

The *Schur function* associated to λ is defined by

$$s_\lambda(\mathbf{x}) \overset{\text{def}}{=} \sum_T x_1^{m_1(T)} x_2^{m_2(T)} \cdots,$$

where T runs over all the column strict plane partitions of shape λ. It is a remarkable fact that $s_\lambda(\mathbf{x})$ is always a symmetric function.

Fact A4.2.1 *Let λ be a partition; then, $s_\lambda(\mathbf{x})$ is a symmetric function, homogeneous of degree $|\lambda|$.*

In fact, much more is true. It is clear that the symmetric functions that are homogeneous of a given degree n form a vector space, and it turns out that the set $\{s_\lambda(\mathbf{x})\}_{\lambda \vdash n}$ is a basis for it.

Let $p \in \mathbb{P}$ and $S \subseteq [p-1]$. A sequence $(a_1, \ldots, a_p) \in \mathbb{P}^p$ is *compatible* with S if the following hold:

(i) $a_1 \leq a_2 \leq \cdots \leq a_p$.

(ii) $a_i < a_{i+1}$ if $i \in S$.

Denote by C_S the set of all the sequences compatible with S. The *fundamental quasi-symmetric function* $Q_{S,p}(\mathbf{x})$ is

$$Q_{S,p}(\mathbf{x}) \overset{\text{def}}{=} \sum_{(a_1, \ldots, a_p) \in C_S} x_{a_1} \cdots x_{a_p}.$$

We then have the following result.

Fact A4.2.2 *Let λ be a partition. Then,*

$$\sum_{T \in \mathrm{SYT}_\lambda} Q_{D(T), |\lambda|}(\mathbf{x}) = s_\lambda(\mathbf{x}),$$

where $D(T)$ is the descent set of T (see Section A3.4).

For example, if $\lambda = (2, 1)$, then there are two tableaux of shape λ, namely

$$\begin{array}{|c|c|} \hline 1 & 2 \\ \hline 3 \\ \cline{1-1} \end{array} \quad \text{and} \quad \begin{array}{|c|c|} \hline 1 & 3 \\ \hline 2 \\ \cline{1-1} \end{array}$$

so

$$s_{(2,1)}(\mathbf{x}) = Q_{\{2\},3}(\mathbf{x}) + Q_{\{1\},3}(\mathbf{x}).$$

Bibliography

[1] R. Adin, F. Brenti, Y. Roichman, Descent numbers and major indices for the hyperoctahedral group, *Adv. Appl. Math.* **27** (2001), 210–224. [243]

[2] R. Adin, F. Brenti, Y. Roichman, Descent representations and multivariate statistics, *Trans. Amer. Math. Soc.*, to appear.

[3] R. Adin, F. Brenti, Y. Roichman, Equi-distribution over descent classes of the hyperoctahedral group, preprint.

[4] R. Adin, A. Postnikov, Y. Roichman, On characters of Weyl groups, *Discrete Math.* **226** (2001), 355–358.

[5] M. Aguiar, N. Bergeron, K. Nyman, The peak algebra and the descent algebras of types B and D, *Trans. Amer. Math. Soc.* **356** (2004), 2781–2824.

[6] D. Alvis, The left cells of the Coxeter group of type H_4, *J. Algebra* **107** (1987), 160–168.

[7] H. H. Andersen, An inversion formula for the Kazhdan-Lusztig polynomials for affine Weyl groups, *Adv. Math.* **60** (1986), 125–153.

[8] H. H. Andersen, The irreducible characters for semi-simple algebraic groups and for quantum groups, *Proceedings of the International Congress of Mathematicians, Zürich, 1994*, Birkhäuser, Basel, Switzerland, 1995, pp. 732–743. [171]

[9] D. André, Solution directe du probléme résolu par M. Bertrand, *C. R. Acad. Sci. Paris* **105** (1887), 436–437. [170]

[10] K. I. Appel, P. E. Schupp, Artin groups and infinite Coxeter groups, *Invent. Math.* **72** (1983), 201–220.

[11] V. I. Arnol´d, Snake calculus and the combinatorics of the Bernoulli, Euler and Springer numbers of Coxeter groups. (Russian) *Uspekhi Mat.*

Nauk **47** (1992), no. 1(283), 3–45, 240; transl. *Russian Math. Surveys* **47** (1992), 1–51. [130]

[12] C. A. Athanasiadis, Generalized Catalan numbers, Weyl groups and arrangements of hyperplanes, *Bull. London Math. Soc.* **36** (2004), 294–302.

[13] C. A. Athanasiadis, On a refinement of the generalized Catalan numbers for Weyl groups, *Trans. Amer. Math. Soc.* **357** (2005), 179–196.

[14] C. A. Athanasiadis, V. Reiner, Noncrossing partitions for the group D_n, *SIAM J. Discrete Math.* **18** (2004), 397–417.

[15] A. Avasjö, *Automata and growth functions of Coxeter groups*, Lic. Thesis, KTH, Stockholm, Sweden, 2004. [123]

[16] L. Balcza, Sum of lengths of inversions in permutations, *Discrete Math.* **111** (1993), 41–48.

[17] H. Barcelo, A. Goupil, Combinatorial aspects of the Poincaré polynomial associated with a reflection group, *Jerusalem Combinatorics '93*, Contemp. Math. **178**, American Mathematical Society, Providence, RI, 1994, pp. 21–44.

[18] H. Barcelo, A. Goupil, Non-broken circuits of reflection groups and factorization in D_n. *Israel J. Math.* **91** (1995), 285–306. [242]

[19] R. Bédard, Cells for two Coxeter groups, *Commun. Algebra* **14** (1986), 1253–1286. [293]

[20] R. Bédard, The lowest two-sided cell for an affine Weyl group, *Commun. Algebra* **16** (1988), 1113–1132.

[21] R. Bédard, Left V-cells for hyperbolic Coxeter groups, *Commun. Algebra* **17** (1989), 2971–2997.

[22] M. K. Bennett, G. Birkhoff, Two families of Newman lattices, *Algebra Universalis* **32** (1994), 115–144.

[23] C. Bennett, R. Blok, Partial orders generalizing the weak order on Coxeter groups, *J. Combin. Theory Ser. A* **102** (2003), 331–346.

[24] A. Berenstein, A. Kirillov, Groups generated by involutions, Gel´fand-Tsetlin patterns, and combinatorics of Young tableaux, *Algebra Analiz* **7** (1995), no. 1, 92–152; translation in *St. Petersburg Math. J.* **7** (1996), 77–127

[25] C. Berge, *Principles of Combinatorics*, Academic Press, New York, 1971. [87]

[26] F. Bergeron, N. Bergeron, A decomposition of the descent algebra of the hyperoctahedral group. I, *J. Algebra* **148** (1992), 86–97.

[27] F. Bergeron, N. Bergeron, R. B. Howlett, D. E. Taylor, A decomposition of the descent algebra of a finite Coxeter group, *J. Algebraic Combin.* **1** (1992), 23–44.

[28] F. Bergeron, A. Garsia, C. Reutenauer, Homomorphisms between Solomon's descent algebras, *J. Algebra* **150** (1992), 503–519.

[29] N. Bergeron, A combinatorial construction of the Schubert polynomials, *J. Combin. Theory Ser. A* **60** (1992), 168–182.

[30] N. Bergeron, S. Billey, RC-graphs and Schubert polynomials, *Experiment. Math.* **2** (1993), 257–269.

[31] N. Bergeron, F. Hivert, J.-Y. Thibon, The peak algebra and the Hecke-Clifford algebras at $q = 0$, *J. Combin. Theory Ser. A* **107** (2004), 1–19.

[32] N. Bergeron, F. Sottile, Schubert polynomials, the Bruhat order, and the geometry of flag manifolds, *Duke Math. J.* **95** (1998), 373–423.

[33] N. Bergeron, F. Sottile, A monoid for the Grassmannian Bruhat order, *European J. Combin.* **20** (1999), 197–211.

[34] N. Bergeron, S. van Willigenburg, A multiplication rule for the descent algebra of type D. *J. Algebra* **206** (1998), 699–705.

[35] I. N. Bernstein, I. M. Gel´fand, S. I. Gel´fand, Schubert cells and the cohomology of the spaces G/P, *Russian Math. Surv.* **28** (1973), 1–26. [63]

[36] I. N. Bernstein, I. M. Gel´fand, S. I. Gel´fand, Differential operators on the base affine space and a study of g-modules, *Lie Groups and Their Representations*, Halsted, New York, 1975, pp. 21–64.

[37] J. Berstel, C. Reutenauer, *Rational Series and Their Languages*, EATCS Monographs on Theoretical Computer Science, **12**. Springer-Verlag, Berlin-New York, 1988.

[38] R. Biagioli, Major and descent statistics for the even-signed permutation group, *Adv. Appl. Math.* **31** (2003), 163–179.

[39] R. Biagioli, F. Caselli, Invariant algebras and major indices for classical Weyl groups, *Proc. London Math. Soc.* **88** (2004), 603–631.

[40] R. Biagioli, F. Caselli, A descent basis for the coinvariant algebra of type D, *J. Algebra* **275** (2004), 517–539.

[41] S. Billey, Pattern avoidance and rational smoothness of Schubert varieties, *Adv. Math.* **139** (1998), 141–156.

[42] S. Billey, T. Braden, Lower bounds for Kazhdan-Lusztig polynomials from patterns, *Transform. Groups* **8** (2003), 321–332.

[43] S. Billey, C. Fan, J. Losonczy, The parabolic map, *J. Algebra* **214** (1999), 1–7.

[44] S. Billey, M. Haiman, Schubert polynomials for the classical groups, *J. Amer. Math. Soc.* **8** (1995), 443–482. [232, 244]

[45] S. Billey, W. Jockusch, R. P. Stanley, Some combinatorial properties of Schubert polynomials, *J. Algebraic Combin.* **2** (1993), 345–374.

[46] S. Billey, V. Lakshmibai, On the singular locus of a Schubert variety, *J. Ramanujan Math. Soc.* **15**, (2000), 155–223.

[47] S. Billey, V. Lakshmibai, *Singular Loci of Schubert Varieties*, Progress in Mathematics, **182**, Birkhäuser Boston Inc., Boston, MA, 2000, xii+251. [24, 171]

[48] S. Billey, T. K. Lam, Vexillary elements in the hyperoctahedral group, *J. Algebraic Combin.* **8** (1998), 139–152. [243]

[49] S. Billey, A. Postnikov, Patterns in root systems and Schubert varieties, preprint.

[50] S. Billey, G. Warrington, Kazhdan-Lusztig polynomials for 321-hexagon-avoiding permutations, *J. Algebraic Combin.* **13** (2001), 111–136.

[51] S. Billey, G. Warrington, Maximal singular loci of Schubert varieties in SL$(n)/B$, *Trans. Amer. Math. Soc.* **355** (2003), 3915–3945.

[52] G. Birkhoff, *Lattice Theory*, Amer. Math. Soc. Colloq. Publ. No. 25, American Mathematical Society, Providence, RI, 1967. [71, 72]

[53] A Björner, The weak ordering of a Coxeter group, unpublished notes, 1981. [87]

[54] A. Björner, Orderings of Coxeter groups, in *Combinatorics and Algebra, Boulder 1983* (ed. C. Greene), Contemp. Math., Vol. 34, American Mathematical Society, Providence, RI, 1984, pp. 175–195. [63, 64, 87]

[55] A. Björner, Posets, regular CW complexes and Bruhat order, *European J. Combin.* **5** (1984), 7–16. [64]

[56] A. Björner, Some combinatorial and algebraic properties of Coxeter complexes and Tits buildings, *Adv. Math.* **52** (1984), 173–212. [88, 200]

[57] A. Björner, Lecture Notes, MIT, 1985. [200]

[58] A. Björner, Essential chains and homotopy type of posets, *Proc. Amer. Math. Soc.* **116** (1992), 1179–1181. [88]

[59] A. Björner, The Möbius function of factor order, *Theoret. Computer Sci.* **117** (1993), 91–98. [88]

[60] A. Björner, Topological Methods, in *Handbook of Combinatorics* (eds. R. Graham, M. Grötschel, and L. Lovász), North-Holland, Amsterdam, 1994, pp. 1819–1872. [303]

[61] A. Björner, F. Brenti, An improved tableau criterion for Bruhat order, *Electron. J. Combin.* **3** (1996), no. 1, #R 22.

[62] A. Björner, F. Brenti, Affine permutations of type A, *Electron. J. Combin.* **3** (1996), no. 2, #R 18. [242, 293, 294]

[63] A. Björner, A. M. Garsia, R. P. Stanley, An introduction to Cohen-Macaulay partially ordered sets, in *Ordered Sets* (ed. I. Rival), Reidel, Dordrecht, 1982, pp. 583–615.

[64] A. Björner, M. LasVergnas, B. Sturmfels, N. White, G. M. Ziegler, *Oriented Matroids*, Cambridge University Press, Cambridge, 1993. [243, 299, 306]

[65] A. Björner, M. Wachs, Bruhat order of Coxeter groups and shellability, *Adv. Math.* **43** (1982), 87–100. [64]

[66] A. Björner, M. Wachs, On lexicographically shellable posets, *Trans. Amer. Math. Soc.* **277** (1983), 323–341.

[67] A. Björner, M. Wachs, Generalized quotients in Coxeter groups, *Trans. Amer. Math. Soc.* **308** (1988), 1–37. [64, 242]

[68] B. D. Boe, Kazhdan-Lusztig polynonomials for Hermitian symmetric spaces, *Trans. Amer. Math. Soc.* **309** (1988), 279–294.

[69] B. D. Boe, A counterexample to the Gabber-Joseph conjecture, in *Kazhdan-Lusztig Theory and Related Topics*, Contemp. Math. Vol. 139, American Mathematical Society, Providence, RI, 1992, pp. 1–3. [170]

[70] B. Boe, W. Graham, A lookup conjecture for rational smoothness, *Amer. J. Math.* **125** (2003), 317–356.

[71] M. Bona, *Combinatorics of Permutations*, Chapman-Hall/CRC Press, New York, 2004.

[72] A. Borel, Foreword to Chevalley's paper [134]. [63]

[73] A. Borel, J. Tits, Compléments à l'article "Groupes réductifs", *I. H. E. S. Publ. Math.* **41** (1972), 253–276.

[74] A. Borovik, *Matroids and Coxeter Groups*, Surveys in Combinatorics, 2003 (Bangor), London Math. Soc. Lecture Note Ser., **307**, Cambridge University Press, Cambridge, 2003, pp. 79–114.

[75] A. Borovik, I. M. Gel′fand, *WP*-matroids and thin Schubert cells on Tits systems, *Adv. Math.* **103** (1994), no. 2, 162–179.

[76] A. Borovik, I. M. Gelfand, N. White, *Coxeter Matroids*, Progress in Mathematics, **216**, Birkhäuser, Boston, MA, 2003.

[77] R. Bott, An application of the Morse theory to the topology of Lie-groups, *Bull. Soc. Math. France* **84** (1956), 251–281. [242]

[78] R. Bott, The geometry and representation theory of compact Lie groups, *Representation Theory of Lie Groups*, London Math. Soc. Lecture Note Ser., 34, Cambridge University Press, Cambridge, 1979, pp. 65–90. [208]

[79] N. Bourbaki, *Groupes et algèbres de Lie*, Ch. 4–6, Éléments de Mathématique, Fasc. XXXIV, Hermann, Paris, 1968; Masson, Paris, 1981. [24, 88, 127, 129, 130, 170]

[80] M. Bousquet-Mélou, K. Eriksson, Lecture hall partitions. *Ramanujan J.* **1** (1997), 101–111. [242]

[81] M. Bousquet-Mélou, K. Eriksson, Lecture hall partitions. II. *Ramanujan J.* **1** (1997), 165–185.

[82] M. Bousquet-Mélou, K. Eriksson, A refinement of the lecture hall theorem, *J. Combin. Theory Ser. A* **86** (1999), 63–84.

[83] T. Braden, R. MacPherson, From moment graphs to intersection cohomology, *Math. Ann.* **321** (2001), 533–551. [171]

[84] N. Brady, J. McCammond, B. Muhlherr, W. Neumann, Rigidity of Coxeter groups and Artin groups, *Geom. Dedicata* **94** (2002), 91–109.

[85] K. Bremke, C. K. Fan, Comparison of a-functions, *J. Algebra* **203** (1998), 355–360.

[86] F. Brenti, q-Eulerian polynomials arising from Coxeter groups, *European J. Combin.* **15** (1994), 417–441. [242, 243]

[87] F. Brenti, A combinatorial formula for Kazhdan-Lusztig polynomials, *Invent. Math.* **118** (1994), 371–394. [170]

[88] F. Brenti, Combinatorial properties of the Kazhdan-Lusztig R-polynomials for S_n, *Adv. Math.* **126** (1997), 21–51. [170]

[89] F. Brenti, Combinatorial expansions of Kazhdan-Lusztig polynomials, *J. London Math. Soc.* **55** (1997), 448–472. [170, 172]

[90] F. Brenti, Kazhdan-Lusztig and R-polynomials from a combinatorial point of view, *Discrete Math.* **193** (1998), 93–116.

[91] F. Brenti, Upper and lower bounds for Kazhdan-Lusztig polynomials, *European J. Combin.* **19** (1998), 283–297.

[92] F. Brenti, Lattice paths and Kazhdan-Lusztig polynomials, *J. Amer. Math. Soc.* **11** (1998), 229–259. [170]

[93] F. Brenti, Twisted incidence algebras and Kazhdan-Lusztig-Stanley functions, *Adv. Math.* **148** (1999), 44–74.

[94] F. Brenti, Approximation results for Kazhdan-Lusztig polynomials, *Adv. Studies Pure Math.* **28** (2000), 57–81.

[95] F. Brenti, Enumerative and combinatorial properties of Dyck partitions, *J. Combin. Theory Ser. A* **99** (2002), 51–74.

[96] F. Brenti, Kazhdan-Lusztig and *R*-polynomials, Young's lattice, and Dyck partitions, *Pacific J. Math.* **207** (2002), 257–286.

[97] F. Brenti, Kazhdan-Lusztig polynomials: history, problems, and combinatorial invariance, *Sém. Lothar. Combin.* **49** (2002), Art. B49b, 30 pp. (electronic).

[98] F. Brenti, *P*-kernels, IC bases and Kazhdan-Lusztig polynomials, *J. Algebra* **259** (2003), 613–627.

[99] F. Brenti, The intersection cohomology of Schubert varities is a combinatorial invariant, *European J. Combin.* **25** (2004), 1151–1167. [159]

[100] F. Brenti, F. Caselli, M. Marietti, Special matchings and Kazhdan-Lusztig polynomials, *Adv. Math.*, to appear. [64, 159, 161]

[101] F. Brenti, S. Fomin, A. Postnikov, Mixed Bruhat operators and Yang-Baxter equations for Weyl groups, *Internat. Math. Res. Notices*, 1999, no. 8, 419–441.

[102] F. Brenti, R. Simion, Enumerative aspects of Kazhdan-Lusztig polynomials, *J. Algebraic Combin.* **11** (2000), 187–196. [170]

[103] B. Brink, The set of dominance-minimal roots, *J. Algebra* **206** (1998), 371–412. [130]

[104] B. Brink, R. Howlett, A finiteness property and an automatic structure for Coxeter groups, *Math. Ann.* **296** (1993), 179–190. [130]

[105] P. Bromwich, *Variations on a Theme of Solomon*, Ph.D. thesis, University of Warwick, 1975. [200]

[106] K. S. Brown, *Buildings*, Springer, New York, 1989. [24, 88]

[107] J. Brundan, Kazhdan-Lusztig polynomials and character formulae for the Lie superalgebra *gl(m|n)*, *J. Amer. Math. Soc.* **16** (2003), 185–231.

[108] J. Brundan, Kazhdan-Lusztig polynomials and character formulae for the Lie superalgebra *q(n)*, *Adv. Math.* **182** (2004), 28–77.

[109] J. B. Carrell, The Bruhat graph of a Coxeter group, a conjecture of Deodhar, and rational smoothness of Schubert varieties, *Algebraic Groups and Their Generalizations: Classical Methods* (University Park, 1991), 53–61, Proc. Sympos. Pure Math. **56**, American Mathematical Society, Providence, RI, 1994. [171]

[110] R. W. Carter, *Simple Groups of Lie Type*, J. Wiley & Sons, London, 1972. [24]

[111] R. W. Carter, *Finite Groups of Lie Type: Conjugacy Classes and Complex Characters*, Wiley Interscience, London, 1985. [24]

[112] F. Caselli, Proof of two conjectures of Brenti-Simion on Kazhdan-Lusztig polynomials, *J. Algebraic Combin.* **18** (2003), 171–187.

[113] F. Caselli, A simple combinatorial proof of a generalization of a result of Polo, *Represent. Theory* **8** (2004), 479–486. [170, 172]

[114] F. Caselli, Non-negativity properties of R-polynomials, preprint.

[115] F. Caselli, M. Marietti, Combinatorial interpretations of certain classes of Kazhdan-Lusztig polynomials, preprint.

[116] N. Caspard, C. Le Conte de Poly-Barbut, M. Morvan, Cayley lattices of finite Coxeter groups are bounded. *Adv. Appl. Math.* **33** (2004), 71–94.

[117] W. Casselman, Machine calculations in Weyl groups, *Invent. Math.* **116** (1994), 95–108. [130]

[118] W. Casselman, Automata to perform basic calculations in Coxeter groups, *Representations of Groups* (Banff, AB, 1994), 35–58, CMS Conf. Proc., 16, American Mathematical Society, Providence, RI, 1995. [130]

[119] W. Casselman, Computation in Coxeter groups. I. Multiplication. *Electron. J. Combin.* **9** (2002), RP 25, 22 pp. [130]

[120] P. Cellini, A characterization of total reflection orders, *Proc. Amer. Math. Soc.* **128** (2000), 1633–1639.

[121] P. Cellini, T-increasing paths on the Bruhat graph of affine Weyl groups are self-avoiding, *J. Algebra* **228** (2000), 107–118.

[122] P. Cellini, P. Papi, The structure of total reflection orders in affine root systems, *J. Algebra* **205** (1998), 207–226.

[123] C. Chameni-Nembua, B. Monjardet, Les treillis pseudocomplementés finis, *European J. Combin.* **13** (1992), 89–107.

[124] C. Chameni-Nembua, B. Monjardet, Finite pseudocomplemented lattices and "permutoèdre," *Discrete Math.* **111** (1993), 105–112.

[125] R. Charney, M. Davis, Reciprocity of growth functions of Coxeter groups, *Geom. Dedicata* **39** (1991), 373–378. [242]

[126] R. Charney, M. Davis, When is a Coxeter system determined by its Coxeter group?, *J. London Math. Soc.* **61** (2000), 441–461.

[127] Y. Chen, Left cells in the Weyl group of type E_8, *J. Algebra* **231** (2000), 805–830.

[128] C. Chen, The left cells of the affine Weyl group of type $\tilde{D}_5$, *Commun. Algebra* **29** (2001), 11–30.

[129] C. Chen, L.-J. Chun, The distinguished involutions with a-value $n^2 - 3n + 3$ in the Weyl group of type D_n, *J. Algebra* **265** (2003), 211–220.

[130] C.-D. Chen, L. Feng, Expression of certain Kazhdan-Lusztig basis elements C_w over the Hecke algebra of type A_n, *J. Algebra* **255** (2002), 174–181.

[131] Y. Chen, J.-Y. Shi, Left cells in the Weyl group of type E_7, *Commun. Algebra* **26** (1998), 3837–3852.

[132] I. Cherednik, Y. Markov, R. Howe, G. Lusztig, *Iwahori-Hecke Algebras and Their Representation Theory*, Lecture Notes in Mathematics, **1804**, Springer-Verlag, Berlin, 2002.

[133] C. Chevalley, Sur certains groupes simples, *Tohoku Math. J.* **7** (1955), 14–66. [242]

[134] C. Chevalley, Sur les décompositions cellulaires des éspaces G/B, manuscript, 1958. *Algebraic Groups and Their Generalizations: Classical Methods* (University Park, 1991), 1–23, Proc. Sympos. Pure Math. **56**, American Mathematical Society, Providence, RI, 1994. [63, 327]

[135] R. Chirivì, Deformation and Cohen-Macaulayness of the multicone over the flag variety, *Comment. Math. Helv.* **76** (2001), 436–466.

[136] R. Chirivì, A relation on minimal representatives and the path model, Special issue in celebration of Claudio Procesi's 60th birthday. *J. Algebra* **258** (2002), 362–385.

[137] C.-O. Chow, On the Eulerian polynomials of type D, *European J. Combin.* **24** (2003), 391–408.

[138] C.-O. Chow, Counting involutory, unimodal, and alternating permutations, preprint.

[139] C.-O. Chow, I. Gessel, On the descent numbers and major indices for the hyperoctahedral group, preprint.

[140] R. J. Clarke, D. Foata, Eulerian calculus. I. Univariable statistics, *European J. Combin.* **15** (1994), 345–362. [243]

[141] R. J. Clarke, D. Foata, Eulerian calculus. II. An extension of Han's fundamental transformation, *European J. Combin.* **16** (1995), 221–252. [243]

[142] R. J. Clarke, D. Foata, Eulerian calculus. III. The ubiquitous Cauchy formula, *European J. Combin.* **16** (1995), 329–355. [243]

[143] F. du Cloux, Un algorithme de forme normal pour les groupes de Coxeter, preprint, École Polytechn. Palaiseau, 1990. [82, 87]

[144] F. du Cloux, The state of the art in the computation of Kazhdan-Lusztig polynomials. *Appl. Algebra Eng. Commun. Comput.* **7** (1996), 211–219.

[145] F. du Cloux, A transducer approach to Coxeter groups, *J. Symb. Comput.* **27** (1999), 311–324. [82, 87]

[146] F. du Cloux, Some Open Problems in the Theory of Kazhdan-Lusztig Polynomials and Coxeter Groups, *Computational methods for representations of groups and algebras* (Essen, 1997), Progr. Math., **173**, Birkhäuser, Basel, 1999, 201–210.

[147] F. du Cloux, An abstract model for Bruhat intervals, *European J. Combin.* **21** (2000), 197–222.

[148] F. du Cloux, Computing Kazhdan-Lusztig polynomials for arbitrary Coxeter groups, *Experiment. Math.* **11** (2002), 371–381.

[149] F. du Cloux, Rigidity of Schubert closures and invariance of Kazhdan-Lusztig polynomials, *Adv. Math.* **180** (2003), 146–175.

[150] A. M. Cohen, Coxeter groups and three related topics. *Generators and Relations in Groups and Geometries* (Lucca, 1990), 235–278, NATO Adv. Sci. Inst. Ser. C **333**, Kluwer, Dordrecht, 1991.

[151] A. M. Cohen, Recent results on Coxeter groups. *Polytopes: Abstract, Convex and Computational* (Scarborough, ON, 1993), 1–19, NATO Adv. Sci. Inst. Ser. C **440**, Kluwer, Dordrecht, 1994.

[152] A. Cohen, H. Cuypers, R. Riebeek, Explorations with the Icosahedral Group. Some Tapas of Computer Algebra, *Algorithms Comput. Math.*, **4**, Springer, Berlin, 1999, pp. 315–322.

[153] A. Cohen, S. Murray, D. Taylor, Computing in groups of Lie type, *Math. Comp.* **73** (2004), 1477–1498 (electronic).

[154] D. Collingwood, Orbits and characters associated to highest weight representations, *Proc. Amer. Math. Soc.* **114** (1992), 1157–1165

[155] L. Comtet, *Advanced Combinatorics*, Reidel, Dordrecht, 1974.

[156] J. H. Conway, T. R. Curtis, S. P. Norton, R. A. Parker, R. A. Wilson, *Atlas of Finite Groups*, Clarendon Press, Oxford, 1985.

[157] J. H. Conway, N. J. A. Sloane, A. R. Wilks, Gray codes for reflection groups, *Graphs Combin.* **5** (1989), 315–325. [88]

[158] G. E. Cooke, R. L. Finney, *Homology of Cell Complexes*, Princeton University Press, Princeton, NJ, 1967. [305]

[159] A. Cortez, Singularités génériques des variétés de Schubert covexillaires, *Ann. Inst. Fourier (Grenoble)* **51** (2001), 375–393.

[160] M. Couillens, Algèbres de Hecke, *Séminaire sur les groupes finis II*, Publ. Math. de l' Université Paris VII, 1983, pp. 77–94.

[161] H. S. M. Coxeter, *Regular Polytopes*, 3rd ed., Dover, New York, 1973.

[162] H. S. M. Coxeter, Discrete groups generated by reflections, *Ann. Math.* **35** (1934), 588–621. [24]

[163] H. S. M. Coxeter, The complete enumeration of finite groups of the form $R_i^2 = (R_i R_j)^{k_{ij}} = 1$, *J. London Math. Soc.* **10** (1935), 21–25. [24]

[164] H. S. M. Coxeter, The product of the generators of a finite group generated by reflections, *Duke Math. J.* **18** (1951), 765–782.

[165] H. S. M. Coxeter, The evolution of Coxeter-Dynkin diagrams. *Polytopes: Abstract, Convex and Computational* (Scarborough, ON, 1993), 21–42, NATO Adv. Sci. Inst. Ser. C, **440**, Kluwer, Dordrecht, 1994.

[166] H. S. M. Coxeter, W. O. J. Moser, *Generators and Relations for Discrete Groups*, 4th revised ed., Ergebnisse der Mathematik und ihrer Grenzgebiete **14**, Springer-Verlag, Berlin, 1980. [24]

[167] C. W. Curtis, Representations of finite groups of Lie type, *Bull. Amer. Math. Soc. (N.S.)* **1** (1979), 721–757.

[168] C. W. Curtis, The Hecke algebra of a finite Coxeter group, *The Arcata Conference on Representations of Finite Groups*, Proc. Symp. Pure Math. 47, part 1, American Mathematical Society, Providence, RI, 1987, pp. 51–60. [200]

[169] C. W. Curtis, Representations of Hecke algebras, *Astérisque* **168** (1988), 13–60.

[170] C. W. Curtis, G. I. Lehrer, Generic chain complexes and finite Coxeter groups, *J. Reine Angew. Math.* **363** (1985), 146–173.

[171] M. W. Davis, M. D. Shapiro, Coxeter groups are almost convex, *Geom. Dedicata* **39** (1991), 55–57. [130]

[172] C. De Concini, D. Eisenbud, C. Procesi, Hodge algebras, *Astérisque* **91** (1982).

[173] C. De Concini, V. Lakshmibai, Arithmetic Cohen-Macaulayness and arithmetic normality for Schubert varieties, *Amer. J. Math.* **103** (1981), 835–850

[174] C. De Concini, C. Procesi, Hodge algebras: a survey, *Astérisque* **87–88** (1981), 79–83.

[175] M. Demazure, Désingularisation des variétés de Schubert généralisées, *Ann. Sci. École Norm. Sup.* **7** (1974), 53–88.

[176] V. V. Deodhar, Some characterizations of Bruhat ordering on a Coxeter group and determination of the relative Möbius function, *Invent. Math.* **39** (1977), 187–198. [63, 64]

[177] V. V. Deodhar, On Bruhat ordering and weight-lattice ordering for a Weyl group, *Indag. Math.* **40** (1978), 423–435.

[178] V. V. Deodhar, On the Kazhdan-Lusztig conjectures, *Indag. Math.* **44** (1982), 1–17.

[179] V. V. Deodhar, On the root system of a Coxeter group, *Commun. Algebra* **10** (1982), 611–630. [64, 130]

[180] V. V. Deodhar, On some geometric aspects of Bruhat orderings. I. A finer decomposition of Bruhat cells, *Invent. Math.* **79** (1985), 499–511. [170]

[181] V. V. Deodhar, Local Poincaré duality and non-singularity of Schubert varieties, *Commun. Algebra* **13** (1985), 1379–1388. [171]

[182] V. V. Deodhar, Some characterizations of Coxeter groups, *Enseign. Math.* **32** (1986), 111–120. [25]

[183] V. V. Deodhar, On some geometric aspects of Bruhat orderings. II. The parabolic analogue of Kazhdan-Lusztig polynomials, *J. Algebra* **111** (1987), 483–506.

[184] V. V. Deodhar, A splitting criterion for the Bruhat orderings on Coxeter groups, *Commun. Algebra* **15** (1987), 1889–1894.

[185] V. V. Deodhar, A note on subgroups generated by reflections in Coxeter groups, *Arch. Math.* **53** (1989), 543–546. [25]

[186] V. V. Deodhar, A combinatorial setting for questions in Kazhdan-Lusztig Theory, *Geom. Dedicata* **36** (1990), 95–119.

[187] V. V. Deodhar, Duality in parabolic set up for questions in Kazhdan-Lusztig theory, *J. Algebra* **142** (1991), 201–209.

[188] V. V. Deodhar, A brief survey of Kazhdan-Lusztig theory and related topics, *Algebraic Groups and Their Generalizations: Classical Methods* (University Park, 1991), 105–124, Proc. Sympos. Pure Math. **56**, American Mathematical Society, Providence, RI, 1994.

[189] V. V. Deodhar, J-chains and multichains, duality of Hecke modules, and formulas for parabolic Kazhdan-Lusztig polynomials, *J. Algebra* **190** (1997), 214–225.

[190] D. I. Deriziotis, D. F. Holt, The Möbius function of the lattice of closed subsystems of a root system, *Commun. Algebra* **21** (1993), 1543–1570.

[191] I. Dolgachev, V. Lunts, A character formula for the representation of a Weyl group in the cohomology of the associated toric variety, *J. Algebra*, to appear. [243]

[192] W. F. Doran, On the homology of distributive lattices, *European J. Combin.* **19** (1998), 441–450.

[193] J. M. Douglass, An inversion formula for relative Kazhdan-Lusztig polynomials, *Commun. Algebra* **18** (1990), 371–387.

[194] B. Drake, S. Gerrish, M. Skandera, Two new criteria for comparison in the Bruhat order, *Electron. J. Combin.* **11** (2004), N6, 4 pp.

[195] C. Droms, H. Servatius, The Cayley graphs of Coxeter and Artin groups, *Proc. Amer. Math. Soc.* **118** (1993), 693–698.

[196] J. Du, The decomposition into cells of the affine Weyl group of type $\tilde{B}_3$, *Commun. Algebra* **16** (1988), 1383–1409.

[197] J. Du, Two-sided cells of the affine Weyl group of type $\tilde{C}_3$, *J. London Math. Soc.* **38** (1988), 87–98.

[198] J. Du, Cells in the affine Weyl group of type $\tilde{D}_4$, *J. Algebra* **128** (1990), 384–404.

[199] J. Du, Sign types and Kazhdan-Lusztig cells, *Chin. Ann. Math. Ser. B* **12** (1991), 33–39.

[200] M. J. Dyer, *Hecke algebras and reflections in Coxeter groups*, Ph.D. Thesis, University of Sydney, 1987. [130, 161, 170]

[201] M. J. Dyer, On some generalisations of the Kazhdan-Lusztig polynomials for "universal" Coxeter groups, *J. Algebra* **116** (1988), 353–371. [170]

[202] M. J. Dyer, Reflection subgroups of Coxeter systems, *J. Algebra* **135** (1990), 57–73. [25]

[203] M. J. Dyer, On the "Bruhat graph" of a Coxeter system, *Compos. Math.* **78** (1991), 185–191. [55, 64, 162]

[204] M. J. Dyer, Hecke algebras and shellings of Bruhat intervals II: twisted Bruhat orders, *Kazhdan-Lusztig Theory and Related Topics* (Chicago, 1989), Contemp. Math., **139**, American Mathematical Society, Providence, RI, 1992, pp. 141–165.

[205] M. J. Dyer, Hecke algebras and shellings of Bruhat intervals, *Compos. Math.*, **89** (1993), 91–115. [170]

[206] M. J. Dyer, The nil Hecke ring and Deodhar's conjecture on Bruhat intervals, *Invent. Math.* **111** (1993), 571–574. [64]

[207] M. J. Dyer, Bruhat intervals, polyhedral cones and Kazhdan-Lusztig-Stanley polynomials, *Math. Z.* **215** (1994), 223–236.

[208] M. J. Dyer, Quotients of twisted Bruhat orders, *J. Algebra* **163** (1994), 861–879.

[209] M. J. Dyer, Algebras associated to Bruhat intervals and polyhedral cones, *Finite-Dimensional Algebras and Related Topics* (Ottawa, 1992), NATO Adv. Sci. Inst. Ser. C, **424**, Kluwer, Dordrecht, 1994, pp. 95–121.

[210] M. J. Dyer, On coefficients of q in Kazhdan-Lusztig polynomials, *Algebraic groups and Lie groups* Austral. Math. Soc. Lect. Ser., **9**, Cambridge University Press, Cambridge, 1997, pp. 189–194. [170]

[211] M. Dyer, On minimal lengths of expressions of Coxeter group elements as products of reflections, *Proc. Amer. Math. Soc.* **129** (2001), 2591–2595. [242]

[212] M. J. Dyer, G. I. Lehrer, On positivity in Hecke algebras, *Geom. Dedicata* **35** (1990), 115–125.

[213] P. H. Edelman, Meet-distributive lattices and the anti-exchange closure, *Algebra Universalis* **10** (1980), 290–299.

[214] P. H. Edelman, The Bruhat order of the symmetric group is lexicographically shellable, *Proc. Amer. Math. Soc.* **82** (1981), 355–358.

[215] P. H. Edelman, A partial order on the regions of R^n dissected by hyperplanes, *Trans. Amer. Math. Soc.* **283** (1984), 617–631.

[216] P. H. Edelman, Lexicographically first reduced words, *Discrete Math.* **147** (1995), 95–106.

[217] P. ·H. Edelman, C. Greene, Combinatorial correspondences for Young tableaux, balanced tableaux, and maximal chains in the weak Bruhat order of S_n, *Combinatorics and Algebra* (Boulder, 1983), Contemp. Math. **34**, American Mathematical Society, Providence, RI, 1984, pp. 155–162. [243]

[218] P. H. Edelman, C. Greene, Balanced tableaux, *Adv. Math.* **63** (1987), 42–99. [243]

[219] P. H. Edelman, J. W. Walker, The homotopy type of hyperplane posets, *Proc. Amer. Math. Soc.* **94** (1985), 221–225.

[220] C. Ehresmann, Sur la topologie de certains éspaces homogènes, *Ann. Math.* **35** (1934), 396–443. [63]

[221] E. Ellers, B. Grünbaum, P. McMullen, A. Weiss, H. S. M. Coxeter (1907–2003), *Notices Amer. Math. Soc.* **50** (2003), 1234–1240.

[222] S. Elnitsky, Rhombic tilings of polygons and classes of reduced words in Coxeter groups, *J. Combin. Theory Ser. A* **77** (1997), 193–221.

[223] H. Eriksson, *Computational and combinatorial aspects of Coxeter groups*, Ph.D. Thesis, KTH, Stockholm, Sweden, 1994. [88, 125, 130, 294]

[224] H. Eriksson, K. Eriksson, Affine Weyl groups as infinite permutations. *Electron. J. Combin.* **5** (1998), no. 1, #R 18, 32 pp. [208, 242, 294]

[225] H. Eriksson, K. Eriksson, J. Sjöstrand, Expected number of inversions after a sequence of random adjacent transpositions, *Formal Power Series and Algebraic Combinatorics (Moscow, 2000)*, Springer, Berlin, 2000, pp. 677–685.

[226] K. Eriksson, Convergence of Mozes's game of numbers. *Linear Algebra Appl.* **166** (1992), 151–165.

[227] K. Eriksson, *Strongly convergent games and Coxeter groups*, Ph.D. Thesis, KTH, Stockholm, Sweden, 1993. [130]

[228] K. Eriksson, Reachability is decidable in the numbers game, *Theoret. Comput. Sci.* **131** (1994), 431–439.

[229] K. Eriksson, A combinatorial proof of the existence of the generic Hecke algebra and R-polynomials, *Math. Scand.* **75** (1994), 169–177.

[230] K. Eriksson, Polygon posets and the weak order of Coxeter groups. *J. Algebraic Combin.* **4** (1995), 233–252. [88]

[231] K. Eriksson, The numbers game and Coxeter groups. *Discrete Math.* **139** (1995), 155–166.

[232] K. Eriksson, Strong convergence and a game of numbers, *European J. Combin.* **17** (1996), 379–390.

[233] K. Eriksson, S. Linusson, A combinatorial theory of higher-dimensional permutation arrays, *Adv. Appl. Math.* **25** (2000), 194–211.

[234] K. Eriksson, S. Linusson, A decomposition of $\mathrm{Fl}(n)^d$ indexed by permutation arrays, *Adv. Appl. Math.* **25** (2000), 212–227.

[235] C. K. Fan, Schubert varieties and short braidedness, *Transform. Groups* **3** (1998), 51–56. [244]

[236] C. K. Fan, J. R. Stembridge, Nilpotent orbits and commutative elements, *J. Algebra* **196** (1997), 490–498. [244]

[237] W. Feller, *An Introduction to Probability Theory and Its Applications*, Vol. 1, Wiley, New York, 1950. [170]

[238] S. Felsner, The skeleton of a reduced word and a correspondence of Edelman and Greene, *Electron. J. Combin.* **8** (2001), RP10, 21 pp.

[239] S. Flath, The order dimension of multinomial lattices, *Order* **10** (1993), 201–219.

[240] P. Fleischmann, On pointwise conjugacy of distinguished coset representatives in Coxeter groups, *J. Group Theory* **5** (2002), 269–283.

[241] P. Fleischmann, I. Janiszczak, The lattices and Mbius functions of stable closed subrootsystems and hyperplane complements for classical Weyl groups, *Manuscr. Math.* **72** (1991), 375–403.

[242] P. Fleischmann, I. Janiszczak, The number of regular semisimple elements for Chevalley groups of classical type, *J. Algebra* **155** (1993), 482–528.

[243] D. Foata, G. N. Han, Calcul basique des permutations signées. I. Longueur et nombre d'inversions, *Adv. Applied Math.* **18** (1997), 489–509. [243]

[244] S. Fomin, C. Greene, V. Reiner, M. Shimozono, Balanced labellings and Schubert polynomials, *European J. Combin.* **18** (1997), 373–389.

[245] S. Fomin, A. Kirillov, Combinatorial B_n-analogues of Schubert polynomials, *Trans. Amer. Math. Soc.* **348** (1996), 3591–3620.

[246] S. Fomin, A. Kirillov, Reduced words and plane partitions. *J. Algebraic Combin.* **6** (1997), 311–319. [242]

[247] S. Fomin, R. P. Stanley, Schubert polynomials and the nil-Coxeter algebra, *Adv. Math.* **103** (1994), 196–207. [242]

[248] W. Fulton, *Young tableaux. With Applications to Representation Theory and Geometry.* London Mathematical Society Student Texts **35**, Cambridge University Press, Cambridge, 1997. [24, 234, 307]

[249] D. Garfinkle, On the classification of primitive ideals for complex classical Lie algebras. I, *Compos. Math.* **75** (1990), 135–169.

[250] D. Garfinkle, On the classification of primitive ideals for complex classical Lie algebras. II, *Compos. Math.* **81** (1992), 307–336.

[251] D. Garfinkle, On the classification of primitive ideals for complex classical Lie algebras. III, *Compos. Math.* **88** (1993), 187–234.

[252] D. Garfinkle, D. Vogan, On the structure of Kazhdan-Lusztig cells for branched Dynkin diagrams, *J. Algebra* **153** (1992), 91–120.

[253] A. M. Garsia, *The saga of reduced factorizations of elements of the symmetric group*, Publ. du LACIM **29**, Univ. du Québec, Montréal, 2002.

[254] A. M. Garsia, T. J. McLarnan, Relations between Young's natural and the Kazhdan-Lusztig representations of S_n, *Adv. Math.* **69** (1988), 32–92. [200]

[255] A. M. Garsia, C. Reutenauer, A decomposition of Solomon's descent algebra, *Adv. Math.* **77** (1989), 189–262.

[256] V. Gasharov, Factoring the Poincaré polynomials for the Bruhat order on S_n, *J. Combin. Theory Ser. A* **83** (1998), 159–164.

[257] V. Gasharov, Sufficiency of Lakshmibai-Sandhya singularity conditions for Schubert varieties, *Compos. Math.* **126** (2001), 47–56.

[258] G. Gasper, M. Rahman, *Basic Hypergeometric Series*, Cambridge University Press, Cambridge, 1990. [319, 320]

[259] M. Geck, On the induction of Kazhdan-Lusztig cells, *Bull. London Math. Soc.* **35** (2003), 608–614.

[260] M. Geck, S. Kim, Bases for the Bruhat-Chevalley order on all finite Coxeter groups, *J. Algebra* **197** (1997), 278–310.

[261] M. Geck, G. Pfeiffer, On the irreducible characters of Hecke algebras, *Adv. Math.* **102** (1993), 79–94. [25]

[262] M. Geck, G. Pfeiffer, *Characters of Finite Coxeter Groups and Iwahori-Hecke Algebras*, London Mathematical Society Monographs, New Series, **21**, The Clarendon Press/Oxford University Press, New York, 2000.

[263] I. M. Gel´fand, V. Serganova, Combinatorial geometries and the strata of a torus on homogeneous compact manifolds, (Russian), *Usp. Mat. Nauk* **42** (1987), no. 2(254), 107–134.

[264] I. M. Gel´fand, V. Serganova, On the general definition of a matroid and a greedoid, (Russian), *Dokl. Akad. Nauk SSSR* **292** (1987), no. 1, 15–20.

[265] S. I. Gel´fand, R. MacPherson, *Verma modules and Schubert cells: a dictionary*, Lect. Notes in Math. **925**, Springer, Berlin, 1982, pp. 1–50.

[266] R. Gill, On posets from conjugacy classes of Coxeter groups, *Discrete Math.* **216** (2000), 139–152. [25]

[267] M. Goresky, *Kazhdan-Lusztig polynomials for classical groups*, preprint, Northeastern University, Boston, 1981.

[268] I. P. Goulden, D. M. Jackson, *Combinatorial Enumeration*, Wiley-Interscience, New York, 1983. [170]

[269] A. Goupil, The poset of conjugacy classes and decomposition of products in the symmetric group. *Can. Math. Bull.* **35** (1992), 152–160.

[270] A. Goupil, Reflection decompositions in the classical Weyl groups, *Discrete Math.* **137** (1995), 195–209.

[271] D. Grabiner, Random walk in an alcove of an affine Weyl group, and non-colliding random walks on an interval, *J. Combin. Theory Ser. A* **97** (2002), 285–306.

[272] R. Green, On 321-avoiding permutations in affine Weyl groups, *J. Algebraic Combin.* **15** (2002), 241–252.

[273] R. Green, J. Losonczy, Fully commutative Kazhdan-Lusztig cells, *Ann. Inst. Fourier (Grenoble)* **51** (2001), 1025–1045.

[274] R. Green, J. Losonczy, Freely braided elements of Coxeter groups, *Ann. Combin.* **6** (2002), 337–348.

[275] R. Green, J. Losonczy, Freely braided elements in Coxeter groups. II, *Adv. Appl. Math.* **33** (2004), 26–39.

[276] J. Griggs, M. Wachs, Towers of powers and Bruhat order, *European J. Combin.* **13** (1992), 367–370.

[277] L. C. Grove, C. T. Benson, *Finite Reflection Groups*, 2nd ed., Springer, New York, 1985.

[278] G. Th. Guilbaud, P. Rosenstiehl, Analyse algébrique d'un scrutin, in *Ordres totaux finis*, Gauthiers-Villars et Mouton, Paris, 1971, pp. 71–100.

[279] E. Gutkin, Geometry and combinatorics of groups generated by reflections, *Enseign. Math.* **32** (1986), 95–110.

[280] A. Gyoja, A generalized Poincaré series associated to a Hecke algebra of a finite or *p*-adic Chevalley group, *Japan J. Math. (N.S.)* **9** (1983), 87–111.

[281] A. Gyoja, On the existence of a *W*-graph for an irreducible representation of a Coxeter group, *J. Algebra* **86** (1984), 422–438.

[282] M. Hagiwara, Minuscule elements of affine Weyl groups (Japanese), *Combinatorial representation theory and related topics* (Japanese) (Kyoto, 2002), Sūrikaisekikenkyūsho Kōkyūroku, **1310** (2003), 1–15.

[283] M. Haiman, On mixed insertion, symmetry, and shifted Young tableaux, *J. Combin. Theory Ser. A* **50** (1989), 196–225.

[284] M. Haiman, Dual equivalence with applications, including a conjecture of Proctor, *Discrete Math.* **99** (1992), 79–113. [232, 242, 244]

[285] M. Haiman, D. Kim, A characterization of generalized staircases, *Discrete Math.* **99** (1992), 115–122.

[286] M. Haiman, Hecke algebra characters and immanant conjectures, *J. Amer. Math. Soc.* **6** (1993), 569–595.

[287] P. de la Harpe, An invitation to Coxeter groups, *Group Theory from a Geometrical Viewpoint* (Trieste, 1990), World Scientific, Singapore, 1991.

[288] L. H. Harper, Stabilization and the edgesum problem, *Ars Combinatoria* **4** (1977), 225–270.

[289] M. Hazewinkel, W. Hesselink, D. Siersma, F. D. Veldkamp, The ubiquity of Coxeter-Dynkin diagrams (an introduction to the A-D-E problem), *Nieuw Arch. Wisk.* **25** (1977), 257–307. [24]

[290] P. Headley, *Reduced expressions in infinite Coxeter groups*, Ph.D. Thesis, University of Michigan, 1994. [130]

[291] P. Headley, On a family of hyperplane arrangements related to the affine Weyl groups, *J. Algebraic Combin.* **6** (1997), 331–338.

[292] A. Heck, A criterion for triple (X, I, μ) to be a W-graph of a Coxeter group, *Commun. Algebra* **16** (1988), 2083–2102.

[293] S. Hermiller, Rewriting systems for Coxeter groups, *J. Pure Appl. Algebra* **92** (1994), 137–148.

[294] G. Higman, Ordering by divisibility in abstract algebras, *Proc. London Math. Soc.* **2** (1952), 326–336. [64]

[295] H. L. Hiller, *Geometry of Coxeter groups*, Pitman, Boston, 1982. [24, 208]

[296] H. L. Hiller, Combinatorics and intersections of Schubert varieties, *Comment. Math. Helv.* **57** (1982), 41–59.

[297] C. Hohlweg, M. Schocker, On a parabolic symmetry of finite Coxeter groups, *Bull. London Math. Soc.* **36** (2004), 289–293.

[298] A. van den Hombergh, About the automorphisms of the Bruhat-ordering in a Coxeter group, *Indag. math.* **36** (1974), 125–131. [38, 64]

[299] R. Howlett, J.-Y. Shi, On regularity of finite reflection groups, *Manuscr. Math.* **102** (2000), 325–333.

[300] R. Howlett, Y. Yin, Inducing W-graphs, *Math. Z.* **244** (2003), 415–431.

[301] A. Hultman, *Combinatorial complexes, Bruhat intervals and reflection distances*, Ph.D. Thesis, KTH, Stockholm, Sweden, 2003. [56]

[302] A. Hultman, Bruhat intervals of length 4 in Weyl groups, *J. Combin. Theory Ser. A* **102** (2003), 163–178. [56]

[303] A. Hultman, Fixed points of involutive automorphisms of the Bruhat order, *Adv. Math.*, to appear. [64]

[304] J. E. Humphreys, *Introduction to Lie Algebras and Representation Theory*, Springer, New York, 1972. [24]

[305] J. E. Humphreys, *Linear Algebraic Groups*, Springer, New York, 1975. [24]

[306] J. E. Humphreys, *Reflection Groups and Coxeter Groups*, Cambridge University Press, Cambridge, 1990. [4, 24, 123, 124, 126, 130, 132, 134, 136, 174, 175, 200, 205, 240]

[307] C. Huneke, V. Lakshmibai, A characterization of Kempf varieties by means of standard monomials and the geometric consequences, *J. Algebra* **94** (1985), 52–105.

[308] L. Iancu, Cellules de Kazhdan-Lusztig et correspondance de Robinson-Schensted, *C. R. Math. Acad. Sci. Paris* **336** (2003), 791–794.

[309] F. Incitti, The Bruhat order on the involutions of the hyperoctahedral group, *European J. Combin.* **24** (2003), 825–848. [64]

[310] F. Incitti, The Bruhat order on the involutions of the symmetric group, *J. Algebraic Combin.* **20** (2004), 243–261. [64]

[311] F. Incitti, Bruhat order on classical Weyl groups: minimal chains and covering relation, *European J. Combin.* **26** (2005), 729–753. [64, 294]

[312] F. Incitti, Bruhat order on the involutions of classical Weyl groups, preprint. [64]

[313] N. Iwahori, H. Matsumoto, On some Bruhat decomposition and the structure of the Hecke rings of *p*-adic Chevalley groups, *Inst. Hautes Études Sci. Publ. Math.* **25** (1965), 5–48.

[314] G. D. James, *The Representation Theory of Symmetric Groups*, Lecture Notes in Math., Vol. 682, Springer-Verlag, Berlin, 1978. [188]

[315] G. D. James, A. Kerber, *The Representation Theory of the Symmetric Group*, Addison-Wesley, Reading, MA, 1981. [188]

[316] J. Jantzen, *Moduln mit Einem Höchsten Gewicht*, Lecture Notes in Math., Vol. 750, Springer-Verlag, Berlin, 1979. [56]

[317] V. G. Kac, *Infinite-Dimensional Lie Algebras*, 3rd ed., Cambridge University Press, Cambridge, 1990. [24]

[318] M. Kaneda, On the inverse Kazhdan-Lusztig polynomials for affine Weyl groups, *J. Reine Angew. Math.* **381** (1987), 116–135.

[319] C. Kassel, A. Lascoux, C. Reutenauer, Factorizations in Schubert cells, *Adv. Math.* **150** (2000), 1–35.

[320] C. Kassel, A. Lascoux, C. Reutenauer, The singular locus of a Schubert variety, *J. Algebra* **269** (2003), 74–108.

[321] S. I. Kato, On the Kazhdan-Lusztig polynomials for affine Weyl groups, *Adv. Math.* **55** (1985), 103–130.

[322] D. Kazhdan, G. Lusztig, Representations of Coxeter groups and Hecke algebras, *Invent. Math.* **53** (1979), 165–184. [131, 170, 171, 173, 175, 188, 196, 200]

[323] D. Kazhdan, G. Lusztig, Schubert varieties and Poincaré duality, in *Geometry of the Laplace operator*, Proc. Sympos. Pure Math. 34, American Mathematical Society, Providence, RI, 1980, pp. 185–203. [170, 171]

[324] A. Kerber, A. Kohnert, A. Lascoux, SYMMETRICA, an object oriented computer-algebra system for the symmetric group, *J. Symbolic Comput.* **14** (1992), 195–203.

[325] S. V. Kerov, *W*-graphs of representations of symmetric groups, *J. Sov. Math.* **28** (1985), 596–605. [64, 193, 200]

[326] A. Kirillov, A. Lascoux, Factorization of Kazhdan-Lusztig elements for Grassmanians, *Adv. Studies Pure Math.* **28** (2000), 143–154.

[327] D. E. Knuth, Permutations, matrices and generalized Young tableaux, *Pacific J. Math.* **34** (1970), 709–727.

[328] D. E. Knuth, *The Art of Computer Programming, Vol. 3, Sorting and Searching*, Addison-Wesley, Reading, MA, 1973. [307]

[329] A. Knutson, E. Miller, Subword complexes in Coxeter groups, *Adv. Math.* **184** (2004), 161–176. [88]

340   Bibliography

[330] W. Kraśkiewicz, Reduced decompositions in hyperoctahedral groups. *C. R. Acad. Sci. Paris, Ser. I Math.* **309** (1989), 903–907. [244]

[331] W. Kraśkiewicz, Reduced decompositions in Weyl groups, *European J. Combin.* **16** (1995), 293–313. [234]

[332] W. Kraśkiewicz, J. Weyman, Algebra of coinvariants and the action of a Coxeter element, *Bayreuth. Math. Schr.* **63** (2001), 265–284.

[333] C. Krattenthaler, L. Orsina, P. Papi, Enumeration of ad-nilpotent *b*-ideals for simple Lie algebras, Special issue in memory of Rodica Simion, *Adv. Appl. Math.* **28** (2002), 478–522.

[334] S. Kumar, *Kac-Moody Groups, Their Flag Varieties and Representation Theory*, Progress in Mathematics, **204**, Birkhäuser, Boston, MA, 2002. [24]

[335] J. Kung, D. Sutherland, The automorphism group of the strong order of the symmetric group, *J. London Math. Soc.* **37** (1988), 193–202.

[336] V. Lakshmibai, C. Musili, C. S. Seshadri, Geometry of G/B, *Bull. Amer. Math. Soc.* **1** (1979), 432–435.

[337] V. Lakshmibai, B. Sandhya, Criterion for smoothness of Schubert varieties in $Sl(n)/B$, *Proc. Indian Acad. Sci. Math. Sci.* **100** (1990), 45–52. [171]

[338] V. Lakshmibai, M. Song, A criterion for smoothness of Schubert varieties in $Sp(2n)/B$, *J. Algebra* **189** (1997), 332–352.

[339] T. K. Lam, *B and D analogues of stable Schubert polynomials and related insertion algorithms*, Ph.D. Thesis, Massachusetts Institute of Technology, 1994. [244]

[340] F. Lannér, On complexes with transitive groups of automorphisms, *Commun. Semin. Math. Univ. Lund* **11** (1950), 71 pp.

[341] A. Lascoux, Polynômes de Kazhdan-Lusztig pour les variétés de Schubert vexillaires. *C. R. Acad. Sci. Paris Sr. I Math.* **321** (1995), 667–670. [171]

[342] A. Lascoux, Polynômes de Schubert: une approche historique. *Discrete Math.* **139** (1995), 303–317.

[343] A. Lascoux, Ordonner le groupe symétrique: pourquoi utiliser l'älgèbre de Iwahori-Hecke? *Proceedings of the International Congress of Mathematicians, Vol. III* (Berlin, 1998). *Doc. Math.* (electronic) (1998), Extra Vol. III, 355–364.

[344] A. Lascoux, Ordering the affine symmetric group, *Algebraic Combinatorics and Applications* (Gösweinstein, 1999), Springer, Berlin, 2001, 219–231.

[345] A. Lascoux, Chern and Yang through ice, preprint.

[346] A. Lascoux, M.-P. Schützenberger, Le monoïde plaxique. *Noncommutative Structures in Algebra and Geometric Combinatorics* (Naples, 1978), Quad. "Ricerca Sci.", **109**, CNR, Rome, 1981, pp. 129–156.

[347] A. Lascoux, M.-P. Schützenberger, Polynômes de Kazhdan & Lusztig pour les grassmanniennes, *Astérisque* **87–88** (1981), 249–266. [193, 200]

[348] A. Lascoux, M.-P. Schützenberger, Polynômes de Schubert. *C. R. Acad. Sci. Paris Sér. I Math.* **294** (1982), 447–450.

[349] A. Lascoux, M.-P. Schützenberger, Structure de Hopf de l'anneau de co-homologie et de l'anneau de Grothendieck d'une variété de drapeaux. *C. R. Acad. Sci. Paris Sér. I Math.* **295** (1982), 629–633. [243]

[350] A. Lascoux, M.-P. Schützenberger, *Symmetry and Flag Manifolds*, Lecture Notes in Mathematics, **996**, Springer, Berlin, 1983, pp. 118–144.

[351] A. Lascoux, M.-P. Schützenberger, Treillis et bases des groupes de Coxeter, *Electron. J. Combin.* **3** (1996), no. 2, #R 27, 35 pp.

[352] G. Lawton, Two-sided cells in the affine Weyl group of type $\tilde{A}_{n-1}$, *J. Algebra* **120** (1989), 74–89.

[353] P. Le Chenadec, Canonical forms in finitely presented algebras. *7th International Conference on Automated Deduction* (Napa, Calif., 1984), Lecture Notes in Comput. Sci., **170**, Springer, Berlin, 1984, pp. 142–165.

[354] B. Leclerc, A finite Coxeter group the weak Bruhat order of which is not symmetric chain, *European J. Combin.* **15** (1994), 181–185.

[355] L. Leclerc, J.-Y. Thibon, Littlewood-Richardson coefficients and Kazhdan-Lusztig polynomials, *Adv. Studies Pure Math.* **28** (2000), 155–220.

[356] C. Le Conte de Poly-Barbut, Le diagramme du treillis permutoèdre est intersection des diagrammes de deux produits directs d'ordres totaux. *Math. Inform. Sci. Humaines* **112** (1990), 49–53.

[357] C. Le Conte de Poly-Barbut, Sur les treillis de Coxeter finis. *Math. Inform. Sci. Humaines* **125** (1994), 41–57.

[358] W. Ledermann, *Introduction to Group Characters*, 2nd ed., Cambridge University Press, Cambridge, 1987. [175, 181, 182, 183]

[359] E. L. Lehmann, Some concepts of dependence, *Ann. Math. Statist.* **37** (1966), 1137–1153. [87]

[360] G. I. Lehrer, On hyperoctahedral hyperplane complements, *The Arcata Conference on Representations of Finite Groups* (Arcata, CA, 1986), Proc. Sympos. Pure Math., **47**, Part 2, American Mathematical Society, Providence, RI, 1987, pp. 219–234.

[361] G. I. Lehrer, On the Poincaré series associated with Coxeter group actions on the complements of hyperplanes, *J. London Math. Soc.* **36** (1987), 275–294. [242]

[362] G. I. Lehrer, A survey of Hecke algebras and the Artin braid groups, *Braids* (Santa Cruz, CA, 1986), Contemp. Math. **78**, American Mathematical Society, Providence, RI, 1988, pp. 365–385. [200]

[363] A. T. Lundell, S. Weingram, *The Topology of CW Complexes*, Van Nostrand, New York, 1969. [305]

[364] G. Lusztig, personal communication, 1980. [161]

[365] G. Lusztig, Some examples of square integrable representations of semisimple p-adic groups, *Trans. Amer. Math. Soc.* **277** (1983), 623–653. [293]

[366] G. Lusztig, Left cells in Weyl groups, *Lie Group Representations I*, Lect. Notes in Math. 1024, Springer, Berlin, 1983, pp. 99–111.

[367] G. Lusztig, Cells in affine Weyl groups, *Algebraic Groups and Related Topics*, Adv. Studies in Pure. Math. **6**, North-Holland, Amsterdam, 1985, pp. 225–287. [198, 200]

[368] G. Lusztig, The two-sided cells of the affine Weyl group of type $\tilde{A}_n$, *Infinite Dimensional Groups with Applications* (Berkeley, 1984), 275–283, Math. Sci. Res. Inst. Publ. **4**, Springer, New York, 1985.

[369] G. Lusztig, Sur les cellules gauches des groupes de Weyl, *C.R. Acad. Sci. Paris. Sér. I Math.* **302** (1986), 5–8.

[370] G. Lusztig, Cells in affine Weyl groups II, *J. Algebra* **109** (1987), 536–548. [198, 200]

[371] G. Lusztig, Cells in affine Weyl groups III, *J. Fac. Sci. Univ. Tokyo Sect. IA Math.* **34** (1987), 223–243.

[372] G. Lusztig, Cells in affine Weyl groups IV, *J. Fac. Sci. Univ. Tokyo Sect. IA Math.* **36** (1989), 297–328.

[373] G. Lusztig, Intersection cohomology methods in representation theory, *Proceedings of the International Congress of Mathematicians* (Kyoto, 1990), Mathematical Society of Japan, Tokyo, 1991, pp. 155–174.

[374] G. Lusztig, Affine Weyl groups and conjugacy classes in Weyl groups, *Transform. Groups* **1** (1996), 83–97.

[375] G. Lusztig, Nonlocal finiteness of a W-graph, *Represent. Theory* **1** (1997), 25–30 (electronic).

[376] G. Lusztig, Periodic W-graphs, *Represent. Theory* **1** (1997), 207–279 (electronic).

[377] G. Lusztig, *Notes on Affine Hecke Algebras*, 71–103, Lecture Notes in Math., **1804**, Springer, Berlin, 2002.

[378] G. Lusztig, *Hecke Algebras with Unequal Parameters*, CRM Monograph Series, **18**, American Mathematical Society, Providence, RI, 2003.

[379] G. Lusztig, N. H. Xi, Canonical left cells in affine Weyl groups, *Adv. Math.* **72** (1988), 284–288.

[380] I. G. Macdonald, The Poincaré series of a Coxeter group, *Math. Ann.* **199** (1972), 161–174.

[381] I. G. Macdonald, *Notes on Schubert Polynomials*, Publ. du LACIM **6**, University du Québec, Montréal, 1991. [234, 242]

[382] M. Mamagani, Rewriting systems and complete growth series for triangular Coxeter groups, (Russian) *Mat. Zametki* **71** (2002), 431–439; translation in *Math. Notes* **71** (2002), 392–399.

[383] Y. Manin, V. Shekhtman, Higher Bruhat orderings connected with the symmetric group, (Russian) *Funkt. Anal. Prilozhen.* **20** (1986), 74–75.

[384] L. Manivel, *Fonctions symétriques, polynômes de Schubert et lieux de dégénérescence*, Cours Spécialisés **3**, Société Mathematic de France, Paris, 1998.

[385] M. Marietti, Closed product formulas for certain R-polynomials, *European J. Combin.* **23** (2002), 57–62.

[386] M. Marietti, Boolean elements in Kazhdan-Lusztig theory, preprint.

[387] M. Marietti, Parabolic Kazhdan-Lusztig polynomials in the symmetric group, preprint.

[388] G. Markowsky, Permutation lattices revisited, *Math. Social Sci.* **27** (1994), 59–72. [88]

[389] A. Mathas, Some generic representations, *W*-graphs, and duality, *J. Algebra* **170** (1994), 322–353. [200]

[390] A. Mathas, A *q*-analogue of the Coxeter complex, *J. Algebra* **164** (1994), 831–848. [200]

[391] A. Mathas, On the left cell representations of Iwahori-Hecke algebras of finite Coxeter groups, *J. London Math. Soc.* **54** (1996), 475–488. [200]

[392] H. Matsumoto, Générateurs et relations des groupes de Weyl généralisés, *C.R. Acad. Sci. Paris* **258** (1964), 3419–3422. [25]

[393] T. McLarnan, G. Warrington, Counterexamples to the 0–1 conjecture, *Represent. Theory* **7** (2003), 181–195.

[394] P. McMullen, The order of a finite Coxeter group. *Elem. Math.* **46** (1991), 121–130.

[395] P. McMullen, Modern developments in regular polytopes. *Polytopes: Abstract, Convex and Computational* (Scarborough, ON, 1993), NATO Adv. Sci. Inst. Ser. C, **440**, Kluwer, Dordrecht, 1994, pp. 97–124.

[396] P. McMullen, E. Schulte, *Abstract Regular Polytopes*, Encyclopedia of Mathematics and its Applications, **92**, Cambridge University Press, Cambridge 2002. [24]

[397] P. Moszkowski, Généralisation d'une formule de Solomon relative à l'anneau de groupe d'un groupe de Coxeter, *C. R. Acad. Sci. Paris Sr. I Math.* **309** (1989), 539–541.

[398] P. Moszkowski, Longueur des involutions et classification des groupes de Coxeter finis, *Sémin. Lotharingien Combin.* **33** (1994), 9 pp. (electronic).

[399] G. Moussong, Hyperbolic Coxeter groups, Ph.D. thesis, Ohio State University, 1988.

[400] S. Mozes, Reflection processes on graphs and Weyl groups, *J. Combin. Theory, Ser. A* **53** (1990), 128–142. [130]

[401] J. R. Munkres, *Elements of Algebraic Topology*, Addison-Wesley, Menlo Park, CA, 1984. [302]

[402] H. Naruse, A combinatorial description of the Grassman-type parabolic Kazhdan-Lusztig polynomial Q^I, *Topics in Combinatorial Representation Theory*, (Japanese) (Kyoto, 2000). Sūrikaisekikenkyūsho Kōkyūroku, **1190** (2001), 126–135.

[403] W. Neidhardt, General Kac-Moody algebras and the Kazhdan-Lusztig conjecture, *Pacific J. Math.* **159** (1993), 87–126.

[404] T. Oda, *Convex Bodies and Algebraic Geometry*, Ergebnisse der Mathematik und ihrer Grenzgebiete **15**, Springer-Verlag, Berlin Heidelberg, 1988. [243]

[405] P. Orlik, L. Solomon, Complexes for reflection groups, *Algebraic Geometry*, Lect. Notes in Math. 862, Springer, Berlin, 1981, pp. 193–207.

[406] P. Orlik, L. Solomon, Coxeter arrangements, *Singularities*, Part 2, Proc. Sympos. Pure Math. **40**, American Mathematical Society, Providence, RI, 1983, pp. 269–291.

[407] P. Orlik, L. Solomon, H. Terao, On Coxeter arrangements and the Coxeter number, *Complex Analytic Singularities*, Adv. Studies Pure Math. **8**, North-Holland, Amsterdam, 1987, pp. 461–477.

[408] M. Pagliacci, A simple product formula for certain Kazhdan-Lusztig R-polynomials, preprint.

[409] W. Parry, Growth series of Coxeter groups and Salem numbers. *J. Algebra* **154** (1993), 406–415.

[410] P. Papi, Convex orderings and symmetric group, *Commun. Algebra* **22** (1994), 4089–4094.

[411] P. Papi, A characterization of a special ordering in a root system, *Proc. Amer. Math. Soc.* **120** (1994), 661–665.

[412] P. Papi, Convex orderings in affine root systems, *J. Algebra* **172** (1995), 613–623.

[413] P. Papi, Convex orderings in affine root systems. II, *J. Algebra* **186** (1996), 72–91.

[414] P. Papi, Affine permutations and inversion multigraphs, *Electron. J. Combin.* **4** (1997), no. 1, #R 5, approx. 9 pp.

[415] L. Paris, Growth series of Coxeter groups, *Group theory from a Geometrical Viewpoint* (Trieste, 1990), World Scientific, Singapore, 1991.

[416] L. Paris, Complex growth series of Coxeter systems, *Enseign. Math.* **38** (1992), 95–102.

[417] L. Paris, Minimal nonstandard Coxeter trees, *J. Algebra* **156** (1993), 76–107.

[418] L. Paris, Commensurators of parabolic subgroups of Coxeter groups, *Proc. Amer. Math. Soc.* **125** (1997), 731–738.

[419] S. Perkins, P. Rowley, Minimal and maximal length involutions in finite Coxeter groups, *Commun. Algebra* **30** (2002), 1273–1292.

[420] G. Pfeiffer, G. Rohrle, Distributive coset graphs of finite Coxeter groups, *J. Group Theory* **6** (2003), 311–320.

[421] P. Polo, Construction of arbitrary Kazhdan-Lusztig polynomials in symmetric groups, *Represent. Theory* **3** (1999), 90–104. [170, 172]

[422] R. A. Proctor, *Interactions between combinatorics, Lie theory and algebraic geometry via the Bruhat orders*, Ph.D. Thesis, Massachusetts Institute of Technology, 1981.

[423] R. A. Proctor, Classical Bruhat orders and lexicographic shellability, *J. Algebra* **77** (1982), 104–126. [294]

[424] R. A. Proctor, Bruhat lattices, plane partition generating functions, and minuscule representations, *European J. Combin.* **5** (1984), 331–350. [64]

[425] R. A. Proctor, A Dynkin diagram classification theorem arising from a combinatorial problem, *Adv. Math.* **62** (1986), 103–117.

[426] R. A. Proctor, Two amusing Dynkin diagram graph classifications, *Amer. Math. Monthly* **100** (1993), 937–941.

[427] R. Proctor, Minuscule elements of Weyl groups, the numbers game, and d-complete posets, *J. Algebra* **213** (1999), 272–303.

[428] R. Proctor, Dynkin diagram classification of λ-minuscule Bruhat lattices and of d-complete posets, *J. Algebraic Combin.* **9** (1999), 61–94.

[429] A. Ram, Standard Young tableaux for finite root systems, preprint. [243]

[430] N. Reading, Order dimension, strong Bruhat order and lattice properties for posets, *Order* **19** (2002), 73–100. [64]

[431] N. Reading, The order dimension of the poset of regions in a hyperplane arrangement, *J. Combin. Theory Ser. A* **104** (2003), 265–285. [88]

[432] N. Reading, The cd-index of Bruhat intervals, *Electron. J. Combin.* **11** (2004), #R 74, 25 pp. (electronic).

[433] V. Reiner, Quotients of Coxeter complexes and P-partitions, *Mem. Amer. Math. Soc.* **95** (1992), no. 460, vi+134 pp. [243]

[434] V. Reiner, Signed posets, *J. Combin. Theory Ser. A* **62** (1993), 324–360. [243]

[435] V. Reiner, Signed permutation statistics, *European J. Combin.* **14** (1993), 553–567. [243]

[436] V. Reiner, Upper binomial posets and signed permutation statistics, *European J. Combin.* **14** (1993), 581–588. [243]

[437] V. Reiner, Signed permutation statistics and cycle-type, *European J. Combin.* **14** (1993), 569–579. [243]

[438] V. Reiner, Descents and one-dimensional characters for classical Weyl groups, *Discrete Math.* **140** (1995), 129–140.

[439] V. Reiner, The distribution of descent and length in a Coxeter group, *Electron. J. Combin.* **2** (1995), R25. [242]

[440] V. Reiner, Non-crossing partitions for classical reflection groups, *Discrete Math.* **177** (1997), 195–222. [243]

[441] V. Reiner, Note on the expected number of Yang-Baxter moves applicable to reduced decompositions, preprint.

[442] V. Reiner, M. Shimozono, Plactification, *J. Algebraic Combin.* **4** (1995), 331–351.

[443] V. Reiner, G. Ziegler, Coxeter associahedra, *Mathematika* **41** (1994), 364–393. [243]

[444] R. W. Richardson, Conjugacy classes of involutions in Coxeter groups. *Bull. Austral. Math. Soc.* **26** (1982), 1–15.

[445] R. Richardson, T. Springer, The Bruhat order on symmetric varieties, *Geom. Dedicata* **35** (1990), 389–436; and **49** (1994), 231–238. [64]

[446] M. Ronan, *Lectures on Buildings*, Academic Press, San Diego, 1989. [88]

[447] I. G. Rosenberg, A semilattice on the set of permutations on an infinite set, *Math. Nachrichten* **60** (1974), 191–199.

[448] J. Rosenboom, On the computation of Kazhdan-Lusztig polynomials and representations of Hecke algebras, *Arch. Math.* **66** (1996), 35–50.

[449] G.-C. Rota, On the foundations of Combinatorial Theory I: Theory of Möbius functions, *Z. Wahrsch. Verw. Gebiete* **2** (1964), 340–368.

[450] B. Sagan, *The symmetric group. Representations, Combinatorial Algorithms, and Symmetric Functions*, 2nd ed., Graduate Texts in Mathematics, **203**, Springer-Verlag, New York, 2001. [175, 188, 307, 319]

[451] I. R. Savage, Contributions to the theory of rank order statistics — the "trend case", *Ann. Math. Statist.* **28** (1957), 968–977.

[452] I. R. Savage, Contributions to the theory of rank order statistics: Applications of lattice theory, *Rev Int. Statist. Inst.* **32** (1964), 52–64. [87]

[453] J. P. Serre, Cohomologie des groupes discrets, *Prospects in Mathematics*, Ann. of Math. Studies, No. 70, Princeton University Press, Princeton, NJ, 1971, pp. 77–169. [242]

[454] B. Shapiro, M. Shapiro, A. Vainshtein, Kazhdan-Lusztig polynomials for certain varieties of incomplete flags, *Discrete Math.* **180** (1998), 345–355. [170, 171]

[455] J.-Y. Shi, *The Kazhdan-Lusztig cells in certain affine Weyl groups*, Lect. Notes in Math. **1179**, Springer, Berlin, 1986. [200, 293]

[456] J.-Y. Shi, Sign types corresponding to an affine Weyl group, *J. London Math. Soc.* **35** (1987), 56–74. [124]

[457] J.-Y. Shi, Alcoves corresponding to an affine Weyl group, *J. London Math. Soc.* **35** (1987), 42–55.

[458] J.-Y. Shi, A two-sided cell in an affine Weyl group, *J. London Math. Soc.* **36** (1987), 407–420.

[459] J.-Y. Shi, A two-sided cell in an affine Weyl group. II, *J. London Math. Soc.* **37** (1988), 253–264.

[460] J.-Y. Shi, A result on the Bruhat order of a Coxeter group, *J. Algebra* **128** (1990), 510–516.

[461] J.-Y. Shi, The joint relations and the set $\mathcal{D}_1$ in certain crystallographic groups, *Adv. Math.* **81**, (1990), 66–89.

[462] J.-Y. Shi, The generalized Robinson-Schensted algorithm on the affine Weyl group of type $\tilde{A}_{n-1}$, *J. Algebra* **139** (1991), 364–394.

[463] J.-Y. Shi, Some numeric results on root systems, *Pacific J. Math.* **160** (1993), 155–164. [130]

[464] J.-Y. Shi, Left cells in affine Weyl groups, *Tohoku Math. J.* **46** (1994), 105–124.

[465] J.-Y. Shi, Some results relating two presentations of certain affine Weyl groups, *J. Algebra* **163** (1994), 235–257.

[466] J.-Y. Shi, Left cells in the affine Weyl group $W_a(\tilde{D}_4)$, *Osaka J. Math.* **31** (1994), 27–50.

[467] J.-Y. Shi, The verification of a conjecture on left cells of certain Coxeter groups, *Hiroshima Math. J.* **24** (1994), 627–646. [293]

[468] J.-Y. Shi, The partial order on two-sided cells of certain affine Weyl groups, *J. Algebra* **179** (1996), 607–621.

[469] J.-Y. Shi, Left cells in certain Coxeter groups, *Group theory in China*, Math. Appl. (China Ser.), **365**, Kluwer, Dordrecht, 1996, pp. 130–148.

[470] J.-Y. Shi, The enumeration of Coxeter elements, *J. Algebraic Combin.* **6** (1997), 161–171.

[471] J.-Y. Shi, The number of $\oplus$-sign types, *Quart. J. Math. Oxford* **48** (1997), 93–105.

[472] J.-Y. Shi, Left cells in the affine Weyl group of type $\widetilde{F}_4$, *J. Algebra* **200** (1998), 173–206.

[473] J.-Y. Shi, Left cells in the affine Weyl group of type $\widetilde{C}_4$, *J. Algebra* **202** (1998), 745–776.

[474] J.-Y. Shi, On two presentations of the affine Weyl groups of classical types, *J. Algebra* **221** (1999), 360–383.

[475] J.-Y. Shi, Conjugacy relation on Coxeter elements, *Adv. Math.* **161** (2001), 1–19.

[476] J.-Y. Shi, Coxeter elements and Kazhdan-Lusztig cells, *J. Algebra* **250** (2002), 229–251.

[477] J.-Y. Shi, Explicit formulae for the Brenti's polynomials $\gamma_{a_1,\dots,a_r}$, *Adv. Math.* **177** (2003), 181–207.

[478] J.-Y. Shi, Fully commutative elements and Kazhdan-Lusztig cells in the finite and affine Coxeter groups, *Proc. Amer. Math. Soc.* **131** (2003), 3371–3378.

[479] J.-Y. Shi, Yang-Baxter bases for Coxeter groups, *J. London Math. Soc.* **69** (2004), 349–362.

[480] L. Solomon, The orders of the finite Chevalley groups, *J. Algebra* **3** (1966), 376–393. [242]

[481] L. Solomon, A decomposition of the group algebra of a finite Coxeter group, *J. Algebra* **9** (1968), 220–239. [200]

[482] L. Solomon, A Mackey formula in the group ring of a Coxeter group, *J. Algebra* **41** (1976), 255–268.

[483] L. Solomon, The number of irreducible representations of a finite Coxeter group, *Lie Algebra and Related Topics* (Madison, 1988), Contemp. Math. **110**, American Mathematical Society, Providence, RI, 1990, pp. 241–263.

[484] L. Solomon, The Bruhat decomposition, Tits system and Iwahori ring for the monoid of matrices over a finite field, *Geom. Dedicata* **36** (1990), 15–49.

[485] L. Solomon, Presenting the symmetric group with transpositions, *J. Algebra* **168** (1994), 521–524.

[486] L. Solomon, H. Terao, The double Coxeter arrangement, *Comment. Math. Helv.* **73** (1998), 237–258.

[487] T. A. Springer, Quelques applications de la cohomologie d'intersection, *Astérisque* **92–93** (1982), 249–273.

[488] T. A. Springer, Some remarks on involutions in Coxeter groups, *Commun. Algebra* **10** (1982), 631–636.

[489] T. A. Springer, A combinatorial result on K-orbits on a flag manifold, *The Sophus Lie Memorial Conference* (Oslo, 1992), Scandinavian University Press, Oslo, 1994, pp. 363–370.

[490] R. P. Stanley, Binomial posets, Möbius inversion, and permutation enumeration, *J. Combin. Theory Ser. A* **20** (1976), 336–356. [210, 242]

[491] R. P. Stanley, Weyl groups, the hard Lefschetz theorem, and the Sperner property, *SIAM J. Algebra Discrete Methods* **1** (1980), 168–184. [239, 294]

[492] R. P. Stanley, Some aspects of groups acting on finite posets, *J. Combin. Theory Ser. A* **32** (1982), 132–161. [200]

[493] R. P. Stanley, On the number of reduced decompositions of elements of Coxeter groups, *European J. Combin.* **5** (1984), 359–372. [234, 242, 243, 244]

[494] R. P. Stanley, Generalized h-vectors, intersection cohomology of toric varieties, and related results, *Adv. Studies Pure Math.* **11** (1987), 187–213.

[495] R. P. Stanley, Subdivisions and local h-vectors, *J. Amer. Math. Soc.* **5** (1992), 805–851.

[496] R. P. Stanley, *Combinatorics and Commutative Algebra*, 2nd ed., Birkhäuser, Boston, 1996. [305]

[497] R. P. Stanley, *Enumerative Combinatorics, Vol. 1*, Wadsworth and Brooks/Cole, Monterey, CA, 1986; and Cambridge Studies in Advanced Mathematics **49**, Cambridge University Press, Cambridge, 1997. [204, 299, 302, 303, 319, 320]

[498] R. P. Stanley, *Enumerative Combinatorics, Vol. 2*, Cambridge Studies in Advanced Mathematics **62**, Cambridge University Press, Cambridge, 1999. [230, 232, 234, 307, 319]

[499] R. Stanley, Positivity problems and conjectures in algebraic combinatorics, *Mathematics: Frontiers and Perspectives*, American Mathematical Society, Providence, RI, 2000, pp. 295–319.

[500] R. Steinberg, *Lectures on Chevalley groups*, Notes, Yale University, 1967.

[501] R. Steinberg, Endomorphisms of Linear Algebraic Groups, *Mem. Amer. Math. Soc.* Nr. 80 (1968). [242]

[502] E. Steingrimsson, Permutation statistics of indexed permutations, *European J. Combin.* **15** (1994), 187–205. [243]

[503] J. R. Stembridge, Eulerian numbers, tableaux, and the Betti numbers of a toric variety, *Discrete Math.* **99**(1992), 307–320. [243]

[504] J. R. Stembridge, On the action of a Weyl group on the cohomology of its associated toric variety, *Adv. Math.* **106** (1994), 244–301. [243]

[505] J. R. Stembridge, On the fully commutative elements of Coxeter groups, *J. Algebraic Combin.* **5** (1996), 353–385. [244]

[506] J. R. Stembridge, Some combinatorial aspects of reduced words in finite Coxeter groups, *Trans. Amer. Math. Soc.* **349** (1997), 1285–1332. [244]

[507] J. R. Stembridge, On the Poincaré series and cardinalities of finite reflection groups, *Proc. Amer. Math. Soc.* **126** (1998), 3177–3181.

[508] J. R. Stembridge, The enumeration of fully commutative elements of Coxeter groups, *J. Algebraic Combin.* **7** (1998), 291–320. [244]

[509] J. R. Stembridge, The partial order of dominant weights, *Adv. Math.* **136** (1998), 340–364.

[510] J. Stembridge, A construction of H_4 without miracles, *Discrete Comput. Geom.* **22** (1999), 425–427.

[511] J. Stembridge, Minuscule elements of Weyl groups, *J. Algebra* **235** (2001), 722–743.

[512] J. Stembridge, Quasi-minuscule quotients and reduced words for reflections, *J. Algebraic Combin.* **13** (2001), 275–293.

[513] J. Stembridge, Computational aspects of root systems, Coxeter groups, and Weyl characters, *Interaction of Combinatorics and Representation Theory*, MSJ Mem., **11**, Mathematical Society of Japan, Tokyo, 2001, pp. 1–38.

[514] J. Stembridge, A weighted enumeration of maximal chains in the Bruhat order, *J. Algebraic Combin.* **15** (2002), 291–301.

[515] J. Stembridge, Combinatorial models for Weyl characters, *Adv. Math.* **168** (2002), 96–131.

[516] J. Stembridge, Tight quotients and double quotients in the Bruhat order, *Electron. J. Combin.*, to appear.

[517] J. Stembridge, D. Waugh, A Weyl group generating function that ought to be better known, *Indag. Math.* (N.S.) **9** (1998), 451–457.

[518] J. Stillwell, *Classical Topology and Combinatorial Group Theory*, 2nd ed., Graduate Texts in Mathematics, **72**. Springer-Verlag, New York, 1993. [24]

[519] D. B. Surowski, Reflection compound representations of groups of type A_n, B_n or C_n, *J. Algebra* **78** (1982), 239–247.

[520] R. Szwarc, Structure of geodesics in the Cayley graph of infinite Coxeter groups, *Colloq. Math.* **95** (2003), 79–90.

[521] H. Tagawa, On the first coefficients in q of the Kazhdan-Lusztig polynomials, *Tokyo J. Math.* **17** (1994), 219–228.

[522] H. Tagawa, On the maximum value of the first coefficients of Kazhdan-Lusztig polynomials for symmetric groups, *J. Math. Sci. Univ. Tokyo* **1** (1994), 461–469.

[523] H. Tagawa, On the non-negativity of the first coefficient of Kazhdan-Lusztig polynomials. *J. Algebra* **177** (1995), 698–707. [170]

[524] H. Tagawa, A decomposition of R-polynomials and Kazhdan-Lusztig polynomials, *Proc. Japan Acad. Ser. A Math. Sci.* **71** (1995), 107–108.

[525] H. Tagawa, Kazhdan-Lusztig polynomials of parabolic type, *J. Algebra* **200** (1998), 258–278.

[526] H. Tagawa, A construction of weighted parabolic Kazhdan-Lusztig polynomials, *J. Algebra* **216** (1999), 566–599.

[527] H. Tagawa, A recursion formula of the weighted parabolic Kazhdan-Lusztig polynomials, *Adv. Studies Pure Math.* **28** (2000), 373–389.

[528] H. Tagawa, Some properties of inverse weighted parabolic Kazhdan-Lusztig polynomials, *J. Algebra* **239** (2001), 298–326.

[529] K. Takahashi, The left cells and their W-graphs of Weyl group of type F_4, *Tokyo J. Math.* **13** (1990), 327–340.

[530] T. Takebayashi, Kazhdan-Lusztig polynomials of the central extension of the elliptic Hecke algebras, *J. Algebra* **269** (2003), 275–284.

[531] L. Tan, On the distinguished coset representatives of the parabolic subgroups in finite Coxeter groups, *Commun. Algebra* **22** (1994), 1049–1061. [242]

[532] I. Terada, Brauer diagrams, updown tableaux and nilpotent matrices, *J. Algebraic Combin.* **14** (2001), 229–267.

[533] H. Terao, Generalized exponents of a free arrangement of hyperplanes and Shephard-Todd-Brieskorn formula, *Invent. Math.* **63** (1981), 159–179.

[534] J. Tits, Groupes et géométries de Coxeter, preprint, I.H.E.S., Paris, 1961. [24]

[535] J. Tits, Géométries polyédriques finies, *Rend. Mat. Appl.* **23** (1964), 156–165.

[536] J. Tits, Structures et groupes de Weyl, *Séminaire Bourbaki (1964/65)*, Exp. 288, Secrétariat Mathématique, Paris, 1966. Reprinted in *Séminaire Bourbaki* **9** Exp. No. 288, 169–183, Société Mathématiques de France, Paris, 1995.

[537] J. Tits, Le problème des mots dans les groupes de Coxeter, *Symposia Mathematica (INDAM, Rome, 1967/68)*, Academic Press, London, 1969, vol. 1, pp. 175–185. [87]

[538] J. Tits, *Buildings of Spherical Type and Finite BN-pairs*, Lecture Notes in Math. No. 386, Springer, Berlin, 1974. [24, 88]

[539] J. Tits, Two properties of Coxeter complexes, *J. Algebra* **41** (1976), 265–268.

[540] W. T. Trotter, *Combinatorics and Partially Ordered Sets, Dimension Theory*, Johns Hopkins University Press, 1992. [59]

[541] P. Ungar, $2N$ noncollinear points determine at least $2N$ directions, *J. Combin. Theory, Ser. A* **33** (1982), 343–347. [88]

[542] T. Uzawa, Finite Coxeter groups and their subgroup lattices, *J. Algebra* **101** (1986), 82–94.

[543] D.-N. Verma, Structure of certain induced representations of complex semisimple Lie algebras, *Bull. Amer. Math. Soc.* **74** (1968), 160–166. [63]

[544] D.-N. Verma, Möbius inversion for the Bruhat order on a Weyl group, *Ann. Sci. École Norm. Sup.* **4** (1971), 393–398. [63, 64]

[545] D.-N. Verma, A strengthening of the exchange property of Coxeter groups, preprint, 1972. [24]

[546] E. B. Vinberg, Discrete reflection groups in Lobachevsky spaces, *Proc. Intern. Congress Math. (Warsaw, 1983)*, PWN, Warsaw, 1984, pp. 593–601. [24]

[547] E. B. Vinberg, Hyperbolic reflection groups, *Russian Math. Surveys* **40** (1985), 31–75. [24]

[548] D. A. Vogan, Jr. A generalized τ-invariant for the primitive spectrum of a semisimple Lie algebra, *Math. Ann.* **242** (1979), 209–224.

[549] M. L. Wachs, Quotients of Coxeter complexes and buildings with linear diagram, *European J. Combin.* **7** (1986), 75–92.

[550] G. Warrington, A formula for certain inverse Kazhdan-Lusztig polynomials in S_n, *J. Combin. Theory Ser. A* **104** (2003), 301–316.

[551] W. C. Waterhouse, Automorphisms of the Bruhat order on Coxeter groups, *Bull. London Math. Soc.* **21** (1989), 243–248. [38]

[552] D. Waugh, Upper bounds in affine Weyl groups under the weak order, *Order* **16** (1999), 77–87.

[553] D. Waugh, On quotients of Coxeter groups under the weak order, *Adv. Appl. Math.* **30** (2003), 369–384.

[554] R. Winkel, A combinatorial derivation of the Poincaré polynomials of the finite irreducible Coxeter groups, *Discrete Math.* **239** (2001), 83–99.

[555] E. Witt, Spiegelungsgruppen und Aufzählung halbeinfacher Liescher Ringe, *Abh. Math. Sem. Univ. Hamburg* **14** (1941), 289–322. [24]

[556] N. Xi, An approach to the connectedness of the left cells in affine Weyl groups, *Bull. London Math. Soc.* **21** (1989), 557–561.

[557] T. Yanagimoto, M. Okamoto, Partial orderings of permutations and monotonicity of a rank correlation statistics, *Ann. Inst. Statist. Math.* **21** (1969), 489–506. [87]

[558] A. Zelevinsky, Small resolutions of singularities of Schubert varieties, *Funct. Anal. Appl.* **17** (1983), 142–144.

[559] Y. M. Zou, D-sets and BG-functors in Kazhdan-Lusztig theory. *Proc. Amer. Math. Soc.* **123** (1995), 935–943.

Index of notation

We collect here the main notation used in the book, with references to the pages where definitions can be found. For general notational conventions, see the beginning of the book.

Index